P9-CMW-869

IPv6 is the 21st Century Internet Protocol. Marc spent a good part of his life pioneering with IPv6 as co-founder of the IPv6 Forum and board member and his book is what every engineer on this planet is mandated to read to become the most advanced New Generation Internet engineer.

Latif Ladid
President, IPv6 Forum
Chair, European IPv6 Task Force

This book is a thorough and logical engineer's walk through the universe of IPv6, ideal for those who need a good technical understanding of the future Internet protocol.

Brian E. Carpenter
IETF Chair
Distinguished Engineer, Internet Standards & Technology, IBM

There are many books on IPv6, but this is certainly the best I've read so far. Marc has managed to take a wealth of material describing all aspects of the next generation of IP technology and turn it into a thoroughly readable book. His approach in this book has both breadth of coverage and also careful attention to detail that any network engineer or system administrator will appreciate. If Marc's aim was to produce both a learning resource and a reference book, then he has succeeded on both counts here. This book will be an invaluable and reliable assistant if your task is to make IPv6 work in your network.

Geoff Huston
Author of the "ISP Survival Guide"
Senior Internet Researcher, Asia Pacific Network Information Centre

Readable, comprehensive, and well documented, *Migrating to IPv6* is destined to be the premier IPv6 text. Marc Blanchet's long involvement in the IPv6 community, active contribution to standards, and extensive practical experience implementing IPv6 bring an authority seldom found in other IPv6 books.

Jeff Doyle
Senior Network Architect
Juniper Networks

This is an excellent book for hands-on users such as system implementers, researchers, and network administrators who need an in-depth understanding of the new IPv6 protocol. I intend to buy copies of this book for our researchers, implementers, and customers working with IPv6.

David Green
CERDEC Site Manager
SRI International

A must have companion to take the first steps on the road to an IP converged networking world of fixed and mobile voice, data, gaming, radio, music, television, location based services etc. An era of sensors, RFID, home networks, plug and play, ad-hoc networks, home networks, Networks in Motion, multicast IPTV, Grid computing, end to end secure VPN's, IP based QoS and many others. The on-line complement to the book should add a further dimension to this companion on the way to IPv6 and even more importantly, the resulting new service and revenue opportunities.

Yves Poppe
Director IP Strategy
Teleglobe

An excellent balance between theory and applied knowledge, this book serves as both a guide and a reference for people looking to bring their networks and systems into the IPv6 world.

William Fernando Maton Sotomayor
Senior Project Manager and Network Engineer, Advanced Infrastructures.
National Research Council of Canada

Internet Protocol version 6 (IPv6) is ready, it works, and there is great momentum in migrating to IPv6. Finally there is a book written by a world class engineer, Marc Blanchet, who has been in the IPv6 trenches over the past 15 years - designing, building, deploying, utilizing, and managing IPv6 networks. It is that breadth of experience that makes Marc Blanchet's *Migrating to IPv6* an excellent and insightful guide to IPv6. *Migrating to IPv6* contains a wide variety of deployment examples that detail how to apply IPv6 in real-world networks–fixed and mobile, as well as a skillful primer focused on "how-to" deploy, utilize and use IPv6 in day-to-day network activities. There are numerous complexities in bridging IPv6 networks to IPv4 networks. *Migrating to IPv6* focuses on how to use IPv6 in real-world intranet and Internet-wide applications and serves as an excellent resources on managing the co-existence of IPv4 and IPv6 infrastructures.

Carl Williams
Senior Architect and IP Lead
KDDI R&D Labs

Migrating to IPv6 is a timely book because the tipping point of IPv6 usage is imminent as DOD and the Agencies begin deploying network centric applications. With the need for end-to-end services over a wide variety of fixed and mobile subnetworks, IPv6's Flow Label, not found in the existing IPv4 protocol, becomes an indispensable tool for facilitating an end-to-end Quality of Service (QOS). In stark contrast to wide area networks having plenty of unused capacity to accommodate variable traffic flows, the wireless networks, including tactical, must have the means to identify flows in which to apply a QOS policy - and IPv6 offers the only solution. Thus, with growing agility and network-centricity needs driven by DOD Transformation and Agencies' adopting of Enterprise Architectures, the need for IPv6 is evident. Marc Blanchet provides the network engineer a thorough description of implementation issues, configuration options and troubleshooting advice. The rich use of illustrations and configuration examples provides the reader an invaluable benefit. Accordingly, this book will become a staple of networking engineers for years to come.

Bob Collet
Chief Engineer, VP of Engineering Transformation, Training and Logistics Group
Science Applications Corporation Inc

Forget the technical stuff, here is the real deal... IPv6 in all its parts and promise! Kudos to Blanchet for writing a thoroughly IPv6 kind of book!!

Bob Fink, co-conspirator of the 6bone

Appréciation sur le livre Migration to IPv6 de Marc Blanchet

Ayant une longue expérience dans le monde pédagogique (ancien professeur en Université et dans des Ecoles d'ingénieurs de télécommunications à Paris) et une longue expérience aussi bien dans l'industrie des télécommunications que dans le monde des opérateurs, je dois reconnaître que Marc Banchet, à travers cet excellent ouvrage *Migration to IPv6*, a réussi une formidable performance.

Trois ingrédients au moins étaient nécessaires pour réussir cette alchimie et Marc est parmi ces rares spécialistes à en détenir le secret:

- une parfaite connaissance et maîtrise des normes et standards IPv6 pour être en mesure d'en tirer l'essentiel et en proposer une reformulation simple et non simpliste au lecteur non averti. De ce point de vue, Marc en tant que pionnier dans la standardisation IPv6 a toute l'envergure et l'épaisseur scientifique. Car il faut bien reconnaître que les RFCs sont certes des "ouvres d'art technique" cependant conçus par des artistes/experts pour des admirateurs/experts. Ce sont des fresques où le "pinceau pédagogique" n'a presque pas droit de cité. Et pourtant, la "galerie" de l'IETF et la "salle IPv6" en particulier, intéressent toute la société et l'économie basées sur l'information et la connaissance. Dommage, il n'y a ni guide ni mode d'emploi! Marc, à sa manière, vient précisément par son ouvrage combler ce déficit. C'est donc non seulement un énorme service qu'il rend à la communauté des utilisateurs au sens large (étudiants, industriels, ingénieurs, opérateurs, entreprises, développeurs), mais aussi à la communauté des normalisateurs elle-même!

- une expérience dans la mise au banc d'essai de ces RFCs sur IPv6 afin de rapprocher le monde des standards du monde pré-industriel et pré-commercial à travers des implémentations et des développements de produits et services pour des utilisateurs potentiels. On passe ainsi de la théorie à la pratique. Pour cela, une capacité et une expérience dans la connaissance des besoins de ces clients potentiels est primordiale. Là encore, Marc en tant que fondateur de la société Hexago ciblée sur le produits IPv6, a convaincu le marché des opérateurs télécoms et des entreprises que la migration vers IPv6, devenue une nécessité, pouvait se faire via des produits fiables et sécurisés, disponible sur étagère et à des coûts raisonnables. Toujours dans un souci d'aider l'utilisateur à toucher la réalité concrète d'IPv6, Marc expose dans son ouvrage, "les marques de fabrique" et implémentations d'autres constructeurs. C'est une réponse au besoin de diversité et donc d'interopérabilité critères cruciaux en tout cas pour les grands opérateurs et grandes entreprises.

- un souci pointu de la pédagogie et un style. C'est la partie scénario et mise en scène de l'ouvrage. Je dois dire que Marc a réussi cette épreuve oh combien difficile et épineuse et qui demande un investissement considérable. C'est la partie tant redoutée par l'écrivain car elle le met face à face avec son lecteur et donc à son jugement. Quand on lit l'ouvrage, on comprend très vite que Marc a délibérément mis le lecteur au centre de ses préoccupations. C'est l'aboutissement entre autre de l'investissement que Marc a fait pour la communauté IPv6 à travers ses tutoriaux et autres actions de dissémination.

Le produit fini est un chef-d'ouvre. Bravo Marc et merci pour ce beau cadeau!

Tayeb Ben Meriem, August 2005

This book is one of the most comprehensive technology in depth books I have read over the many years regarding IPv6. It provides a new view and perspective, relating the most recent IPv6 features and capabilities to the reader, that will benefit the engineer, architect, technologist, or a network manager, and is an important reference book for anyone that transitions to IPv6.

Jim Bound CTO,
IPv6 Forum/Chair North American IPv6 Task Force/HP Fellow

IPv6 will revolutionize IP networking over the next 10 years, and learning about its features and deployment today will put anyone working in IT at significant advantage. Marc's writing style is clear and concise, and this an excellent book for every IT worker - network guru's through engineering management.

John Spence,
CTO, Native6

Migrating to IPv6

Migrating to IPv6

A Practical Guide to Implementing IPv6 in Mobile and Fixed Networks

Marc Blanchet
Québec, Canada

Telephone (+44) 1243 779777

Email (for orders and customer service enquiries): cs-books@wiley.co.uk
Visit our Home Page on www.wiley.com

Other Wiley Editorial Offices

John Wiley & Sons Inc., 111 River Street, Hoboken, NJ 07030, USA

Jossey-Bass, 989 Market Street, San Francisco, CA 94103-1741, USA

Wiley-VCH Verlag GmbH, Boschstr. 12, D-69469 Weinheim, Germany

John Wiley & Sons Australia Ltd, 42 McDougall Street, Milton, Queensland 4064, Australia

John Wiley & Sons (Asia) Pte Ltd, 2 Clementi Loop #02-01, Jin Xing Distripark, Singapore 129809

John Wiley & Sons Canada Ltd, 22 Worcester Road, Etobicoke, Ontario, Canada M9W 1L1

Library of Congress Cataloging in Publication Data

Blanchet, Marc, 1964–
Migrating to IPv6 : a practical guide to implementing IPv6 in mobile and fixed networks / Marc Blanchet.
p. cm.
Includes bibliographical references and index.
ISBN 0-471-49892-0 (pbk. : alk. paper)
1. TCP/IP (Computer network protocol) 2. Internet. 3. Mobile computing. I. Title.
TK5105.585.B5437 2005
004.6′2—dc22

2005025352

British Library Cataloguing in Publication Data

A catalogue record for this book is available from the British Library

ISBN 0-471-49892-0

Typeset in 10/12pt Times by Integra Software Services Pvt. Ltd, Pondicherry, India
Printed and bound in Great Britain by Antony Rowe Ltd, Chippenham, Wiltshire
This book is printed on acid-free paper responsibly manufactured from sustainable forestry in which at least two trees are planted for each one used for paper production.

Ce livre est dédié aux femmes de ma vie, sans prétention:

À Maman, toujours fière de son fils.
À Julie, ma fille, qui m'apporte tant de joies tous les jours et que j'aime sans mesure.
À Isabelle, Véronique et Monique, Marie, Diane, Josée et les autres, des amies si fidèles.

Et surtout:
À Lucie, mon épouse qui m'appuie dans toutes mes folles aventures et que j'aime sans mesure. Elle s'est dédiée sans mesure pour que je finisse ce projet.

If you like this book, it is because my wife Lucie has been so dedicated to help me finish it. Please thank her, not me.

Contents

Foreword

The premise of this book is that readers already have some idea about the important nature of the Internet Protocol layer in the Internet and that there is some concern that version 4 of the Internet Protocol (IPv4) may need to be succeeded by a newer protocol that has a larger complement of address space and additional features designed to match the evolving requirements of applications of the Internet. What is most valuable about this volume is the clarity and scope of information that the author brings to the table.

The original work on the Internet design began in 1973 and benefited from experience with the predecessor to Internet, the Arpanet. The design of IPv4 took place over the period from 1973 to 1978. It was the product of a recurring series of specifications, implementations and tests that ultimately led to standardization of IPv4 in mid-1978. By the early 1990s it was feared that the rate of consumption of IPv4 address space and the relative inefficiency of its assignment would exhaust the resource within a few years. A crash program was initiated to develop a new version with larger address space and a feature set that benefited from the many years of experience with IPv4. There ensued a great deal of debate and many different proposals. Ultimately, IPv6 was standardized.

The book you are about to read takes you through the concrete technology and issues associated with the implementation of IPv6. What is important to recognize is that not all the issues are fully resolved. Moreover, the use of this new protocol is unlikely to be entirely independent of the existing and spreading IPv4 system. Indeed, it is expected that IPv4 will support IPv6, for example, by providing tunnels through the IPv4 protocol space to link IPv6 layers of protocol.

A number of application programs may need to be revised to cope with IPv6 simply because the syntax of Uniform Resource Locators (URLs), for example, allows for the use of literal IPv4 and IPv6 addresses. This means that any program that processes URLs, including browsers, must be prepared to parse IPv6 addresses rendered as a series of hexadecimal digits. Here is an example of the variability of IPv6 address representations that must be parsed:

2001:0000:1234:0000:0000:C1C0:ABCD:0876

can be represented as:

2001:0000:1234:0000:0000:c1c0:abcd:0876

which can be compressed to:

2001:0:1234:0:0:c1c0:abcd:876

which can be further compressed to:

2001:0:1234::c1c0:abcd:876

All of these representations are equivalent and must be treated that way by the programs that reference them.

IPv6 offers a number of other features including a "flow ID" whose function has not yet been fully determined but that adds the potential to treat sequences of otherwise independent packets with a common priority or service class, even if they belong to different applications.

IPv6 has also been designed to be easier to parse in the sense that the header is kept simple in structure and that contributes to the utility of the new packet format when processing real-time traffic flowing through the routers. It is also anticipated that mobility can be supported more conveniently under IPv6 if the 128 bit address field can be treated as a two-part address. The prefix would be used to route the IPv6 packet to a target network and the rest would be used to deliver to a target within that network. This is not terribly different than the present handling of IPv4 addresses as having two parts, a network part and a host part, the former reflected in the so-called subnet mask that says which part of the IPv4 address references the target network.

IPv6 also specifies that if one party wishes to enter an encrypted mode of communication (with IPSEC) the other party is obliged to support this. That requirement is not mandatory in the IPv4 space.

The major benefit that IPv6 confers is a vastly larger end-to-end address space within which to operate. Many experts predict that there will be billions of terminal devices on the Internet, perhaps as many as 100 per user. The vast IPv6 address space should allow for sufficient space for these "Internet-enabled" devices for many years, if not decades, to come.

None of these benefits will be obtained without a substantial effort requiring clarity of thought, determination, persistence and planning. The transitional introduction of IPv6 will take place over a period of years if not decades. Consequently, it is vital for a wide range of participants to have awareness of the specifics of IPv6 and what it will mean to fully support it and make it inter-work with IPv4 or at least make applications running either or both able to work with each other.

It may prove to be the case that the so-called Network Address Translation (NAT) devices will end up being instruments for the introduction of IPv6 into the normal Internet operation. One could use NAT methods to map from IPv6 to IPv4 at need. It remains to be seen how well this works in practice but there can be little serious disputing of the major premise: that IPv6 will surely have a critical role to play in the expansion of the Internet.

As this book goes to press, IPv6 is in deployment especially in Asia and the Pacific Rim where internauts have taken an aggressive posture with regard to IPv6 implementation and deployment. The Regional Internet Registries, working cooperatively as the Number Resource Organization and in concert with the Internet Corporation for Assigned Names and Numbers, are moving ahead with new global IPv6 allocation rules. Consumer equipment makers are starting to think through the use of IPv6 for mobiles, set top boxes and other devices that are produced in large numbers and need to be on the Internet.

Mobility has become a touchstone for Internet evolution as more users access the Internet in untethered fashion. The ability to support mobility and portability (episodic re-connection with the Internet) will be a premium capability for many users.

ISPs still have much to do to integrate IPv6 into their network management systems, order entry, provisioning, billing and application support services. For the most part, the host and router vendors have implemented IPv6 capability but much testing is needed to assure inter-working of many implementations. The University of New Hampshire has a major testing facility and MCI has recently linked that to its MAE system of Internet exchanges to support easy interconnection and exchange with the UNH site. The IPv6 Forum and its various regional counterparts, such as the North American V6 Task Force (NAV6TF) are hard at work focusing on the tasks ahead to accelerate deployment.

With the first decade of the 21st Century already half over, we need to move quickly if we are to meet deadlines set for 2008 by the US Defense Department, for one. Other government agencies, in the US and elsewhere, are equally keen to see the introduction of IPv6 capability into their networks and generally available in the public Internet as well as in virtual private networks of all kinds.

Vint Cerf
McLean, VA

Preface

Internet Protocol version 6 (IPv6) is a major improvement to IPv4, which currently faces many challenging issues. Designed in the 1970s, IPv4 was initially deployed over a network of few nodes. The 1990s saw its deployment to a large base of end-users, stretching its capabilities. Since then, new applications with additional requirements such as mobility, tiny server nodes, appliances, global reachability and end-to-end communications have (or soon will be) deployed. While IPv6 is an important building block for these new applications, the current deployed IPv4 network and applications will remain dominant for a significant part of the period 2000–2010.

Purpose of this Book

This book attempts to give a comprehensive view of IPv6 and related protocols, from the layers below IPv6 to the application and end-user layers. It takes the perspective that current networks use IPv4 as the dominant networking protocol. The assumption is that IPv6 will be deployed together with current IPv4 networks, so that network engineering will be taking into consideration both protocols.

To make learning for the reader more efficient, the book describes IPv6 by comparing it with IPv4, without reintroducing IPv4. Therefore, the reader is assumed to have minimum familiarity with IPv4. The emphasis is on deploying IPv6 in current IPv4 networks, with case studies and examples.

Hands-on examples on major implementations of hosts and routers are provided to give practical experience for the reader.

In summary, this book provides the reader with a comprehensive view and understanding of IPv6, given that the reader is familiar with IPv4 and that most current networks are IPv4-dominant.

Writing Style

The raw information about IPv6 is in the Internet Engineering Task Force (IETF) Request For Comments (RFC) documents. These documents describe the standards from an implementer point of view. Many (if not most) IPv6 books are written by following the order and the depth of the RFCs. However, because RFCs are numerous and difficult to read for a beginner who is not an IP stack programmer, so are the books based on the RFCs.

Top-down Approach

This book is structured differently and takes a top-down approach. A significant length of time was taken to order the IPv6 topics by their relevance, for the benefit of the reader. The concepts are presented first, and then their implementations, like the bits in the packets. Sometimes, even the bits are not presented in the book. For example, instead of describing the exact location of some option of the neighbor discovery protocol in the IPv6 packet, the book describes why and what the option is for, leaving the bit location in the RFCs for the implementer. However, important bits are shown, such as the packet header and the addressing structure, the foundations of IPv6.

RFCs are exhaustive in describing all bits of a specific protocol. Often, the bits in an RFC describe multiple uses of the protocol. For example, the neighbor discovery protocol is used for finding the neighbors on a link as well as autoconfiguring a node. In this book, these two functions are described in different chapters, so the reader learns one function at a time and more logically.

Saving your Time

A 'direct to the point' writing style is used. I care about the precious time you have invested to read this book. So I have tried to save you time and give you a direct and clear understanding of the topics while not omitting to clarify where needed.

A Learning as well as a Reference Book

One goal is to provide both a learning and reference book simultaneously. After reading the book, my personal goal is that you keep it around as good reference material when you work on IPv6. This is challenging, and I hope to have succeeded.

The reference function of the book is achieved in many ways. The back cover contains a reference card with the key data structures and packet formats. Throughout the book, great effort is made to summarize topics in table or graphical form to help you browse the book at a later date, giving fast access to information. Practical hints as well as key points are also provided in a formatted box. References and Further Reading are put at the end of each chapter.

Modular

Each chapter is as self-contained as possible. Even if the recommended order of reading is to follow the order of chapters, topics are cross-referenced to help you read the chapters in any order while being able to look back for necessary information in previous chapters. Hypertext used on the Web is somewhat duplicated here with a lot of cross-references.

Practical and Hands-on

Theory is always better understood with a practical example. Most chapters contain generic examples of the topics discussed, as well as major implementation configuration examples at the end of the chapter.

Audience

Any computer or networking professional interested in IPv6 is the primary target of this book. Computer and network users who are curious about technology will also find it of interest. Managers involved in networking technologies will get a good understanding of IPv6. The book is written so that details and examples can be skipped while not affecting the overall understanding of IPv6.

IPv4 Knowledge

The reader should have basic IPv4 knowledge. Since IP is horizontal and is used by a wide variety of upper protocols and functions, such as applications, security, mobility and routing, introductory material is given for less common protocols. For example, mobileIP is described in general and then followed by IPv6 mobility specifics. In general, basic IPv4 functionality is described at a minimum level.

Terminology

Accurate definitions of words are vital for complete understanding. Since IPv6 introduces some new words and some additional meanings to existing words, Table P.1 defines some of the words used in this book.

For easier reading, the word 'IPv6' is explicit everywhere in the book unless specifically stated. In which case when 'IP' is used, it means that any version of IP and 'IPv4' is specific to that version.

Implementations

Examples of configurations and commands are given for major implementations of hosts and routers. The versions used in this book are listed in Table P.2. Unless stated, only the out-of-box versions, without any additional packages installed, have been used for the examples.

Since OpenBSD, NetBSD and FreeBSD share the same IPv6 code base, the examples for FreeBSD usually apply to the OpenBSD and NetBSD. MacOS X/Darwin being a derivative of FreeBSD, the examples of FreeBSD often apply to MacOS X. This common IPv6 code comes from the Kame Japanese project, which is described as the best IPv6 stack available. When FreeBSD is used in this book to describe some IPv6 behavior it is, in fact, the Kame code. The behavior should be very similar or identical on the other Kame platforms.

Linux has many distributions. IPv6 is yet to be fully normalized in the Linux distributions. The examples are based on RedHat 8.0 Linux distribution, which is applicable to the other

Table P.1 Definitions

Word	Definition
datagram	'The unit of transmission in the network layer (such as IP). A datagram may be encapsulated in one or more packets passed to the data link layer.' [RFC1661]
packet	'The basic unit of encapsulation, which is passed across the interface between the network layer and the data link layer. A packet is usually mapped to a frame; the exceptions are when data link layer fragmentation is being performed, or when multiple packets are incorporated into a single frame.' [RFC1661]. In most cases, a datagram and a packet are identical. This book uses the word datagram in the generic sense and packet when there is a need to identify specifically the datagram.
frame	'The unit of transmission at the data link layer. A frame may include a header and/or a trailer, along with some number of units of data.' [RFC1661]
node	any IP device: host, appliance, router, server, etc.
IP	Internet Protocol. When the word IP is used, it means IPv4 or IPv6.
Payload	Data part, after the header, of a datagram.
Intermediate nodes	Any IP forwarding device in a path between the source and the destination: router or firewall.
Forwarding node	An IP forwarding device, usually a router.
Link	'A communication facility or medium over which nodes can communicate at the link layer.' [RFC2460]. Similar to subnet, vlan or broadcast domain. In IPv6, many prefixes can be used on the same link, which could be considered different subnets at the IP layer.
Prefix	Leftmost part of an IP address, used to describe a 'subnet mask', a route or an address range.

Table P.2 Versions of implementations used

Implementation	Version
Cisco	IOS 12.2(13)T
FreeBSD	5.2
Hexago	HexOS 3.0
Juniper	JunOS 5.5
Linux	RedHat 8.0
Microsoft	Windows XP-SP1
Sun	Solaris 9
Zebra	0.93b

distributions. The Japanese USAGI project[1] is making a common IPv6 code for all Linux platforms.

The examples are given at the end of each chapter, for the topic discussed in that chapter. The intent is to give a very good start for a user on the implementations. The reader should refer to the implementation documentation for complete information.

[1] Kame means turtle in Japanese. USAGI, a similar project with similar goals, also comes from Japan and means rabbit.

Conventions used

A '<name>' is to be replaced by the actual value, depending on the context.

The `Courier` font is used for typed characters in commands or text files.

RFCs and Internet-drafts are IETF documents and can be found at the IETF Web site (http://www.ietf.org) or at various mirror sites. This book's Web site (http://www.ipv6book.ca) also makes these documents and a search engine available. Internet-drafts have version numbers in the file name (e.g., draft-ietf-ipv6-addr-arch-08.txt) and usually disappear after six months. To find the Internet-draft, one can try to find the most recent version by incrementing the value of the version. Approved drafts are published as RFCs.

Comments and Errata

I've been involved in IPv6 since 1995, when it started to take form. Since then, I've been using it, implementing it, consulting for providers and organizations, writing white papers, and giving tutorials both at conferences and privately. I also co-authored the 'Implementing IPv6 Networks' Cisco course. This book is the result of many years' work on IPv6 and I hope to share this knowledge and practical experience with you in the most efficient way through this book. I hope you will enjoy the book.

Together with the editor, significant time has been spent to guarantee the quality. However, nothing is perfect. I will be very pleased to receive any comments on this book. Please send your comments and suggestions to author@ipv6book.ca. Any future revisions of this book will take into account your comments.

Book Web Site

Since its initial design, IPv6 has evolved; some features have been removed or deprecated. Keeping in mind that some of these features might still be discussed in other literature, the deprecated features are not discussed in this book, to keep the main text flow seamless. Also, some techniques are no longer relevant; some topics are for specific interests only, and new topics will arise subsequent to the publishing of this book. For all these reasons, a Web site has been created to complement this book. Here, you will find additional information such as:

- deprecated or less used protocols or specifications;
- more detailed coverage of some specific topics, such as some migration techniques;
- new features or considerations subsequent to the publishing of this book;
- complete bibliography;
- book errata.

The book Web site is located at http://www.ipv6book.ca.

Acknowledgments

First and most importantly, I would like to thank my family for giving me the support, both in time and encouragement, to write and finish this book. Without the patience and dedication of my wife and my daughter (and by extension to my young son) when I was away writing the book, I would never have completed this book.

Many people contributed to this book as initial readers or reviewers. I'm very honored to have had the following people dedicate time for reviewing the manuscript: Jun-ichiro itojun Hagino, William Fernandez, Maton Sotomayor, John Spence, Hesham Soliman, Robert Fink, Michel Py, David Greene, Jim Bound and Brian Carpenter. Finally, thanks to the team of Wiley, namely Richard Davies, Andy Finch, Birgit Gruber, Sally Mortimore, Joanna Tootill and Julie Ward for their patience, encouragement and support in publishing this book.

1

IPv6 Rationale and Features

Back in the 1970s, the Internet Protocol (IP) was designed upon certain assumptions and key design decisions. After more than 25 years of deployment and usage, the resulting design has been surprisingly appropriate to sustain the growth of the Internet that we have seen and continue to see; not only the increase of the number of devices connected, but also of the kinds of applications and usage we are inventing everyday. This sustainability is a very impressive achievement of engineering excellence.

Despite the extraordinary sustainability of the current version (IPv4), however, it is suffering and the Internet Protocol needs an important revision. This chapter describes why we need a new version of the IP protocol (IPv6), by describing the Internet growth, the use of techniques to temper the consequences of that growth and the trouble experienced in deploying applications in current IPv4 networks. Some architecture considerations are then discussed and new features needed in current and future networks presented.

Next, the work towards IPv6 at the IETF is shown along with the key features of IPv6. Some milestones are also tabled. Finally, the IPv6 return on investment and drivers is discussed.

1.1 Internet Growth

The origin of IPv6 work lay in the imminent exhaustion of address space and global routing table growth; both could be summarized as Internet growth.

1.1.1 IPv4 Addressing

The Internet is a victim of his own success. No one in the 1970s could have predicted this level of penetration into our lives.

Migrating to IPv6: A Practical Guide to Implementing IPv6 in Mobile and Fixed Networks Marc Blanchet

In theory, 32 bits of IPv4 address space enables 4 billion hosts. Studies [RFC1715] have shown that the effectiveness of an address space is far less. For example, RFC1715 defines a H ratio as: H = log (number of objects using the network)/number of bits of the address space. Based on some empirical studies of phone numbers and other addressing schemes, the author concluded that this H ratio usually never reaches the value of 0.3, even with the most efficient addressing schemes. An optimistic H ratio is 0.26 and a pessimistic one (for not very efficient addressing schemes) is 0,14. At H = 0.26, with an addressing of 32 bits, the maximum number of objects, in the case of IPv4 the number of reachable hosts, is 200 000 000.[1] When IPv4 Internet reaches 200 million reachable nodes, the IPv4 addresses will be exhausted.

Moreover, the IPv4 address space was designed with three classes (A, B and C)[2] which makes the address space usage even less efficient than with the optimistic H ratio. In August 1990 at Vancouver IETF, a study [Solensky, 1990] demonstrated the exhaustion of class B address space by March 1994. Figure 1.1 shows the summary slide presented during that IETF. This was an important wakeup call for the whole Internet engineering community.

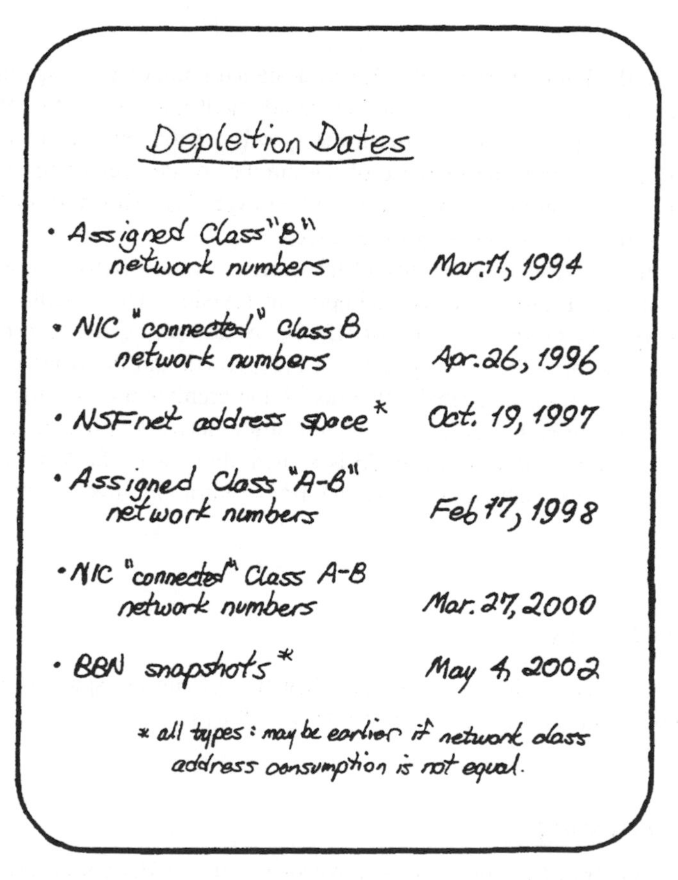

Figure 1.1 Solensky slide on IPv4 address depletion dates

[1] RFC1715 was also used as input to define the IPv6 address length to 128 bits.

[2] D and E classes also exist but are not for unicast generic use.

At that time, most organizations requesting an address space pretty easily obtained a class B address block, since there was plenty of IPv4 address space. Assigning class C address blocks to organizations was the first cure; it decreased the initial address consumption problem but introduced more routes in the global routing table, therefore creating another problem.

1.1.2 IPv4 Address Space Utilization

Let's talk about the current IPv4 address space utilization. The IPv4 address space is 32 bits wide. IANA allocates by 1/256th (0.4%) chunks to regional registries, which corresponds to a /8 prefix length or to the leftmost number in an IPv4 address. Since the 224.X.X.X to 239.X.X.X range is reserved for multicast addressing and the 240.X.X.X to 254.X.X.X range is the experimental class E addressing, the total unicast available address space is of 223 /8 prefixes.

Figure 1.2 shows the cumulative number of /8 prefixes allocated since the beginning of IPv4. At the end of 2004, there are 160 /8 prefixes allocated, representing 71% of the total unicast available address space.

In 2003, 5 /8 prefixes were allocated by IANA to the regional registries. In 2004, 9 /8 prefixes were allocated (80% annual increase). In January 2005 alone, 3 /8 prefixes were allocated. If every year after 2004, we are flattening the annual consumption to the 2004 number (9 /8 prefixes: i.e. 0% annual increase for the next 7 years), then Figure 1.3 shows the exhaustion of IPv4 address space (223 /8 prefixes) by 2011.

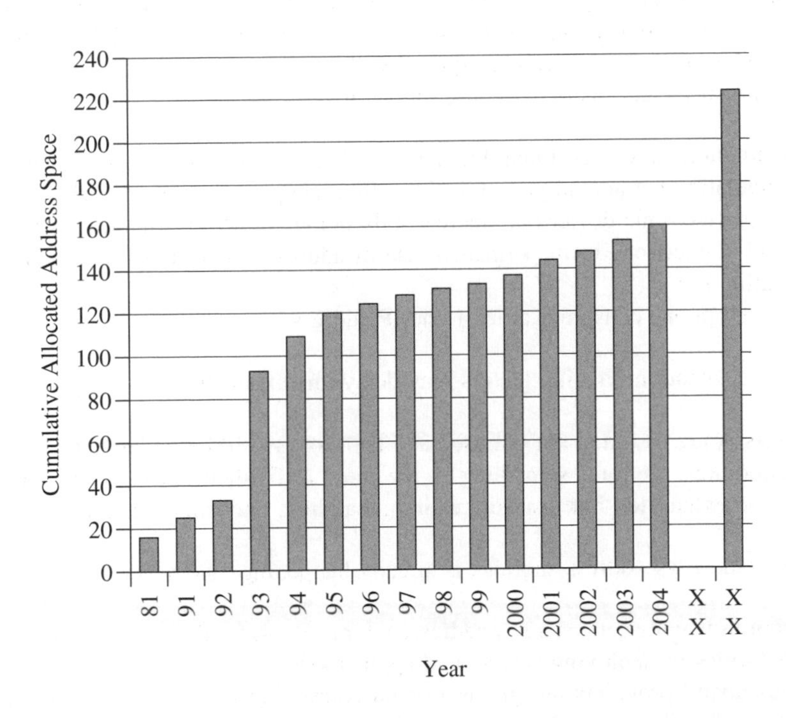

Figure 1.2 IPv4 cumulative allocated address space as of 2004–12

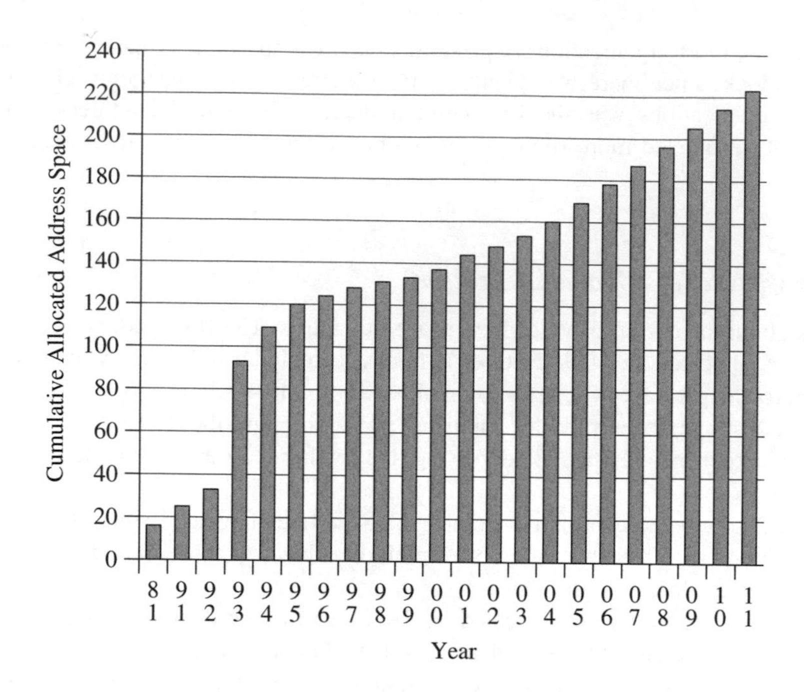

Figure 1.3 Prediction of IPv4 allocated address space with flat annual consumption

If we are slightly more aggressive by increasing the annual consumption by 2 additional /8 prefixes every year after 2004, which results in an annual increase of 22%, then Figure 1.4 shows the exhaustion of IPv4 address space by 2009.

A 20% annual increase is pretty conservative, given that:

- large populations in China, India, Indonesia and Africa are not yet connected;
- world population net annual growth is 77 million people [Charnie, 2004];
- all kinds of electronic devices are increasingly being connected and always on;
- broadband connections incur permanent use of addresses instead of temporary addresses when dialing up;
- each 3G cell phone consumes at least one IP address.

On the other hand, mitigating factors may delay this exhaustion:

- some class A are assigned but not used and therefore could be reclaimed;
- as in economics, the rarer something is, the more difficult it is to get and more it costs, slowing the exhaustion but instead creating an address exchange market.

Despite this, the IPv4 address shortage is already happening, and severely, because

- organizations usually get just a few addresses (typically 4) for their whole network, limiting the possibilities of deploying servers and applications;
- some broadband providers are giving private address spaces to their subscribers, which means the subscriber computers cannot be reached from the Internet.

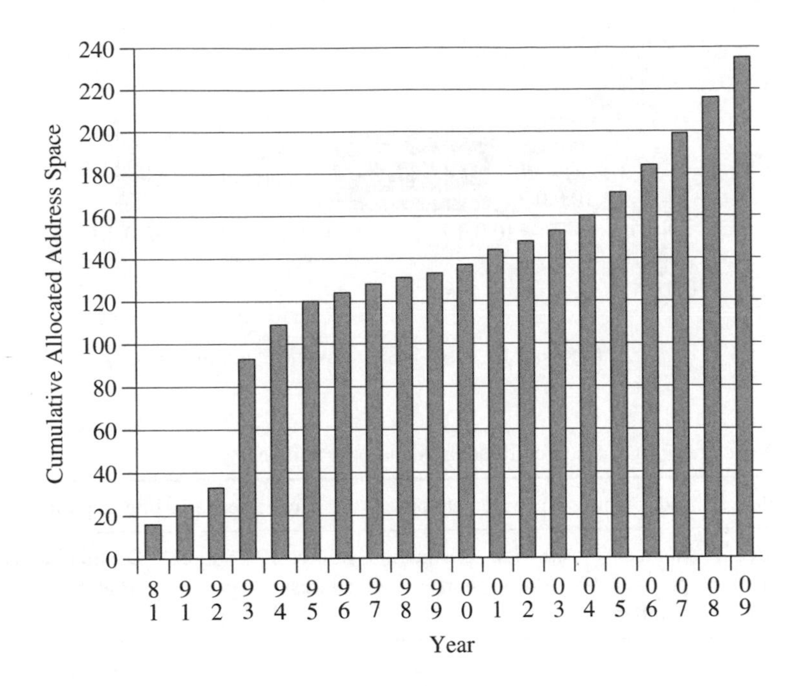

Figure 1.4 Prediction of IPv4 allocated address space with incremented annual consumption

1.1.3 Network Address Translation

The most important change regarding IP addressing is the massive use of Network Address Translation (NAT). The NAT functionality is usually implemented within the edge device of a network, combined with firewalling. For example, most organization networks have a firewall with NAT at the edge of their network and most home networks have a home router which implements firewalling and NAT.

NAT maps multiple internal private IP addresses to a single external IP address.[3] By allocating new external port numbers for each connection, essentially this NAT mapping process extends the address space by adding 16 bits of the port address space.

Figure 1.5 shows a basic network diagram of a private network with 2 computers (N1 and N2) and a public network, such as the Internet with one server (S). The private network uses private address space [RFC1918]. When internal nodes N1 and N2 connect to server S, the source addresses (10.0.0.3, 10.0.0.4) of the packets are translated to the NAT external IP address (192.0.2.2) when the packet is traversing the NAT. Server S receives connections coming from the same single source address (192.0.2.2), as if it comes from one single computer.

Table 1.1 shows how the detailed process works based on Figure 1.5. When the packet traverses the NAT, the source IP address and port are translated to the external IP address of the NAT and a new allocated port, respectively. For example, N1 source IP address 10.0.0.3 is translated to 192.0.2.2 and the source port 11111 is translated to the new allocated

[3] NAT can map multiple internal addresses to more than one external address, but for simplication we are discussing the most current used case: multiple internal to a single external address.

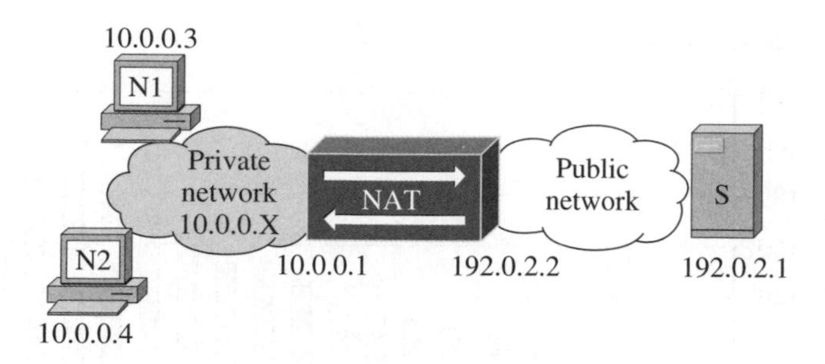

Figure 1.5 NAT basic network diagram

Table 1.1 NAT changing the source IP address and port number

Flow	Packet header while in private network				Packet header while in public network			
	Source IP address	Source port number	Destination IP address	Destination port	Source IP address	Source port number	Destination IP address	Destination port
N1 to S	10.0.0.3	11111	192.0.2.1	80	192.0.2.2	32001	192.0.2.1	80
N2 to S	10.0.0.4	22222	192.0.2.1	80	192.0.2.2	32002	192.0.2.1	80

external port 32001 by the NAT. The mapping is kept inside the NAT for the lifetime of the connection. If the connection is to get a Web page from the HTTP server S, then the mapping will remain for the duration of the GET request. Any new connection, even to the same server, creates a new mapping. Also, Table 1.1 shows another connection from N2. From the S perspective, the two connections have the same source IP address, so they appear to come from the same source.

This translation technique effectively hides the nodes on the private network and conserves the public IP address space. The private address space [RFC1918] enables a very large network having millions of nodes hidden behind a single IP address, given that the total number of simultaneous connections is less than 65 K, since one connection takes one external port and port numbers are 16 bits wide. The proliferation of NATs enabled the whole Internet to continue growing at a much higher rate than the actual consumption rate of the IPv4 address space.

However, NAT does not come free. When internal nodes are using application protocols that negotiate the IP address and/or port numbers within the application protocol, then the application in server S will receive the private address of the node, not the public translated address by the NAT. The server application will then reply to the private address which is not reachable and routable from the public network. Therefore, the application does not work. For example, FTP [RFC959] with its separate control and data connections does not work if a NAT is in the path between the server and the client.

To overcome this limitation, the NAT must understand each application protocol that traverses it, inspect each application payload and modify the application payload to replace the private source address and port number by the external source IP address and port number.

This processing at the application layer is called an application level gateway (ALG). Every NAT implementation includes a FTP ALG to enable this widely used protocol to traverse NATs. An ALG does not work if the application payload is encrypted or integrity protected by the application protocol or by a layer below such as IPsec.

Moreover, when the IP header itself is integrity protected, for instance with the IPsec AH mode, then the translation of the source IP address and port number destroys the integrity protection.

NAT and its side effects are discussed more throughout this book.

1.1.4 HTTP Version 1.1 Virtual Hosting

The simplicity of HTML and Web servers generated a lot of interest in the 1990s when everyone wanted to have their own Web server. This resulted in a very rapid growth of Web servers. Version 1.0 of the HyperText Transfert Protocol (HTTP) [RFC1945] required each Web site to have a specific public IP address. To aggregate resources, many Web sites are hosted on the same server, requiring the operating system to support multiple IPv4 address on the same interface, usually named secondary IP addresses.[4] This increased the consumption rate of IP addresses.

Version 1.1 [RFC2068] of HTTP supports virtual hosting, where multiple Web sites with different domain names (http://www.example1.com, http://www.example2.com) are served by the same IP address. A version 1.0 HTTP client sends only the path at the right of the domain name (for example: path=/a/b.html of the full URL: http://www.example1.com/a/b.html) to the HTTP server. A version 1.1 HTTP client sends the full hostname to the HTTP server (for example: http://www.example1.com/a/b.html), enabling the HTTP server to forward appropriately the request to the proper Web site handler. With version 1.1 of HTTP, the Web server now needs only one IP address to serve a virtually unlimited number of Web sites.

However, with this virtual hosting technique, IP filtering based on the address of the destination Web server is nearly impossible, since all Web sites share the same IP address. The filtering has to be done at the application level, requiring filtering devices to open the packet payload to inspect and parse the HTTP statements in order to identify the target Web server, which creates more burden on security gateways.

Compared to HTTP version 1.0, HTTP version 1.1 conserves the public IP address space by enabling virtual hosting.

In a typical enterprise scenario, the enterprise needs only two IP addresses: the external address of the NAT hiding its internal network and one address for all its Web sites. This created the defacto ISP practice to provide only four IPv4 addresses to organizations. The bad side effect is that the organizations now have to justify the need for more than four IPv4 addresses, moving the burden of allocation and usage of the IPv4 address space to the organization. Does your organization have to justify the need for more than four telephone numbers?

1.1.5 Variable Length Subnet Mask

In organization networks, the original IPv4 design requires a single subnet mask throughout the network. An address plan identifies the subnet mask by finding the largest possible

[4] Secondary IP addresses were not supported on many OS at that time, which gave more headaches to network managers who hosted many Web sites.

number of hosts on a single subnet and using the according bit-level subnet mask. If the largest number of hosts on a single subnet in a network is 65, then the network will have a 7 bits subnet mask ($2^7 = 128 > 65$), which enables 128 hosts on each subnet. A subnet of 3 nodes in that network consumes 128 IPv4 addresses. A single subnet mask for the whole network decreases the efficiency of the IP address utilization.

To make address plans less sparse, resulting in conservation of address space, the variable length subnet mask technique (VLSM) [RFC1812] was introduced for routers and routing protocols. With VLSM, the routing infrastructure can handle a specific subnet mask on each subnet. Routing protocols such as the Routing Information Protocol (RIP) [RFC1058] had to be updated to support VLSM.

1.1.6 Classless IPv4

Introduced to reduce the growth of the global routing table, the Classless Inter-Domain Routing (CIDR) [RFC1519] converts the classful IPv4 address space into a classless address space. In the classless model with CIDR, any network size is possible using any address number. The A class (from 0.0.0.0 to 127.255.255.255 with 24 bits each), the B class (from 128.0.0.0 to 191.255.255.255 with 16 bits each) and C classes (from 192.0.0.0 to 223.255.255.255 with 8 bits each) are no longer relevant. Therefore, the address consumption rate is decreased since allocation will be more effective.

Owing to the variable size of network prefixes, the /n notation (often named CIDR notation) after the prefix was introduced to make IP address writing notation shorter and more efficient. For example, '/25' in 192.0.2.0/25 identifies the number of significant leftmost bits in the address. Therefore, '/25' gives a $32 - 25 = 7$ bits prefix range, which gives $2^7 = 128$ addresses. '192.0.2.0/25' defines the range of addresses between 192.0.2.0 and 192.0.2.127. The CIDR notation is the only notation used to describe ranges or prefixes of IPv6 addresses and is the one used throughout this book for both IPv4 and IPv6.

1.1.7 Provider-based Assignment and Aggregation of IPv4 Network Prefixes

Since the beginning of IPv4, organizations had been requesting IPv4 address space directly from IANA and IANA assigned a IPv4 address range of one class. This range was assigned to this organization permanently. These assignments were not related to the topology of the network, disabling any aggregation of the prefixes by a common provider. This is another cause of the growth of the global routing table.

In 1994, the policy of assignments changed. The new policy enforces provider-based assignments to the organizations. Now, IANA assigns blocks of addresses to regional registries (ARIN, RIPE, APNIC and others). Within their assigned blocks, regional registries assign smaller blocks of addresses to providers, which in turn assign smaller blocks to organizations. In this context, organization's prefixes are aggregated at the provider level, resulting in a more aggregated global routing table and decreased rate of growth of the table. This aggregation also has the benefit of more stability in the routing table, since organization's prefixes are not specifically announced in the global routing table. Since the leaf of the network (such as the link connecting the organization to the provider) is likely to be less stable, leaf announcements result in BGP updates in the routing table. On the other hand, the links connecting providers to exchange points are likely to be more stable.

An important side effect of the provider-based assignments is that the address space assigned to an organization is not owned by that organization but by the provider. If the organization changes provider, the organization receives a different address space from the new provider and will not be able to use the previously assigned address space anymore, since it is owned by the previous provider. This results in a renumbering of its entire network. Since IPv4 was not designed with mechanisms to facilitate renumbering, this situation results in huge trouble and inconvenience for organizations. Since they are locked by the address space of the provider and the cost of renumbering is high, the organization is locked to the provider. To overcome this big issue, organizations started to limit the use of public addresses to enable smoother change of providers. The limitation is accomplished by using a NAT at the edge of the network and limiting the number of servers using global address space.

1.1.8 Constrained Allocation Policy of IPv4 Addresses

To further conserve IPv4 addresss space, successive versions of IPv4 address allocation policies and guidelines [RFC1466, RFC2050] were put in place by the IANA and the registries. These policies had the following goals [RFC2050]:

- conservation of address space to maximize the lifetime of the IPv4 address space;
- hierarchical routing for routing scalability on the public Internet;
- public registry.

Policy RFC2050 allocates IPv4 addresses space in smaller chunks to providers in a slow-start procedure. ISPs are asked to document the address assignments to the end organizations. The result is a slower rate of consumption of the IPv4 address space.

1.1.9 Global Routing

Figure 1.6 [Huston, 2005] shows the size of the IPv4 BGP global routing table.

Despite the use of NAT, HTTP virtual hosting, VLSM, CIDR, provider-based aggregation and constrained address allocations since the mid 1990s, the growth of the global routing table has been mostly linear, while the slope is increasing in recent years. The growth, due mainly to the increase of small prefixes (/24), comes from the growth of the Internet itself, the use of multihoming and traffic engineering techniques using routing [Huston, 2001].

1.1.10 Summary of Internet Growth

Whatever metric one take and despite all the invented solutions mentioned above, Internet growth is heading towards the exhaustion of IPv4 addresses in a few years and to ever increasing large global routing tables. A major fix to these issues must be deployed soon and IPv6 is the only solution currently worked out.

Let's now take another perspective by discussing other current issues in IPv4 networks.

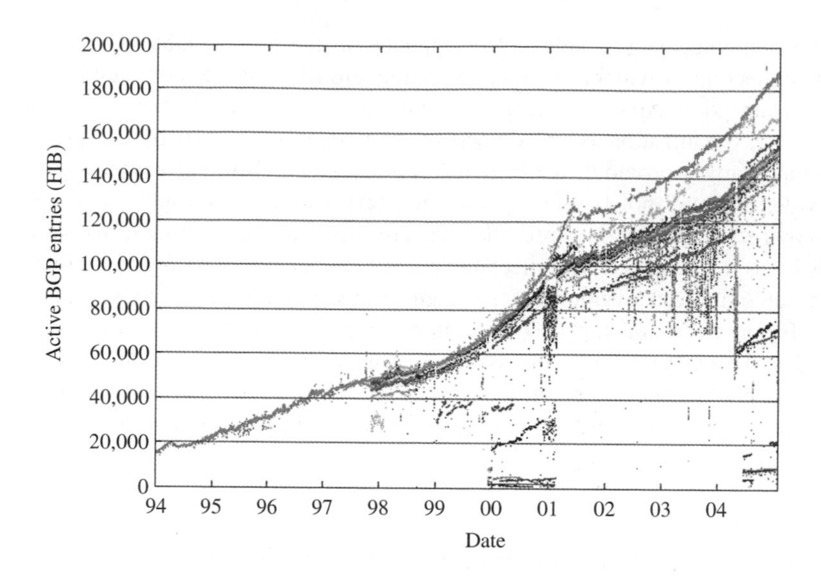

Figure 1.6 IPv4 BGP Global Routing Table Size

1.2 Real Issues and Trouble with IPv4

The shortage of IPv4 addresses is responsible for many issues and trouble in IP deployments today. Real world issues described in this section show the hidden costs of the lack of IPv4 addresses and the lack of functionalities in IPv4 for the current and future use of IP networking.

1.2.1 Deploying Voice over IP

Skype [Skype] is a Peer-to-Peer (P2P) Voice over IP(VoIP) application and network. The Skype designers claim to traverse any NAT or firewalls to achieve P2P. Since the Skype protocol is not publicly disclosed, researchers have analyzed the protocol and described the process to traverse NAT and firewalls [Baset and Schulzrinne, 2004].

In a nutshell, a Skype client knows in advance some Skype gateways, named supernodes, and discovers others that help (the client) to find its external IPv4 address. The authors of the analysis think this technique is similar to STUN [RFC3489] and TURN [Rosenberg, 2004], discussed more in Section 1.3.1. The client tries to connect to the gateways using UDP; if unsuccessful, it tries TCP; if unsuccessful, it then tries TCP on the HTTP port (80); and if unsuccessful, it tries TCP on the HTTPS port (443). Since HTTP ports are usually opened for outgoing connections in most organization networks, Skype uses these ports as a last resort to traverse the firewall. When this last resort does not work, Skype loops again twice more and if still not successful, finally gives up.

In most cases, the voice traffic between the two VoIP peers goes through two other nodes, named supernodes, as shown in Figure 1.17. This figure shows the Skype network where small dots represents VoIP end-users. Supernodes are normal Skype nodes elected to be intermediary nodes, shown as bigger dots in Figure 1.7.

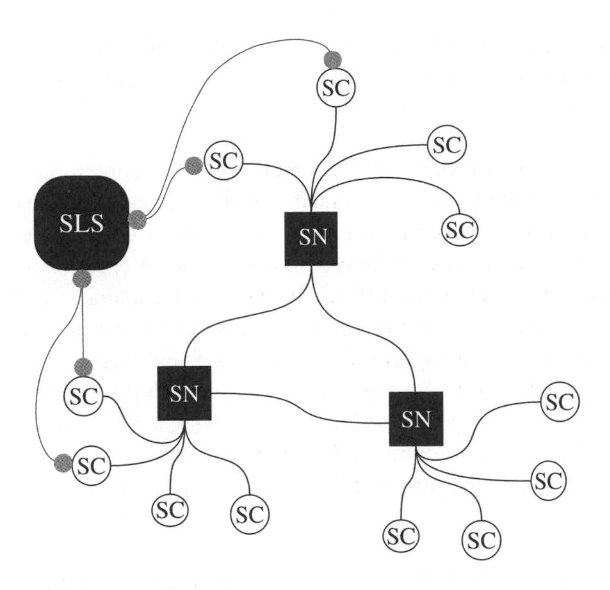

SN: Skype supernode
SC: Skype client
SLS: Skype login server

Figure 1.7 Skype overlay network

The rationale behind electing supernodes in the Skype network is to enable fully automated NAT traversal without a specific network of servers. Any Skype node can be automatically elected as supernode if the node has a public address and some available CPU resources.

True peer-to-peer is almost impossible on IPv4 networks because of the presence of NAT. This Skype 4-nodes routing between the two peers through the supernodes obviously introduces delay, jitter and less performance. Also, since the election process to be a supernode is not managed by the user, a user bandwidth might be filled with the traffic of others, just because its computer was elected as supernode on the Skype overlay network. As a claimed peer-to-peer protocol, Skype is not a peer-to-peer protocol, not because of its authors failing to design a true peer-to-peer protocol, but because the deployed IPv4 networks disable peer-to-peer applications.

This demonstrates how unpredictable it is for an application to get the needed basic IP connections. It also demonstrates how the applications are involved in delivering basic connections, which should be handled by lower layers such as IP and transport.

Owing to the current issues with IPv4, applications have become much more complex just to reach other peers. The IP architecture was based on the end-to-end principle, where the network will be 'dumb' and end nodes can be reachable directly and easily. Current IPv4 networks have too much processing and break the end-to-end direct connectivity. Skype smarts or similar will need to be implemented in all applications requiring end-to-end connections, which makes applications complex, provokes latency in the application and network, makes application fragile to any network change and makes the overall connectivity unpredictable.

For the same reasons, an industry leader and well-known developer of open-source software stopped the development of a VoIP peer-to-peer application named SpeakFreely. Here is a excerpt from his announcement:

> The Internet of the near future will be something never contemplated when Speak Freely was designed, inherently hostile to such peer-to-peer applications. I am not using the phrase 'peer to peer' as a euphemism for 'file sharing' or other related activities, but in its original architectural sense, where all hosts on the Internet were fundamentally equal. Certainly, Internet connections differed in bandwidth, latency, and reliability, but apart from those physical properties any machine connected to the Internet could act as a client, server, or (in the case of datagram traffic such as Speak Freely audio) neither – simply a peer of those with which it communicated. Any Internet host could provide any service to any other and access services provided by them. New kinds of services could be invented as required, subject only to compatibility with the higher level transport protocols (such as TCP and UDP). Unfortunately, this era is coming to an end.
>
> [Walker, 2004]

It is terrible that the well designed IP protocol that offered so much innovation in its first 20 years is now stopping innovation, because of the introduction of NAT in the network.

SIP [RFC3261] is the IETF standard protocol for VoIP. SIP was designed for a pure IP network without NAT. It works fine only when no NAT is present between the peers. The pervasive presence of NAT means that SIP and its related protocols such as RTP are not deployable as in the current IPv4 networks with NAT dominance. These protocols have been augmented by various NAT traversal techniques. However, none of these techniques take care of all cases unless the audio path goes through a gateway, which disables the essence of VoIP performance which is to carry voice over IP on the direct path between the two peers.

I've been using a SIP softphone with a VoIP provider on my laptop. This SIP software implements most of the NAT traversal techniques. In many cases, VoIP calls just do not work. This software shows me the following error message: 'Login timed out! Contact Network Admin.' Very useful message! A few times, the error message was: 'Cannot identify the Cone NAT correctly'. Another very useful message for an end-user.

As a user, I have no clue what kind of NAT, firewall or other network devices are in the path. So I called the VoIP provider technical support. As a technical person, I investigated and discovered the situation prior to the support call, but I took the 'dumb' user hat when calling technical support. After literally one and half hours of support over the phone (not the VoIP one obviously, but a plain old one), I had escalated two levels of technical support. It was suggested that I reinstall the SIP software and change the configuration of my operating system, which I did to the point where they asked me to reinstall the whole operating system, which I refused to do! 90 minutes of technical support, no good answer was given, the service was not restored and I was a frustrated customer. Why? There was a symmetric NAT in the path.[5] This kind of NAT is not supported by the NAT traversal techniques used by the SIP software. If most users of a VoIP deployment started calling the technical support and spent 90 minutes each, the VoIP company would go bankcrupt pretty fast!

These few examples show that VoIP is difficult to deploy over IPv4 networks because of NAT. True peer-to-peer is no longer possible. Innovation is hindered by current IP networks.

[5] Various kind of NATs are discussed in the book Web site: http://www.ipv6book.ca.

We need to restore this network to a good state in order to maintain innovation, user confidence and good experience. IPv6 provides the features needed to deploy applications seamlessly.

1.2.2 Deploying IP Security

The IP protocol did not have any widespread security at the IP layer. Over time, security was added at the application layers, such as the secure socket layer (SSL) for the Web. Right now, we have similar and duplicated security functionality in several application protocols, creating a whole set of new problems, such as multiple, different and incompatible key management functions.

While discussing requirements of IPv6, the IETF decided to work on an IP security layer, named IPsec [RFC2401], to protect the whole IP packet for authentication, integrity protection and confidentiality. IPsec (see Chapter 13) is available for IPv4 as an option and mandatory for IPv6. By protecting the IP layer, the application layers over IP do not usually need additional protection.

However, the deployment of IPsec on current IPv4 networks have shown the difficulty of protecting IP packets when NATs are in the path. IPsec[6] protects the whole packet, so any modification of the packet between the source and the destination violates the security of the packet. NAT modifies addresses and port numbers of IP packets, therefore disabling the full protection of the IP packet, and disabling full security deployment.

Since IPv6 does not need NAT, full end-to-end IP security is deployable without those issues.

1.2.3 Deploying Application Security

An enterprise has setup an e-commerce Web site with connection to its internal SQL database located in its private network. The Web site server is reachable from the internet. As shown in Figure 1.8, the connection from the Web server to the internal SQL database goes through the firewall which also implements NAT.

The SQL connection protocol between the Web server and the database backend negotiates IP addresses and port numbers within the protocol. So, by default, it does not traverse a NAT. The NAT-firewall product supports this protocol by inspecting the exchange and replacing the IP addresses and port numbers by the translated ones, within the application payload. This makes NAT and the SQL connection work. However, the organization then wants to encrypt

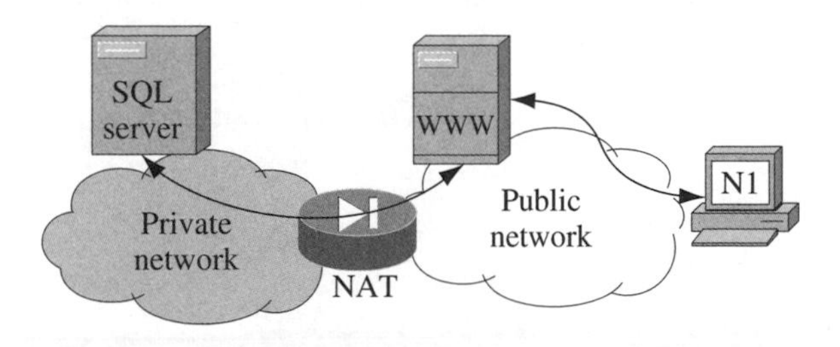

Figure 1.8 Backend database connection to Web site

[6] More specifically the Authentication Header (AH) mode, see Chapter 13.

the data in the Web to SQL connection to prevent any snooping of the confidential data. By turning on encryption in the application protocol, the NAT-firewall is then unable to inspect and replace the IP addresses. The SQL encrypted connection does not work across NATs.

NAT disables the use of security in application protocols.

1.2.4 Videoconferencing

A school board had to deliver a videoconferencing solution to help students in schools in remote communities to have access to professors in cities. Any professor from any school can give the course and any remote school class can attend the videocast. Remote students interact with the professor as if they were in the physical class.

Figure 1.9 shows the network where videoconferencing stations are located in remote networks. Multiple NATs are in the path between any combination of stations. The video feed station can be in any class and all sites are actively participating in the videoconference.

The videoconferencing software did not work by default in this configuration, since all the stations were hidden to the others by private address space. For any class, the teacher had to make a request to the IT department one week in advance, so that the IT department could configure a static mapping of addresses for all the NATs in the path and then configure the videoconferencing stations to use this mapping. This manual process of the IT department hindered the capability of the teachers to use the service as a commodity service. It is not the fault of the IT department, it is the trouble caused by the NATs: the inability to deploy and use applications.

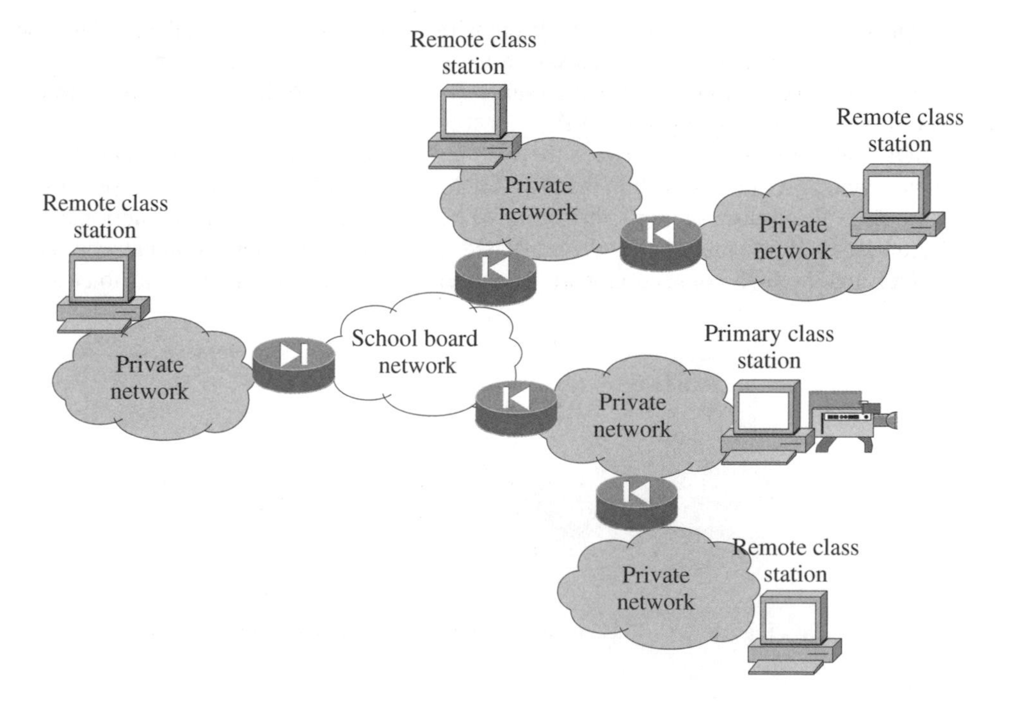

Figure 1.9 Videoconferencing with multiple NAT in remote networks

Deploying IPv6 in this network solves the problem, since all nodes are reachable and videoconferencing is seamlessly working in all cases. A transition mechanism[7] can be used to enable IPv6 in the whole network before a full upgrade becomes possible.

1.2.5 A Simple Web Server at Home

With digital cameras and powerful computers, people have a library of digital pictures on home computers connected to a home network with broadband connectivity. They would like to share these pictures with friends and family that could access them remotely from the Internet. Available broadband bandwidth makes this possible.

Figure 1.10 shows a home network with the computer that has the picture library and runs a Web server.

This Web server is not reachable from the Internet because it has a private address hidden by the NAT function implemented in the home gateway. Therefore, friends cannot access the pictures. To solve the issue, the home gateway is configured with a static mapping of the external address to the internal address of the Web server for the HTTP port. If the home network has multiple computers to be Web servers (such as parents and teenagers Web servers), this mapping works only for a single computer. Moreover, it requires some IP networking knowledge beyond most end-users. There is a good chance that you, the reader, who has good IP networking knowledge, is being requested by the non-techie friends and family to set up these devices! Right? Point taken?

As we can see, the bandwidth, the computer, the data and the software are all available to make this simple application possible. What disables the application from actually working is the NAT.

Home gateways have evolved into very complex devices these days. It is common to receive a 200 page user's manual, discussing a lot of complex IP configurations. A majority of these pages and the overall complexity come from the NAT presence and its related issues.

By restoring reachability, IPv6 in this home network setup will make the application work seamlessly.

1.2.6 Using Remote Procedure Calls

Remote procedure calls (RPC) are used for distributed computing, where applications developers have access to an application programming interface (API) to access services on remote computers by a simple function call. RPC processes in the distributed network talk to each other by exchanging their IP addresses and by dynamically allocating ports.

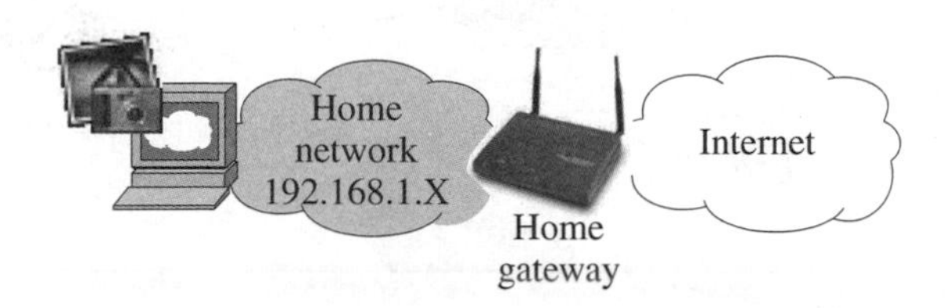

Figure 1.10 Home network

[7] In this case the TSP tunnel broker, see Section 16.2.9.

When a NAT is in the path between the computers, RPC no longer works. Very complex configurations prone to any changes in the network can overcome some of the simple setups, but then do not scale well and maintain states in the network.

Distributed computing networks are difficult to deploy over IPv4: IPv6 deployment solves this RPC problem right away.

1.2.7 Remote Management of Applications and Servers

Many organizations are outsourcing their IT services to a third party. This third party organization usually sets up a network operations center (NOC) to manage remotely the servers, networks and applications of its multiple customers. As shown in Figure 1.11, the NOC is connected to the customer's networks through private networks or the Internet.

However, many organizations have one or many NATs in their internal network. The remote management station in the NOC cannot reach the servers behind NAT. One has to define static translations on all the NATs to make this work, when possible. Even if they make it happen, a lot of static configuration is introduced in the NAT network, where any fault NAT will make the network unreachable by the NOC. The NOC is responsible for troubleshooting and keeping the network running, while it has no tools to manage the network! The support organization can then not deliver its service level agreement conditions such as 99.99% uptime, since the NOC cannot manage the network. This has nothing to do with security and firewalling, but rather lack of address space and the related presence of NAT.

Using IPv6 in this scenario solves the reachability issue. Moreover, it will enhance security since in all paths, end-to-end security and end-to-border security can be established.

1.2.8 VPN Between Same Address Space

Many organizations, subsidiaries within an organization, divisions within an organization or recently merged or acquired organizations have separate IT departements. For that matter,

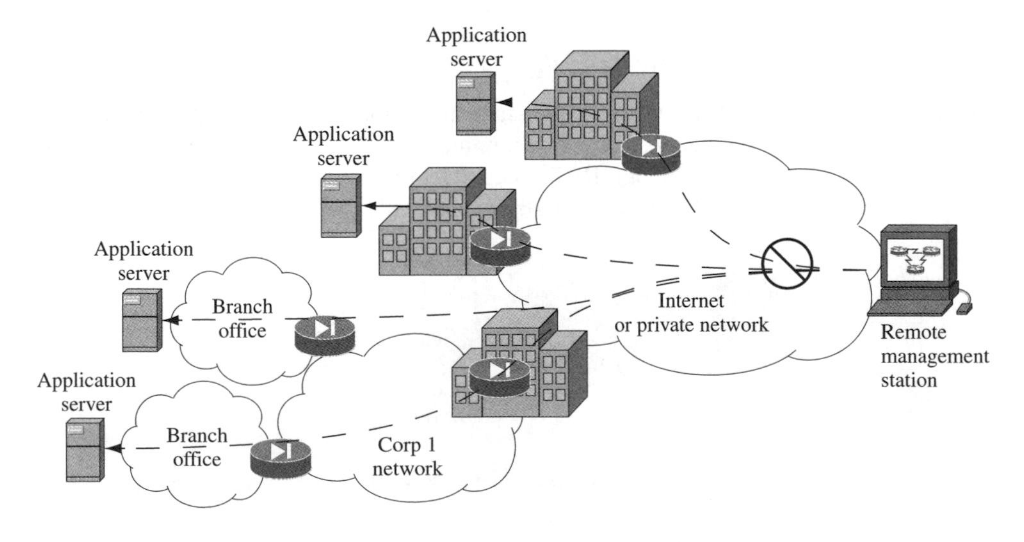

Figure 1.11 Remote management of servers in private networks

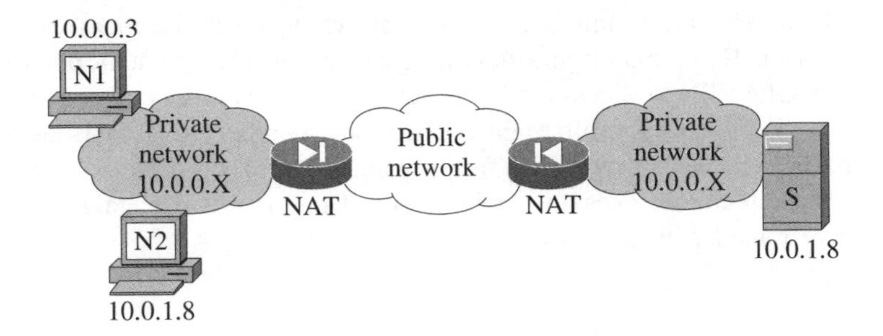

Figure 1.12 VPN between same address spaces

they manage their own address space. In nearly all cases, each network uses the 10.X.X.X private address space. When two or more of these networks are connected together, an address collision happens.

Figure 1.12 shows a simple case of this situation. A VPN is created between the two NATs, at the borders of each network. When N1 needs to reach the server S in the remote network, it sends the packet to S address: 10.0.1.8. However, the packet never reaches S but instead reaches N2 in the same network as N1.

To overcome this situation, one defines address views for each host available to the other network, creating a large static map on both NATs (often called double NAT), not dynamically managed as DNS and routing are good for. When a user calls the IT tech support for a problem reaching the other side, it is very hard to troubleshoot because of this double NAT process. This situation creates network management and support costs and does not scale well. The alternative is to renumber one of the networks, entailing important work and causing downtime on the network.

With its huge address space, IPv6 does not have these address collision issues.

1.2.9 Deploying Services in the Home Network

Figure 1.13 shows an example of a remote monitoring service in the home network, where network cameras are placed in the home. The owner, while out of his home, wants to see what is happening in the home using its PC or its graphical cell phone.

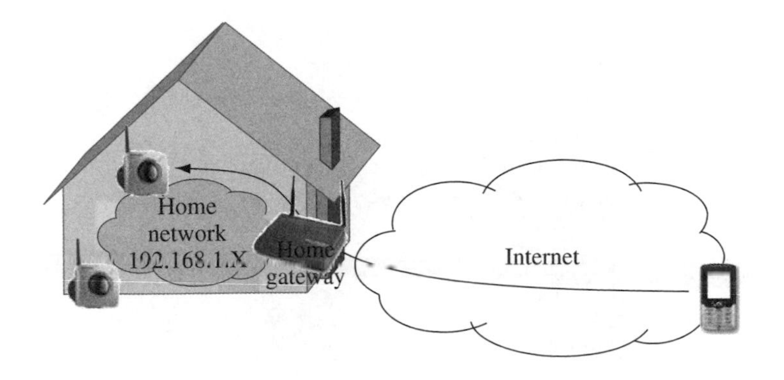

Figure 1.13 Video monitoring in a home network

With multiple cameras in the home, there is no simple way to reach all of them from a device on the Internet. One would have to configure the home gateway with multiple static port translations, or to create a specific VPN to access the home network. The more services one has at home, the more these tricks become painful. Many applications are related to sensors and appliances in the home that are accessed or controlled from outside the home network.

With IPv6, these home services are straightforward to enable, manage and use, because reachability is restored by the addressing.

1.2.10 Merging or Connecting Two Networks Together

When two organizations merge or connect their networks, their address spaces collide because both usually use the same address space 10.X.X.X. Figure 1.14 shows such a situation.

This creates similar problems to the one discussed in Section 1.2.8. With a large address space, IPv6 addressing in networks will not collide when networks connect or merge.

1.2.11 Large Networks

For some large corporate networks, given the non-optimized allocation of address space with subnet masks, the private address space 10.X.X.X is just not sufficient for their numbering [Hain, 2004]. As the networks expand, they need to have more address space. Many organizations are using non-allocated address space such as 1.X.X.X for their additional address space. As we can see, large networks need more private address space than is available in IPv4.

On the other hand, IPv6 has sufficient public and private address space to support these scenarios.

1.2.12 Address Plans and Secondary Addresses

An enterprise address plan identifies a subnet mask for each link, which establishes the maximum number of nodes on that link. When more hosts than the maximum are put on one link, a second prefix is used on that link. Figure 1.15 shows such a situation where the initial prefix is 192.168.1.X/24. Among the computers on that link, N1 has 192.168.1.2 and the router has 192.168.1.1.

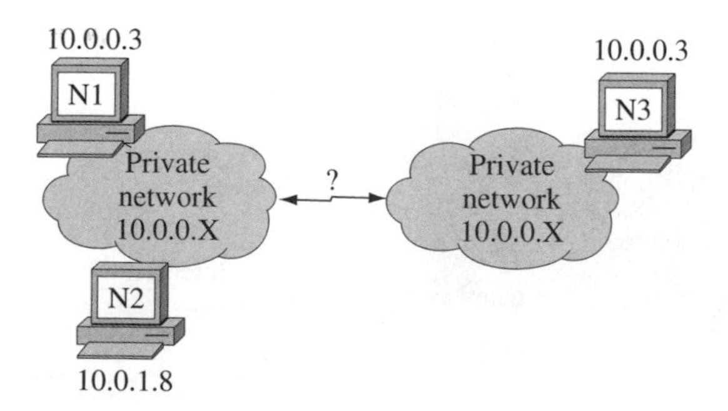

Figure 1.14 Merging two networks with the same address space

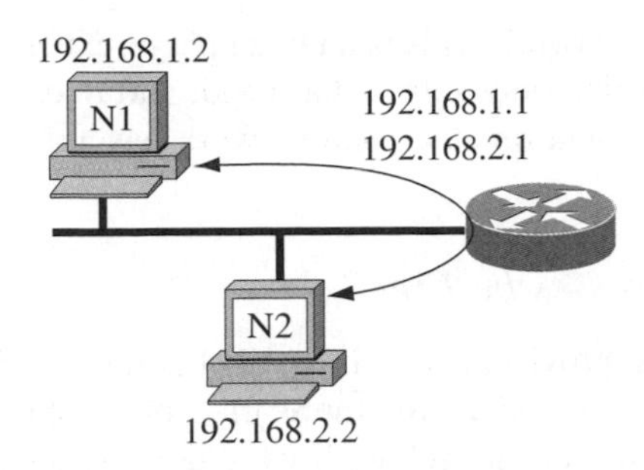

Figure 1.15 Secondary address traffic

When the maximum number of nodes is reached, then another prefix (192.168.2.X/24) is added on the link. The router now has an additional address, 192.168.2.1, and N2 is part of the second prefix (192.168.2.2). For simpler network management, nodes do not participate in routing. To send a packet to N2, N1 finds that N2 is not on the same link, given that it does not have the same prefix. N1 sends the packet to its default router and the router resends the packet on the same link to N2. So all communications between the two nodes are duplicated on the same link, adding delay and decreasing the available bandwidth by half.

With virtually unlimited numbers of addresses for nodes on a link, IPv6 does not suffer from this behavior.

1.2.13 Provider VPN Address Collisions

Nowadays, most organizations are using 10.0.0.0/8 address spaces inside their corporate network. When a provider offers the VPN service to its enterprise customers, the routing inside the VPN core carries many 10.0.0.0/8 routes originating from different networks, as shown in Figure 1.16.

This address collision in the routing table makes the routing incoherent and exposes the organization's networks to others within the provider network. To overcome this problem, a route distinguisher is added to the routing protocols to identify uniquely each organization

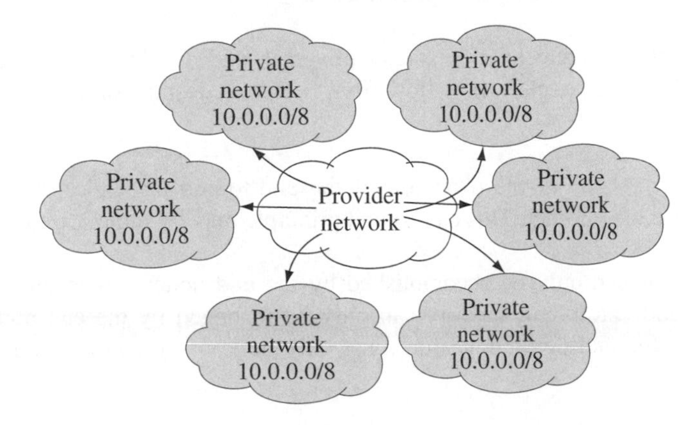

Figure 1.16 Provider VPN address collision

network inside the provider network. This is an example of extending the IPv4 address space for a very specific need, which only solves that need, and does not help with other issues. IPv6 does not have that problem because each network has a unique large address space.

1.2.14 Should IP Addresses be Free?

In many markets nowadays, providers are billing IPv4 addresses. For example, some broadband providers are asking a premium for more than one public IP address to the home networks. This cost is a result of the lack of IPv4 address space and also the lack of functionality in the IP framework to deliver a range of addresses to a large number of networks, such as home networks. By restoring address space and providing prefix delegation methods, IPv6 solves these issues and IPv6 addresses should be free. [8]

1.2.15 Summary

Each example above indicates the costs and issues, often combined together, related to IPv4 networks today. The mitigation techniques used to keep IPv4 up and running were also discussed. This set is just a sample and many other instances exist. For each example, a possible 'workaround' might exist, but these workarounds combined together create an important network management problem.

Moreover, the end-to-end reachability is now lost, disabling innovative applications and security to be deployed.

Many issues are related to the existence of NAT. Either we exacerbate the problem by continuing to procrastinate, incurring more and more costs, or we solve the problem by deploying IPv6. With the right transition tool, the deployment of IPv6 costs less than the current visible and hidden costs of NAT.

1.3 Architectural Considerations

IP architecture considerations are at the core of the issues facing IPv4 today.[9] In architecture terminology, IPv4 has lost transparency, defined as:

> the original Internet concept of a single universal logical addressing scheme, and the mechanisms by which packets may flow from source to destination essentially unaltered. [RFC2775]

Transparency is related to the existence of the end-to-end principle, at the core of the design of IPv4 and the Internet. This end-to-end principle may be summarized as [RFC2775]:

- Certain functions can only be accomplished by the end nodes. For example, failures in transmission and end-to-end security can only be managed by the end nodes. As such, state of the end-to-end communication must only be kept by end nodes and not by the

---

[8] Apart, that is, from some fees paid by the providers to the registries for the registry own operations. However, these fees are near to zero when shared over all the provider's customers.

[9] This section is based on RFC2775 and RFC2993, very good documents to read for a more exhaustive description.

network. The network is enabled to re-route packets transparently and efficiently, since no state is kept in the network.
- Transport protocols are designed to provide the required functions over a non-guaranteed IP network. Enhancements [RFC2581] were also integrated in end-nodes to better manage congestion.
- Packets can flow unaltered throughout the network and IP addresses are used as unique labels for end systems.

Implications of NAT in the network are illustrated by the following issues [RFC2993]:

- NAT is a single point of failure. Since a NAT keeps state, any failure of the NAT requires that all the current connections of all nodes behind the NAT be re-established.
- Application-level gateways(ALG) are complex. ALG are used in NAT devices to inspect application protocol packets to modify them on the fly. Any application requires a synchronization of all ALG in the field to support the deployment of the application.
- NAT violates TCP states. TCP states are defined for end nodes to manage the connections. A device in the network that is assigning transient addresses and ports without managing TCP states will collide with non terminated TCP connections.
- NAT requires symmetric state management. In the event of link flappings, multiple NATs must be fully synchronized in real time in order to keep the state of connections and address and port assignments.
- NAT disables the use of a global name for advertising services. NAT hides devices such that services behind cannot be advertised in the DNS to be accessed from anywhere.
- Private address space used for VPNs are colliding. L2TP tunnels and other VPN technologies enable networks to be connected together. However, the address spaces usually collide since private networks use the same 10.X.X.X address space.
- Correlation in network events is difficult. Since source addresses are changed on the fly, correlation of network events based on IP address becomes a huge problem since it requires the dynamic state of translation of all the NATs in the path being saved and then correlated by some qualitative heuristic.

During an IETF plenary session, Steve Deering, primary author of multicast [RFC1112] and IPv6 [RFC1883], described the initial IP architecture model as an hourglass. The following figures are from his presentation [Deering, 2001]. The initial and true model of the Internet Protocol is shown in Figure 1.17.

The hourglass architecture was based on the following design criteria:

- An internet layer to
  - make a bigger network (than a layer 2 layer such as ATM);
  - provide global addressing (instead of local addressing which makes connecting networks very difficult);
  - virtualize the network to isolate end-to-end protocols from network details/changes.
- A single internet protocol to
  - maximize interoperability;
  - minimize the number of service interfaces.
- A narrow internet protocol that
  - assumes least common network functionality to maximize the number of usable networks.

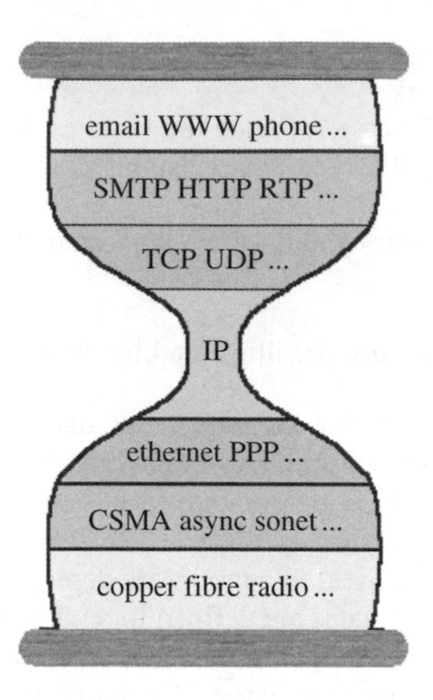

Figure 1.17 IP hourglass model

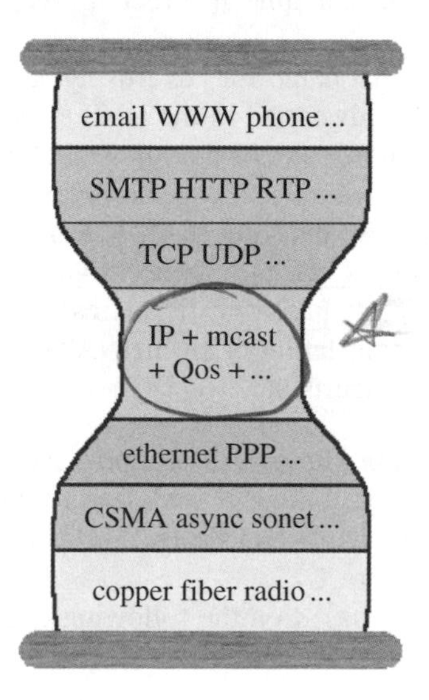

Figure 1.18 IP hourglass architecture fattening

Over time, this model got 'fatter on the waist', as shown in Figure 1.18, by adding additional services to the IP protocol itself, such as multicast, QoS, security, MPLS, L2TP, and others.

This fattening requires additional functionality from the underlying layers, which makes these new functionalities more difficult to deploy. Moreover, the introduction of network address translation (NAT) and application level gateways (ALG) broke the IP model as shown in Figure 1.19. With these middle boxes, state management is introduced in the network and behavior is unpredictable.

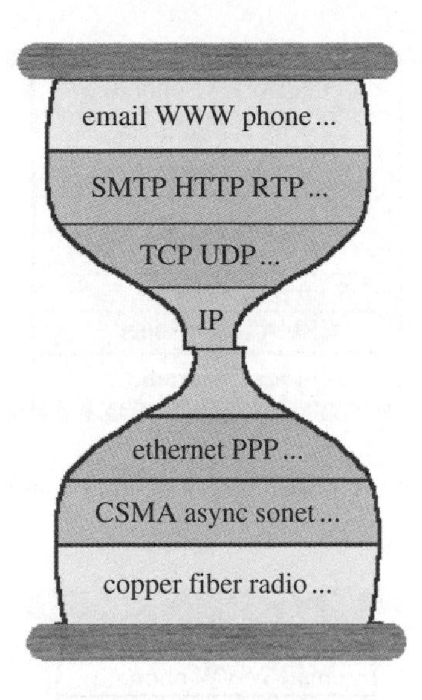

Figure 1.19 NAT and ALG breaking the IP hourglass architecture

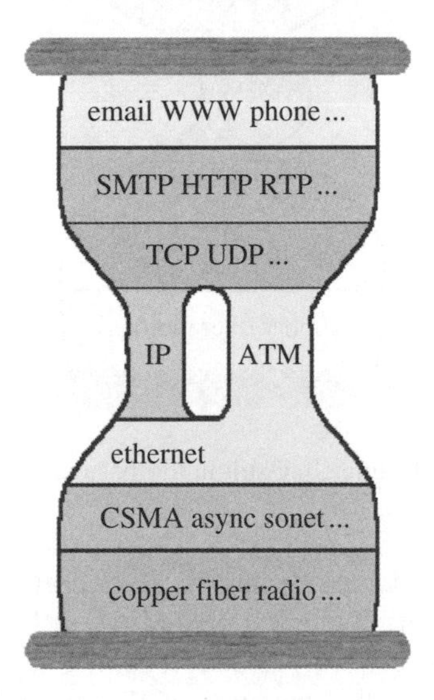

Figure 1.20 ATM replacing IP in the hourglass architecture

Asynchroneous Transfer Mode (ATM) tried to become a layer 3 protocol, as shown in Figure 1.20, but was unsuccessful.

Eventually, the IP layer might become overloaded making the architecture too fat, as shown in Figure 1.21.

On the positive side, for the architecture, we have used IP tunneling to overlay networks, as shown in Figure 1.22.

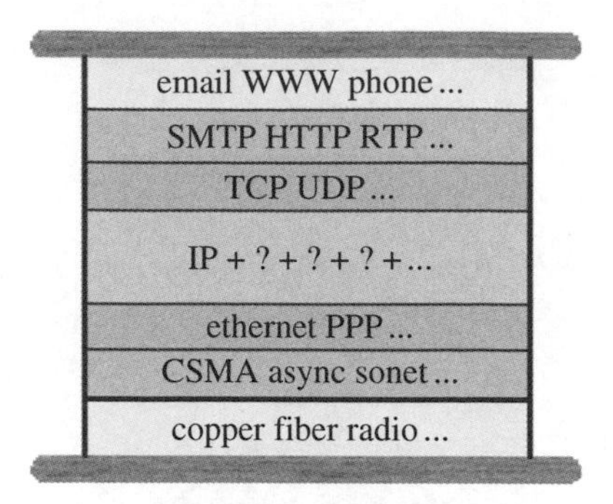

Figure 1.21 IP overloaded and IP hourglass architecture too fat

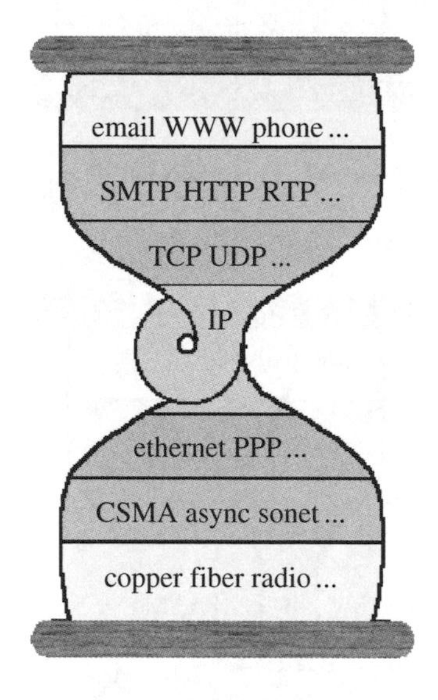

Figure 1.22 IP tunneling still in the IP hourglass architecture

The IP tunneling techniques do not change the IP hourglass model, since the layering is preserved. Such a technique is used to encapsulate IPv6 in IPv4 packets, enabling fast IPv6 deployment over current IPv4 networks, as discussed in Chapter 16.

Figure 1.23 shows the dual-stack integration of IPv4 with IPv6 where the two are used together. This is the main path we are taking towards deploying IPv6, as discussed throughout this book. However, it introduces two service interfaces, requires changes on both upper and lower layers and is not interoperable.

Figure 1.24 shows the target architecture where IPv4 is replaced by IPv6 to restore a thin layer 3 architecture, leaving the lower layer to handle the wires and the upper layers to handle the application requirements.

This final target would fully restore the initial IP architecture, leaving the maximum of flexibility for transport and application layers. It is hoped that, at some point in time, we will drink the wine!

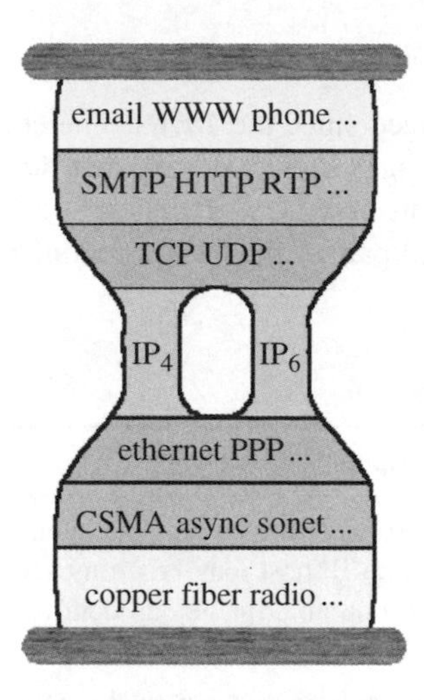

Figure 1.23 Dual stack in the IP hourglass architecture

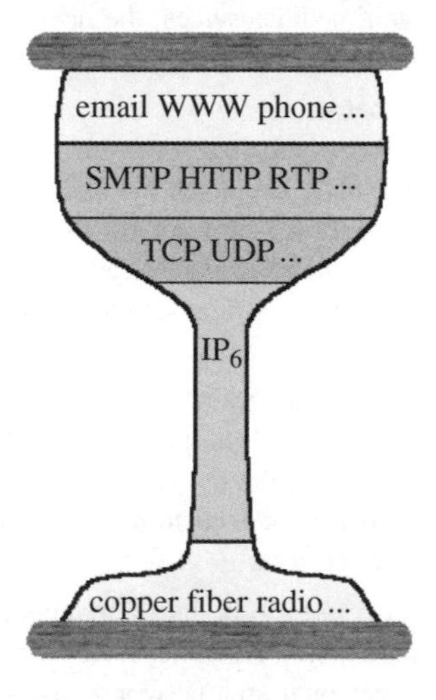

Figure 1.24 End target with IPv6 for the IP hourglass architecture

1.3.1 Network Address Translator Variations

Worst of all, the engineering community have found that NATs have a wide variety of behaviors[RFC3022, RFC3027, RFC3235]. This situation became apparent while application protocol designers were defining techniques to overcome NAT limitations in their respective application protocol. These techniques were based on the early identification of NAT laziness to make the translation of IP addresses and ports as precise as possible. Therefore, various techniques of NAT-traversal [RFC3489, Rosenberg, 2004] have been designed, none of which works in all cases and seamlessly enables the application. This single issue might by itself create sufficient network management costs to justify a full and rapid deployment of IPv6. The book Web site contains a section describing the variety of NATs.

1.4 Paradigm Shift

The networking world has changed since the 1970s. The concept of hosts and networks has changed. Table 1.2 lists the changes in networking that demonstrates the paradigm shift. It also lists the requirements for this new networking.

The listed requirements are all part of the design features of IPv6.

Table 1.2 Changes in networking

Past role	New Role	Description	Requirements
Host	Node or device	A host used to be a big fixed computer. Now, an IP host may be a tiny sensor, a control on an airplane, an identification tag on a cow or a videocamera.	IP efficiency. Autoconfiguration.
Host	Server	With the advent of personal computers and client-server models, the hosts became clients. Nowadays, with VoIP, peer-to-peer and multimedia services, the host is offering services to the network, as an edge device.	IP reachability. Large address space.
Host	Router	Where before a host was present, now it is a router. With personal area networks, one brings a PDA, a cell phone and a laptop: one of them is the router for the others. With broadband access to the home, where once a single host was attached, now it is a router with a network.	Automated routing and router configuration. Network prefix delegation.
Static	Mobile	With Wifi, 3G and other wireless technologies, and the pretty small devices existing today, the devices are mobile and society becomes accustomed to using mobility services.	IP mobility.
Network	Unmanaged small network	Networks used to be large, managed by IP routing experts. Nowadays, networks in the home, personal area networks and sensor networks are all examples of small networks not managed by IP routing experts, but mostly unmanaged.	Automated network deployments.
Friendly	Not friendly	Back in the early Internet, there was a good level of trust between the organizations, mainly universities and research centers, connected to the Internet. The requirements for security were basic. Now, the trust that a user should have when connecting to the Internet is probably near zero.	Security

Note: Well, we (the engineering community) try hard to make the home networks unmanaged, but when one looks at the configuration needed on a typical home gateway, we have still a long way to go to make it easy!

1.5 IETF Work Towards IPv6

As discussed in Section 1.1.1, the Solensky study [Solensky, 1990] that demonstrated the exhaustion of class B address space by March 1994 was the first wakeup call in the IETF. Table 1.3 shows the steps taken by the IETF [RFC1752] towards what was first named IPng and then renamed IPv6, when the new IP version number was assigned to 6.

From 1991 to 1995 and parallel with the IETF requirements work thread, many IPng protocol candidates were designed and discussed. Table 1.4 lists the candidates.

Figure 1.25 shows the generation tree of the proposals. Initially, there were six different protocols, and over time, some merged and some were not considered in the final evaluation step for IPng.

Table 1.3 IETF major steps towards IPv6

Date	Step	Description	Reference
August 1990	Predicted exhaustion of class B addresses by 1994	First wakeup call within the IETF. The class B address space could be exhausted in 4 years!	[Solensky, 1990]
November 1991–November 1992	Routing and addressing(ROAD) working group formed	ROAD wg was formed to address the routing and addressing issues.	[RFC1380]
		Recommendations were: CIDR and new protocol through a request for proposal process.	
June 1992	Internet Architecture Board (IAB) recommendation on IP version 7	IAB recommended to use CLNP instead of a new IP protocol. The proposal was named IP version 7.	[IAB, 1992]
July 1992	IETF rejects IAB proposal	During the IETF meeting, the IETF rejects the IAB proposal for IP version 7 based on CLNP.	[IABREJECT]
September 1993	Recommendation to use Classless Inter-Domain Routing (CIDR)	CIDR removes the class structure, enabling more efficient address assignments and aggregation.	[RFC1519]
December 1993	Solicitation for IPng requirements and selection criteria	The IPng area directors sollicit contributions on requirements and selection criteria for the new IPng protocol.	[RFC1550]
July 1994	IPv4 address exhaustion estimated between 2005 and 2011	The Address Lifetime Expectations (ALE) working group was chartered to estimate the remaining lifetime of IPv4 address space. They concluded that the IPv4 address space end of life is between 2005 and 2011.	[RFC1752]

(*continued overleaf*)

Table 1.3 (*continued*)

Date	Step	Description	Reference
August 1994	20 White Papers on IPng requirements and selection criteria	20 white papers were contributed responding to the solicitation [RFC1550], from the following subjects or industries: cable TV, cellular, electric power, military, ATM, mobility, accounting, routing, security, large corporate networking, transition, market acceptance, host implementations and others.	[RFC1667], [RFC1668], [RFC1669], [RFC1670], [RFC1671], [RFC1672], [RFC1673], [RFC1674], [RFC1675], [RFC1676], [RFC1677], [RFC1678], [RFC1679], [RFC1680], [RFC1681], [RFC1682], [RFC1683], [RFC1686], [RFC1687], [RFC1688], [RFC1753]
December 1994	IPng Area Formed	A new area within IETF is formed to manage the IPng effort. The framework of efforts is also defined.	[RFC1719]
December 1994	Technical criteria to choose IPng	A list of criteria is defined and going to be used against all the IPng protocol proposals.	[RFC1726]
January 1995	Recommendation for IPng	The IPng area directors main recommendation is to use the SIPP 128bits version as the basis of the new IPng protocol. Many working groups are formed.	[RFC1752]
December 1995	IPv6 specification	The first version of the IPv6 specification is published.	[RFC1883]
December 1996	Ngtrans working group first meeting	The Next generation transition (ngtrans) working group is formed to handle the transition to IPv6.	[IETF37ngtrans]
December 1998	New version of IPv6 specification	Based on implementations and additional work, a new version of the IPv6 specification is published. It slightly changes the header format, clarifies many items such as Path MTU, traffic class, flow label and jumbograms.	[RFC2460]

Note: When an RFC document is the reference, the date is the publication date of the RFC. In most cases, the actual step happened many months before the publication date.

SIPP, TUBA and CATNIP were the protocol candidates reviewed more carefully by the IPng directorate [RFC1752]. SIPP was chosen with some additional modifications, such as increasing the address size from 64 bits to 128 bits, after a long debate on the appropriate size of the address space.

Table 1.4 IPng protocol candidates

Protocol Name	Full name	Reference
IP encaps	Internet Protocol Encapsulation	[RFC1955]
SIP	Simple Internet Protocol	
PIP	P Internet Protocol	[RFC1621], [RFC1622]
Simple CLNP	Simple Connectionless-mode Network Layer Protocol	
Nimrod	New IP Routing and Addressing Architecture	[RFC1753], [NIMROD], [RFC1992]
TP/IX		[RFC1475]
IPAE	IP Address Encapsulation	
TUBA	TCP and UDP with Bigger Addresses, TCP/UDP Over CLNP-Addressed Networks	[RFC1347], [RFC1561]
CATNIP	Common Architecture for Next-Generation IP	[RFC1707]
SIPP	Simple Internet Protocol Plus	[RFC1710]
SIPP 128bits ver	Simple Internet Protocol Plus, 128 bits address version	
IPng	Internet Protocol Next Generation	[RFC1883]
IPv6	Internet Protocol version 6	[RFC1883]

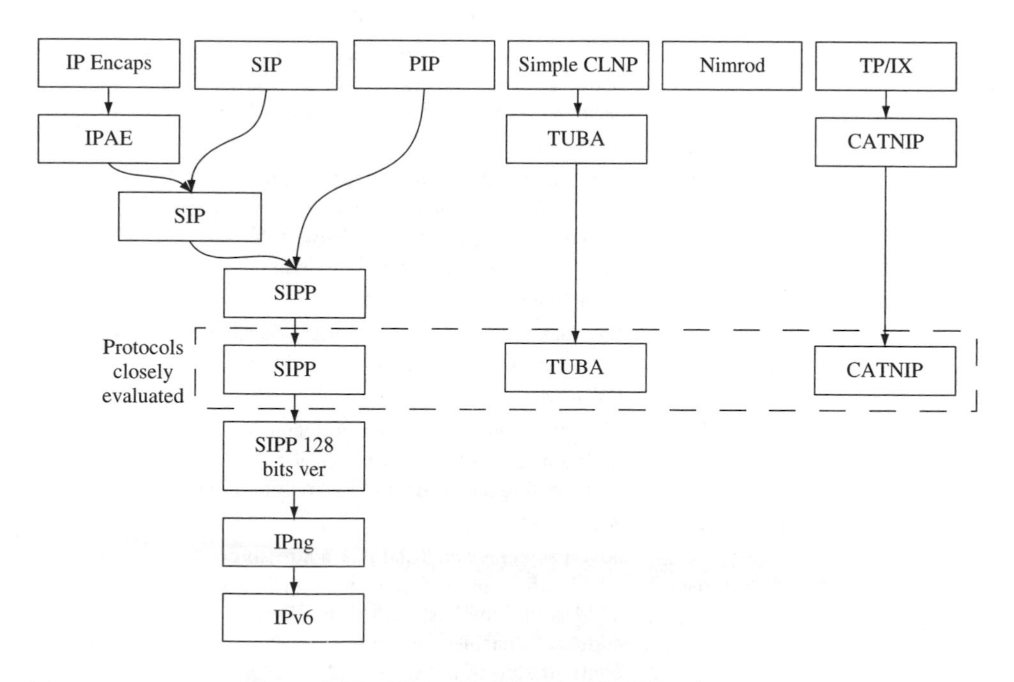

Figure 1.25 IPng protocol proposals generation tree

Main Features

IPv6 is a new version of the Internet Protocol. It has been designed as an evolutionary, rather than revolutionary, step from IPv4. Functions which are generally seen as working in IPv4 were kept in IPv6. Functions which don't work or are infrequently used were removed or made optional. A few new features were added where the functionality was felt to be necessary." [RFC1752]

Table 1.5 lists the main features of IPv6, which introduces many concepts, discussed in the following chapters.

Table 1.5 IPv6 features

Feature	Implementation	Benefit	Book chapter or section
Larger addresses	128 bit addresses	From 32 bit address space in IPv4 to 128 bit address space. It enables all nodes to be addressable and reachable, removing the need for network address translation and restoring the end-to-end model for end-to-end capabilities such as security.	4
More levels of addressing hierarchy	Address architecture	Multiple levels in the addressing hierarchy provide better aggregation of routes, easier allocation of addresses to downstreams and scalability of the global routing table.	4
Scoping in the address	Specific bits in the address	Address scoping enables easy filtering at boundaries, such as link or site and better security against remote attack on link layer protocols.	4
Simple and fixed address architecture	/48 for sites, /64 for a link	Simplified address architecture enables easier addressing plans, which decreases the network management costs. Now, subnet masks are fixed and provide virtually unlimited numbers of nodes on a link.	4
Privacy addresses	Specific bits in the address	Provides privacy for the end-user where the IP address cannot be used for tracking traffic usage.	13.4
Multiple addresses on an interface	IPv6 stack	Multiple addresses on interfaces enables multiple use, virtual hosting, easier renumbering and a method for multihoming.	5
Autoconfiguration of nodes	IPv6 stack, router advertisements	Auto-configuration is based on advertisements about the link addressing sent by the routers. Nodes insert their MAC address into the host part of the IPv6 address. It enables fast and reliable configuration of nodes, as well as easy renumbering.	5

No address conflicts on links	IPv6 stack	Embedding the unique link address (MAC) into the host part of the IPv6 address and a duplicate address detection method guarantee uniqueness of the address on the link.	5
Better reliability in auto configuration	Router advertisements	Each router on a link sends auto-configuration information to nodes, so if one router is dead, others are still sending. The router infrastructure is always nearer to the host and more fault tolerant than DHCP servers.	5
Multicast address scoping	Specific bits in the address	A multicast address now contains a scope. IPv4 multicast had to rely on TTL to manage the reachability of a multicast channel, which makes multicast management complex. IPv6 multicast is easy to manage since the scope of the channel is within the IPv6 multicast address.	4
Simpler and more efficient IP header	Less number of fields, no checksum, 64 bit aligned fields	Routers process the packets faster and more efficiently, which improves the forwarding performance.	3
Extension headers	Options are placed after the base IPv6 header	Options for IPv6 packets are implemented as extension headers and are tagged with processing options. Routers do not have to look at most extension headers which increases their forwarding performance. New headers can be added incrementally without any impact on implementations.	3
Mandatory IP security	IPsec	IPsec is mandatory in IPv6, which makes all nodes in a position to secure their traffic, if they have the necessary underlying key infrastructure.	13
Source routing	Extension header	Source routing is implemented in a way so that routers not directly involved in the source routing can still make policy decisions based on the destination address. This feature makes source routing more deployable.	9.2
Simple and flexible transition	Transition protocols	In the foundation and requirements of IPv6, there was a clear need to make a smooth transition. The requirements were: incremental upgrade, incremental deployment, easy addressing and low start-up costs.	16, 17, 18
Labeling flows for QoS	Flow label header field	A flow label is defined in a specific field in the basic header, enabling the labeling and policing of traffic by the routers, without the need to inspect the application payload by the routers, resulting in more efficient QoS processing.	14

(*continued overleaf*)

(continued)

Feature	Implementation	Benefit	Book chapter or section
Multihoming capabilities	Multiple prefix on the same link and on interfaces	Multiple prefixes can be announced in router advertisements, which creates multiple addresses on interfaces. Lifetimes of prefixes are managed by the nodes which provides an easy way to multihome nodes.	9.12
More efficient use of links	Neighbor discovery	Link scope interactions between nodes and between nodes and routers are optimized.	5.2.2, 6.1, 6.2
Use of Multicast for discovery and link-local interaction	Neighbor discovery	No broadcasts are used in IPv6. In most cases, only relevant nodes receive the requests.	5.2.2, 6.1, 6.2, 6.3
Mobility	MobileIPv6	Mobility is integrated in IPv6 headers, stacks and implementations, making mobility a seamless and deployable feature.	11
Private but unique address space	Unique local address space	Private addresses are used for unconnected networks to the Internet. Different than RFC1918 private IPv4 address space, private IPv6 address space remains unique to the site, which makes it easy to connect private networks together.	4.3.2.3

1.7 IPv6 Milestones

Table 1.6 lists some major milestones of IPv6.

1.8 IPv6 Return on Investment

A study [Pau, 2002a] has established a return on investment framework for IPv6:

- adopters of IPv6 run smaller risks than waiting;
- targeted ratio of approx. 16% of IPv6 creates positive ROI on incremental deployment;
- migration costs can hardly be a deciding factor in deploying IPv6.

Another study [Pau, 2002b] from the same author uses an analytical model to reveal that for the ISP operator, net revenue with IPv6 is intrinsically and systematically higher than for IPv4.

Table 1.6 Some IPv6 milestones

Date	Step	Description	Reference
August 1990	Predicted exhaustion of class B addresses by 1994	First wakeup call within the IETF. The class B address space could be exhausted in 4 years!	[Solensky, 1990]
January 1995	Recommendation for IPng	The IPng area directors main recommendation is to use the SIPP 128 bits version as the basis of the new IPng protocol. Many working groups are formed.	[RFC1752]
December 1995	IPv6 specification	The first version of the IPv6 specification is published.	[RFC1883]
July 1996	First IPv6 test network over Internet (6bone)	The 6bone IPv6 test backbone is started.	[6bonehistory]
February 1999	Freenet6 service started	During an IPng working group interim meeting, the Freenet6 tunnel broker service is announced, providing the world community with easy access to the IPv6 Internet using automated tunnels.	http://www.freenet6.net
July 1999	Registry-based IPv6 address space allocation is started	The regional registries, RIPE, ARIN and APNIC, start allocating IPv6 address space to providers.	[RIPE-196]
July 1999	IPv6Forum	The IPv6Forum body is formed.	http://www.ipv6forum.com
February 2000	Solaris8	The first commercial OS to include IPv6 in the product as standard feature is Sun Solaris 8.	http://www.sun.com/ipv6
March 2000	FreeBSD 4.0	FreeBSD open source operating system now includes IPv6 in its standard distribution.	http://www.freebsd.org
May 2001	Freenet6 second generation	The freenet6 service second generation uses the TSP tunnel broker protocol.	http://www.freenet6.net

1.9 What Happened to IPv5?

IPv5 is the IP protocol number of the Stream Protocol (ST) [RFC1190], an experimental protocol for streaming traffic. Figure 1.26 shows where ST fits in the IP architecture, including some specific streaming transport protocols named PVP and NVP.

To differentiate IPv4 packets from ST IP packets at the link layer, ST requires a specific IP version number. At the time of ST, the next version number available for IP was ‘5’. So IANA [IANA, 2001] allocated 5 to ST, so ST is also known as IPv5. When IPng was designed, the next version number available for IP was ‘6’, so IPng is IPv6.

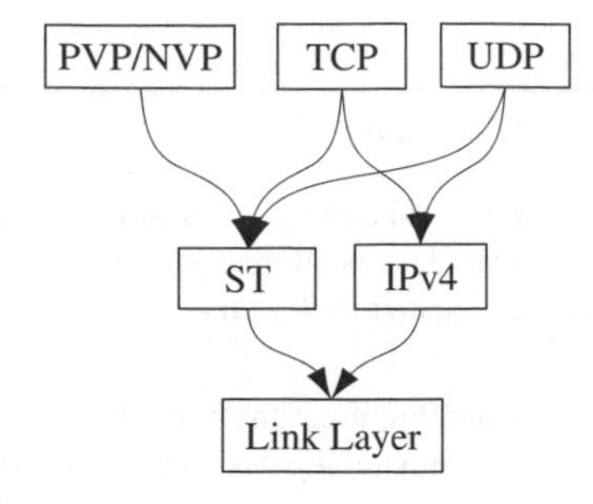

Figure 1.26 Streaming Protocol and IPv4 Architecture

ST is an experimental protocol and is not deployed. Streaming is handled with multicast, RTP and other protocols.

1.10 Summary

IPv4 was deployed mainly in the university networks before the Internet became so pervasive. At that time, the Internet was for information sharing and electronic communications. For information sharing, ftp sites were used, then came Archie to index ftp sites, then Gopher which structures the information, then Veronica to index the gopher sites and then came the Web and the first client Mosaic. Before the Web and Mosaic, there was 'no killer application'. However, there was a playground fertile for future innovations, and now it is part of our daily life.

As we have seen in the first sections of this chapter, the current IPv4 protocol is no longer fertile for innovations and is now a pretty constrained network.

IPv6 restores the fully-fledged network needed to deploy new applications, most of them probably still unknown. In the near future, there will be very little or zero IPv4 address space remaining.

The drivers for IPv6 are multiple, such as mobility, reachability, network management, multimedia, and others described in this chapter. More trouble and higher costs with IPv4, combined with new applications requiring the new IPv6 functionalities, is driving IPv6.

1.11 References

[6bonehistory] Ngtrans historic milestones, http://www.6bone.net/ngtrans/ngtrans_charter.html

Baset S. and Schulzrinne H., 'An Analysis of the Skype Peer-to-Peer Internet Telephony Protocol', Columbia University, September 15, 2004, http://www.cs.columbia.edu/~library/TR-repository/reports/reports-2004/cucs-039-04.pdf

Charnie, J., 'Statement to the commission on population and development', United Nations, March 2004, http://www.un.org/esa/population/cpd/37OPEN.pdf

Deering, S., 'Watching the Waist of the Protocol Hourglass', Proceedings of the 51st Internet Engineering Task Force, August 2001, http://www.ietf.org/proceedings/01aug/slides/plenary-1/index.html

Hain, T., 'Expanded Address Allocation for Private Internets', Internet-Draft draft-hain-1918bis-00, April 2004.

Huston, G., 'Analyzing the Internet's BGP Routing Table', January 2001, http://www.potaroo.net/papers/ipj/4-1-bgp.pdf

Huston, G., 'Growth of the BGP Table', January 2005, http://bgp.potaroo.net/

IAB, Internet Activities Board Meeting Minutes, June 1992, http://www.iab.org/documents/IABmins/IABmins.1992-06-18.html

[IABREJECT] Proceeding of the twenty-fourth Internet Engineering Task Force, http://www.ietf.org/proceedings/prior29/IETF24.pdf

IANA, 'IP protocol numbers', ftp://ftp.iana.org/assignments/version-numbers, November 2001.

[IETF37ngtrans] Proceedings of the 37th Internet Engineering Task Force, CNRI, December 1996, http://www.ietf.org/proceedings/96dec/toc.html

Nimrod: 'A Scalable Routing Architecture for Large, Heterogeneous, and Dynamic Internetworks', http://www.ir.bbn.com/projects/nimrod/nimrod-index.html

Pau, L.-F., (2002a) 'IPv6 Return on Investment (R.O.I) Analysis Framework at a Generic Level, and First Conclusions', Erasmus Research Institute of Management (ERIM), ERIM Report Series Research in Management, Rotterdam, The Netherlands, ERS-2002-78-LIS, September 2002.

Pau, L.-F., (2002b) 'A Business Evaluation of The Next Generation IPv6 Protocol in Fixed And Mobile Communication Services: An Analytical Study and Calculation', Erasmus Research Institute of Management (ERIM), ERIM Report Series Research in Management, Rotterdam, The Netherlands, ERS-2002-78-LIS, September 2002.

[RFC959] Postel, J. and Reynolds, J., 'File Transfer Protocol', STD 9, RFC 959, October 1985.

[RFC1058] Hedrick, C., 'Routing Information Protocol', RFC 1058, June 1988.

[RFC1112] Deering, S., 'Host Extensions for IP multicasting', RFC 1112, August 1989.

[RFC1190] Casner, S., Lynn, C., Park, P., Schroder, K. and Topolcic, C., 'Experimental Internet Stream Protocol: Version 2 (ST-II)', RFC 1190, October 1990.

[RFC1347] Callon, R., 'TCP and UDP with Bigger Addresses (TUBA), A Simple Proposal for Internet Addressing and Routing', RFC 1347, June 1992.

[RFC1380] Gross, P. and Almquist, P., 'IESG Deliberations on Routing and Addressing', RFC 1380, November 1992.

[RFC1466] Gerich, E., 'Guidelines for Management of IP Address Space', RFC 1466, May 1993.

[RFC1475] Ullmann, R. 'TP/IX: The Next Internet', RFC 1475, June 1993.

[RFC1519] Fuller, V., Li, T., Yu, J. and Varadhan K., 'Classless Inter-Domain Routing (CIDR): An Address Assignment and Aggregation Strategy', RFC 1519, September 1993.

[RFC1550] Bradner, S. and Mankin, A., 'IP: Next Generation (IPng) White Paper Solicitation', RFC 1550, December 1993.

[RFC1561] Piscitello, D., 'Use of ISO CLNP in TUBA Environments', RFC 1561, December 1993.

[RFC1621] Francis, P., 'Pip Near-term Architecture', RFC 1621, May 1994.

[RFC1622] Francis, P., 'Pip Header Processing', RFC 1622, May 1994.

[RFC1667] Symington, S., Wood, D. and Pullen, J., 'Modeling and Simulation Requirements for IPng', RFC 1667, August 1994.

[RFC1668] Estrin, D., Li, T. and Rekhter Y., 'Unified Routing Requirements for IPng', RFC 1668, August 1994.

[RFC1669] Curran, J., 'Market Viability as a IPng Criteria', RFC 1669, August 1994.

[RFC1670] Heagerty, D., 'Input to IPng Engineering Considerations', RFC 1670, August 1994.

[RFC1671] Carpenter, B., 'IPng White Paper on Transition and Other Considerations', RFC 1671, August 1994.

[RFC1672] Brownlee, N., 'Accounting Requirements for IPng', RFC 1672, August 1994.

[RFC1673] Skelton, R., 'Electric Power Research Institute Comments on IPng', RFC 1673, August 1994.

[RFC1674] Taylor, M., 'A Cellular Industry View of IPng', RFC 1674, August 1994.

[RFC1675] Bellovin, S., 'Security Concerns for IPng', RFC 1675, August 1994.

[RFC1676] Ghiselli, A., Salomoni, D. and Vistoli, C., 'INFN Requirements for an IPng', RFC 1676, August 1994.

[RFC1677] Adamson, R., 'Tactical Radio Frequency Communication Requirements for IPng', RFC 1677, August 1994.

[RFC1678] Britton, E. and Tavs, J., 'IPng Requirements of Large Corporate Networks', RFC 1678, August 1994.

[RFC1679] Green, D., Irey, P., Marlow, D. and O'Donoghue, K., 'HPN Working Group Input to the IPng Requirements Solicitation', RFC 1679, August 1994.

[RFC1680] Brazdziunas, C., 'IPng Support for ATM Services', RFC 1680, August 1994.

[RFC1681] Bellovin, S., 'On Many Addresses per Host', RFC 1681, August 1994.

[RFC1682] Bound, J., 'IPng BSD Host Implementation Analysis', RFC 1682, August 1994.

[RFC1683] Clark, R., Ammar, M. and Calvert, K., 'Multiprotocol Interoperability In IPng', RFC 1683, August 1994.

[RFC1686] Vecchi, M., 'IPng Requirements: A Cable Television Industry Viewpoint', RFC 1686, August 1994.

[RFC1687] Fleischman, E., 'A Large Corporate User's View of IPng', RFC 1687, August 1994.

[RFC1688] Simpson, W., 'IPng Mobility Considerations', RFC 1688, August 1994.

[RFC1707] McGovern, M. and Ullmann, R., 'CATNIP: Common Architecture for the Internet', RFC 1707, October 1994.

[RFC1710] Hinden, R., 'Simple Internet Protocol Plus White Paper', RFC 1710, October 1994.

[RFC1715] Huitema, C., 'The H Ratio for Address Assignment Efficiency', RFC 1715, November 1994.

[RFC1719] Gross, P., 'A Direction for IPng', RFC 1719, December 1994.

[RFC1726] Partridge, C. and Kastenholz, F., 'Technical Criteria for Choosing IP The Next Generation (IPng)', RFC 1726, December 1994.

[RFC1752] Bradner, S. and Mankin, A., 'The Recommendation for the IP Next Generation Protocol', RFC 1752, January 1995.

[RFC1753] Chiappa, J., 'IPng Technical Requirements of the Nimrod Routing and Addressing Architecture', RFC 1753, December 1994.

[RFC1812] Baker, F., 'Requirements for IP Version 4 Routers', RFC 1812, June 1995.

[RFC1883] Deering, S. and Hinden, R., 'Internet Protocol, Version 6 (IPv6) Specification', RFC 1883, December 1995.

[RFC1918] Rekhter, Y., Moskowitz, R., Karrenberg, D., Groot, G. and Lear, E., 'Address Allocation for Private Internets', BCP 5, RFC 1918, February 1996.

[RFC1945] Berners-Lee, T., Fielding, R. and Nielsen, H., 'Hypertext Transfer Protocol – HTTP/1.0', RFC 1945, May 1996.

[RFC1955] Hinden, R., 'New Scheme for Internet Routing and Addressing (ENCAPS) for IPNG', RFC 1955, June 1996.

[RFC1992] Castineyra, I., Chiappa, N. and Steenstrup, M., 'The Nimrod Routing Architecture', RFC 1992, August 1996.

[RFC2050] Hubbard, K., Kosters, M., Conrad, D., Karrenberg, D. and Postel, J., 'Internet Registry IP Allocation Guidelines', BCP 12, RFC 2050, November 1996.

[RFC2068] Fielding, R., Gettys, J., Mogul, J., Nielsen, H. and Berners-Lee T., 'Hypertext Transfer Protocol – HTTP/1.1', RFC 2068, January 1997.

[RFC2401] Kent, S. and Atkinson, R., 'Security Architecture for the Internet Protocol', IETF RFC 2401, November 1998.

[RFC2460] Deering, S. and Hinden, R., 'Internet Protocol, Version 6 (IPv6) Specification', RFC 2460, December 1998.

[RFC2581] Allman, M., Paxson, V. and Storens, W., 'TCP Congestion Control' RFC 2581, April 1999.

[RFC2775] Carpenter, B., 'Internet Transparency', RFC 2775, February 2000.

[RFC2993] Hain, T., 'Architectural Implications of NAT', RFC 2993, November 2000.

[RFC3022] Srisuresh, P. and Egevang, K., 'Traditional IP Network Address Translator (Traditional NAT)', RFC 3022, January 2001.

[RFC3027] Holdrege, M. and Srisuresh, P., 'Protocol Complications with the IP Network Address Translator', RFC 3027, January 2001.

[RFC3235] Senie, D., 'Network Address Translator (NAT)-Friendly Application Design Guidelines', RFC 3235, January 2002.

[RFC3261] Rosenberg, J., Schulzrinne, H., Camarillo, G., Johnston, A., Peterson, J., Sparks, R., Handley, M. and Schooler E., 'SIP: Session Initiation Protocol', RFC 3261, June 2002.

[RFC3489] Rosenberg, J., Weinberger, J., Huitema, C. and Mahy, R., 'STUN – Simple Traversal of User Datagram Protocol (UDP) Through Network Address Translators (NATs)', RFC 3489, March 2003.

[RIPE-196] RIPE NCC, 'Provisional IPv6 Assignment and Allocation Policy Document', RIPE-196, July 1999.

Rosenberg, J., 'Traversal Using Relay NAT (TURN)', Internet-Draft draft-rosenberg-midcom-turn-04, February 2004.

Solensky F., 'Continued Internet Growth', Proceedings of the 18th Internet Engineering Task Force, August 1990, http://www.ietf.org/proceedings/prior29/IETF18.pdf

[Skype] http://www.skype.com

Walker, J., 'Speak Freely: End of Life Announcement', January 2004, http://www.fourmilab.ch/speakfree/unix

1.12 Further Reading

Walker, J., 'The Digital Imprimatur', November 2003, http://www.fourmilab.ch/documents/digital-imprimatur/

2

I Can't Wait to Get my Hands Dirty!

This chapter is made for the impatient who can't wait to understand everything before trying IPv6. A real network example is used in this chapter. You are encouraged to duplicate the setup in your environment and try it, even if you haven't read the whole book. Skip this chapter if you want to learn IPv6 in a more stepwise way.

2.1 Setup Description

Figure 2.1 shows a home network, with one gateway implementing IPv4 Network Address Translation (NAT) connected to the IPv4 Internet and attaching a subnet with private IPv4 address space and a few hosts. We will take one of the internal hosts to provide IPv6 connectivity to the home LAN.

The network setup is shown in Figure 2.1 and described below:

- one local LAN;
- an IPv4 network address translation (NAT) device (R1), such as typical home gateways, connected to the IPv4 Internet on one side with IPv4 address 192.0.2.1, and to the local LAN on the other side, with IPv4 private address 192.168.1.1;
- one IPv6-enabled node using IPv4 private address 192.168.1.2 that will act as IPv6 router(N2);
- one IPv6-enabled host (N3) (it is also IPv4-enabled, but this fact will be ignored).

The IPv6 connectivity is provided by the Freenet6 service [Freenet6] to the host N2, which will become a router later in this chapter. Freenet6 implements the Tunnel Setup Protocol (TSP) tunnel broker solution, described in Section 16.2.9. It creates automatically IPv6 in IPv4 tunnels between the node and the tunnel broker. TSP is the control protocol

Migrating to IPv6: A Practical Guide to Implementing IPv6 in Mobile and Fixed Networks Marc Blanchet

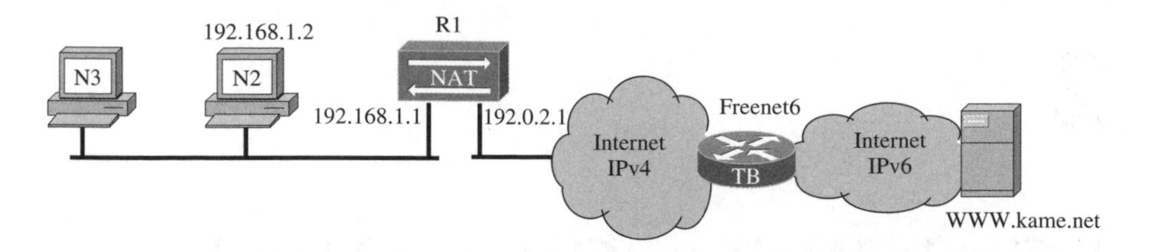

Figure 2.1 Small network setup

establishing the tunnel between the client and the tunnel broker. A thin TSP client is used in N2 to negotiate and establish the tunnel.

Within a few steps, the LAN will be a functional IPv6 network connected to the IPv6 Internet. Try it!

2.2 Steps

Instructions are given for FreeBSD 5.2, Linux RedHat 8 and Windows XP in this chapter. Note the following:

- An IPv6 address has 128 bits and is written in hexadecimal notation with ':' as separator between the 8 fields of 16 bits each.
- The IPv6 address '::1' is the equivalent of the IPv4 127.0.0.1 loopback address.
- IPv6 addresses starting with 'fe80::' are link-local addresses automatically assigned to all IPv6-enabled interfaces. Link-local addresses provide ready to use addresses for subnet scope communications. The last 4 fields of the address contains a slightly modified version of the link-layer (MAC) address of the interface.
- The 'subnet mask' in IPv6 is fixed to /64. For every IPv6 address configured on physical interfaces such as Ethernet, '/64' is appended at the end of the address to show the subnet mask.

2.2.1 Enabling IPv6 on N2 and N3

The first step is to enable IPv6 on the operating system of N2 and N3 and verify it is working. The following sub-sections show how to do this for each operating system.

2.2.1.1 FreeBSD

In FreeBSD, IPv6 is enabled by setting the `ipv6_enable` variable to `yes` in the `/etc/rc.conf` configuration file.

```
# cat /etc/rc.conf
ipv6_enable=yes
```

While not necessary, it is simpler for this exercise to reboot FreeBSD after enabling IPv6.

Pinging to the loopback (::1) is a good way to see that basic IPv6 is working.

```
% ping6 ::1
```

Reply packets should now be seen on the screen.
To verify the IPv6 addresses on an interface, use `ifconfig <interface_name>`.

```
% ifconfig fxp0
```

An IPv6 address starting with fe80:: should be assigned to the interface. If it is, then IPv6 is working.

2.2.1.2 Linux

In most Linux distributions, IPv6 is enabled by setting the `NETWORKING_IPV6` variable to `yes` in the `/etc/sysconfig/network` configuration file.

```
# cat /etc/sysconfig/network
NETWORKING_IPV6=yes
```

Restart network services, enabling IPv6 at the same time, with the '`service`' command:

```
# /sbin/service network restart
```

Pinging to the loopback (::1) is a good way to see that basic IPv6 is working.

```
% ping6 ::1
```

Reply packets should be seen on the screen.
To verify the IPv6 addresses on an interface, use `ifconfig <interface_name>`.

```
% ifconfig eth0
```

An IPv6 address starting with fe80:: should be assigned to the interface. If it is, then IPv6 is working.

2.2.1.3 Windows

In Windows XP, IPv6 modules are loaded on the disk as part of the installation, but are not enabled by default. To enable IPv6, the following command needs to be entered once by an administrator:

```
C> netsh interface ipv6 install
```

Pinging to the loopback (::1) is a good way to see that basic IPv6 is working.

```
C> ping ::1
```

Reply packets should be seen on the screen.

The 'ipconfig' command shows the IPv4 and IPv6 addresses, DNS configuration and default router for all interfaces.

```
C> ipconfig
```

An IPv6 address starting with fe80:: should be assigned to one or several interfaces. Windows automatically enables many transition mechanisms as discussed in Chapter 16. Do not worry for now about all the interfaces.

2.2.2 Two Nodes Talking Already!

Now that you have configured N2 and N3 with IPv6, you could start sending IPv6 packets! A broadcast-like address is defined as ff02::1, a multicast address in IPv6[1]. Pinging to this address will give you replies from all IPv6-enabled nodes on the local LAN. Note that the replies will show the 'fe80::' source address of the neighbors. Try to ping directly to the other neighbor by using its link-local (fe80::) address.

As you can see, without any further configuration, IPv6 traffic can flow between nodes on a same link.

2.2.3 Installing and Configuring the TSP Client on N2

The next step is to get IPv6 connectivity to the IPv6 Internet. If your provider does not offer IPv6 connectivity to your home, then you need to use some IPv6 transition mechanism, as described in Chapter 16. This chapter uses the TSP tunnel broker which works in most cases, including behind a NAT, and has the broadest support of client platforms.

Some open source operating systems include the TSP client as part of their distribution or ports, usually under freenet6 name. If not, then get the TSP client software from the Freenet6 Web site (http://www.freenet6.net). Download and install it based on the instructions provided with the software.

The next subsections describe how to proceed for each operating system.

2.2.3.1 FreeBSD

The TSP tunnel broker client is available on FreeBSD in the ports/net/freenet6, or from the freenet6 Web site (http://www.freenet6.net). If you choose to use the ports, then do install in the freenet6 directory. This will install the TSP client software and a default configuration file.

2.2.3.2 Linux

The TSP tunnel broker client is included in many Linux distributions, or from the freenet6 Web site (http://www.freenet6.net).

2.2.3.3 Windows

The TSP tunnel broker client is available on Windows from the freenet6 Web site (http://www.freenet6.net).

---

[1] IPv6 multicast addresses start with 'ff'.

2.2.4 Creating an IPv6 in IPv4 Tunnel with Freenet6

To get IPv6 connectivity, this setup creates an IPv6 over the IPv4 tunnel on N2 towards the Freenet6 tunnel broker. Freenet6 is connected to the IPv6 Internet. The TSP client execution connects to the Freenet6 tunnel broker, negotiates an IPv6 over the IPv4 tunnel, configures its own tunnel endpoint and creates an IPv6 default route through that tunnel interface. A tunnel interface is created or used for the IPv6 in the IPv4 tunnel and is shown in the list of interfaces.

To run the TSP client, use the administrator privilege (root or Administrator) and type the `tspc` command in a shell on N2:

```
# tspc
```

After execution, you should get a message that the tunnel is setup with freenet6. If such a message does not show, retry with the verbose mode on:

```
# tspc -vvv
```

2.2.5 Testing IPv6 on N2

Now N2 should be connected to the IPv6 Internet. Try IPv6 ping (ping6 on Linux or FreeBSD, ping on Windows) to a well known address, such as www.kame.net.

```
% ping6 www.kame.net
```

Replies should be seen on the screen. You are connected!!!

A well known milestone in the IPv6 community to test basic IPv6 connectivity is to open your browser (most recent browsers support IPv6) and then connect to the Kame Web site (http://www.kame.net). For most browsers, you should see a Web page with a dancing turtle, which is shown only when one is connecting to the Kame Web site using IPv6. If you are connecting over IPv4, the turtle does not dance. Kame is the site of the IPv6 developers for the BSD based distributions.

2.2.6 Requesting an IPv6 Prefix Delegation

N2 is now connected to IPv6. You could redo the same procedure with N3 and both are connected. However, they use a separate tunnel which is not the most efficient way. A better way is to convert N2 to a software router using the same physical interface for the tunnel and for the IPv6 connectivity on the LAN. In this setup, N2 will request an IPv6 address range (a prefix) from the Freenet6 tunnel broker and will announce that prefix on the LAN. In typical IPv6 configuration, no DHCP server is used. Instead, the routers are advertising the prefix on the LAN and the hosts autoconfigure themselves based on that prefix. This is the autoconfiguration feature of IPv6, described in Chapter 5, a key and novel way to manage configuration of hosts.

To request a prefix to the Freenet6 tunnel broker, you must first register to freenet6.net using a Web browser to get a free username and password account. Then modify the tspc.conf file generated at the install of the TSP client software by changing the variables listed in Table 2.1.

Table 2.1 TSP client tspc.conf statements

Variable	New value	Description
`userid`	*the registered username*	The Freenet6 user identification string.
`passwd`	*the received password*	The password for the `userid`.
`host_type`	`router`	Sets the TSP client as a router. In router mode, the client will receive a prefix.
`prefixlen`	`48`	The length of the prefix desired by the client.
`if_prefix`	*interface name*	The interface on the operating system of the client used to send router advertisements with the received prefix from the broker. On linux, set 'eth0'. On FreeBSD, use the name of the physical interface (e.g., 'fxp0'). On Windows, use '1'.

Re-executing the TSP client (tspc) should negotiate a tunnel and an IPv6 prefix with the Freenet6 tunnel broker, create the tunnel and set the router advertisements announcing the received prefix on the local LAN.

If N3 is IPv6-enabled, then it should have received the router advertisements and have configured itself with an IPv6 address based on the received prefix from N2. By looking at the interface (ifconfig or ipconfig), one should see 2 IPv6 addresses on N3, a fe80:: link-local address and a 2001:: IPv6 address. The 2001:: address is the new address configured by concatenating the prefix received in the router advertisements with a slightly modified version of the MAC address of the interface.

By setting the `prefixlen` to 48 in the tspc.conf file, N2 has requested a /48 prefix to the Freenet6 tunnel broker service. A /48 is a very large prefix enabling your home network to have a maximum of 65536 subnets, each having a maximum of 2^{64} hosts. This address space is many many orders of magnitude larger than the whole IPv4 Internet!

Test IPv6 connectivity on N3 by doing ping or by using the Web browser to the Kame Web site, as discussed previously.

Now try an IPv6 application, such as a VoIP softphone application [Linphone]!

2.3 Summary

A home network can be easily IPv6-enabled with a few steps, using the Freenet6 tunnel broker service as the IPv6 provider to get connectivity to the IPv6 Internet. Moreover, one node on the home LAN may be used for providing connectivity to the other nodes acting as a software IPv6 router. Freenet6 provides IPv6 address space for this usage.

2.4 References

[Freenet6] http://www.freenet6.net.
[Linphone] http://www.linphone.org.

2.5 Further Reading

[Kame] http://www.kame.net.

3

IPv6 Datagram

This chapter describes the format and properties of the IP datagram, starting with the IPv4 header used since the early 1980s on the Internet and then following with the IPv6 header. Comparisons between the two protocols are discussed throughout the chapter, but no effort is made to describe the IPv4 header in details.

Since layer 3 protocols like IP are mainly described by their header, this chapter is fundamental to the understanding of IPv6. It first describes the basic IPv4 header and then the basic IPv6 header, with a comparison of fields of the two protocols. Then the IPv6 extension headers are described. Transport layer considerations are the last part of the chapter.

This chapter, like most chapters in this book, is written to be used both for first reading and as a reference. For first reading purposes, many details are skipped in favor of more in-depth description in the next chapters. However, for reference purposes, it tries to be as exhaustive as possible. Therefore, some items might not be complete on the first reading because they are covered in subsequent chapters.

3.1 Description of the IP Datagram

Figure 3.1 shows an IP datagram inside a generic link layer frame which starts with its own link layer header and terminates with a trailer. After the link layer header, the IP header is the first part of the IP datagram followed by the transport header and followed by the application protocol data.

Each part contains a field identifying which protocol is in the inner payload. For example, the Ethernet link layer header contains the Ethernet Type code of 0x0800 identifying that the payload is an IPv4 datagram. The IP header contains the 'Protocol' field identifying that the payload has a TCP header.

Migrating to IPv6: A Practical Guide to Implementing IPv6 in Mobile and Fixed Networks Marc Blanchet

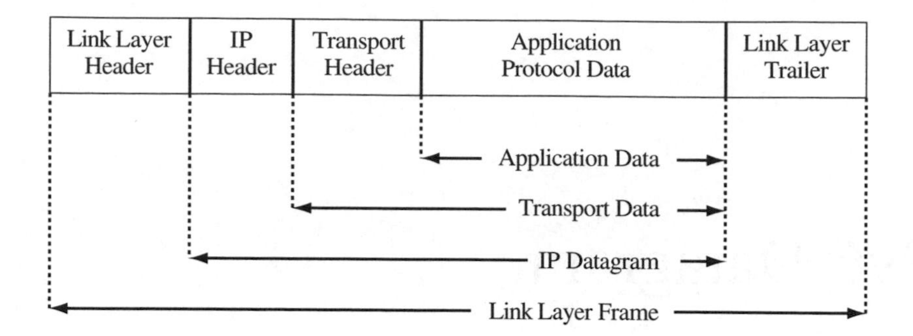

Figure 3.1 IP Datagram inside a link layer frame

3.2 IPv4 Header

Figure 3.2 shows the header of the IPv4 datagram [RFC791]. It has 12 fields and its size is 20 octets, without options. Options are variable-length fields with final padding for 32 bit alignment. The maximum header length with options is 60 octets.

The header fields are briefly described in Table 3.1. The 'IPv6 use' column describes the change in the IPv6 header, explained in the next section.

The data part after the IP header usually starts with the transport header. The transport protocol, like TCP, is identified in the 'Protocol' header field, as illustrated in Figure 3.3.

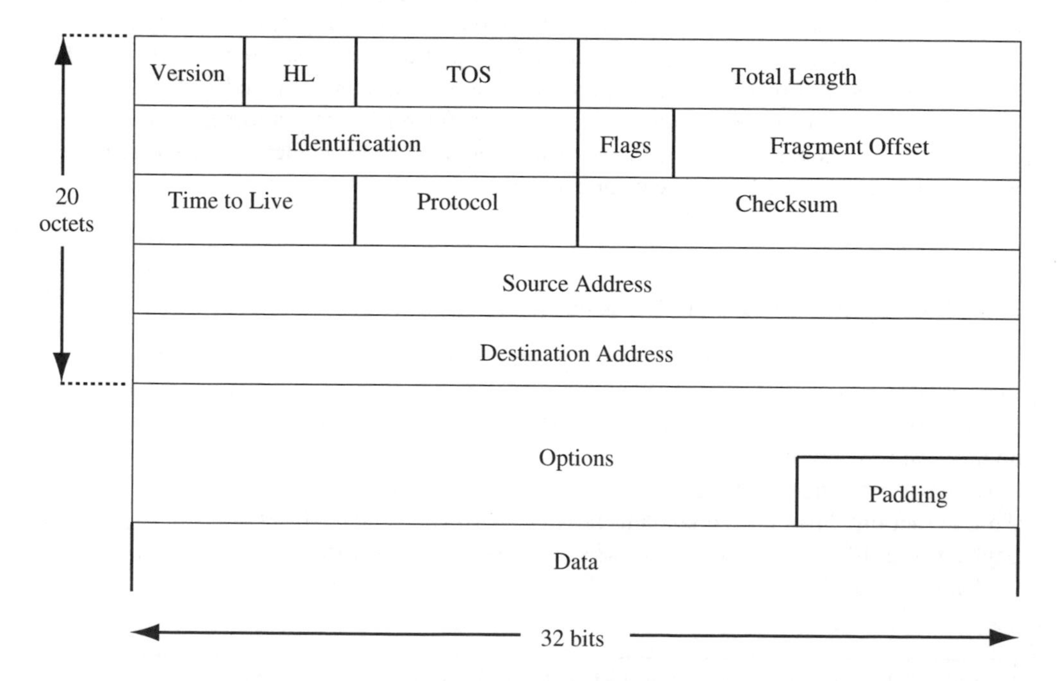

Figure 3.2 IPv4 Header

Table 3.1 IPv4 header fields and their IPv6 use

Field	Size (in bits)	Description	IPv6 use
Version	4	= 4 for IPv4	= 6 for IPv6
Header Length	4	Length of the header	Removed
TOS	8	Class of service [RFC2474].	Same but renamed to 'Traffic Class'
Total Length	16	Length of the datagram	Different: payload length
Identification	16	Fragment Id	In Fragment Extension Header
Flags	3	Fragmentation flags	In Fragment Extension Header
Fragment Offset	13	Pointer	In Fragment Extension Header
Time to Live	8	Decreasing by one on each hop	Same but renamed to 'Hop Limit'
Protocol	8	Transport protocol identification	Same but renamed 'Next Header'
Checksum	16	Checksum of the header	Removed
Source Address	32	IPv4 address	IPv6 address = 128 bits
Destination Address	32	IPv4 address	IPv6 address = 128 bits

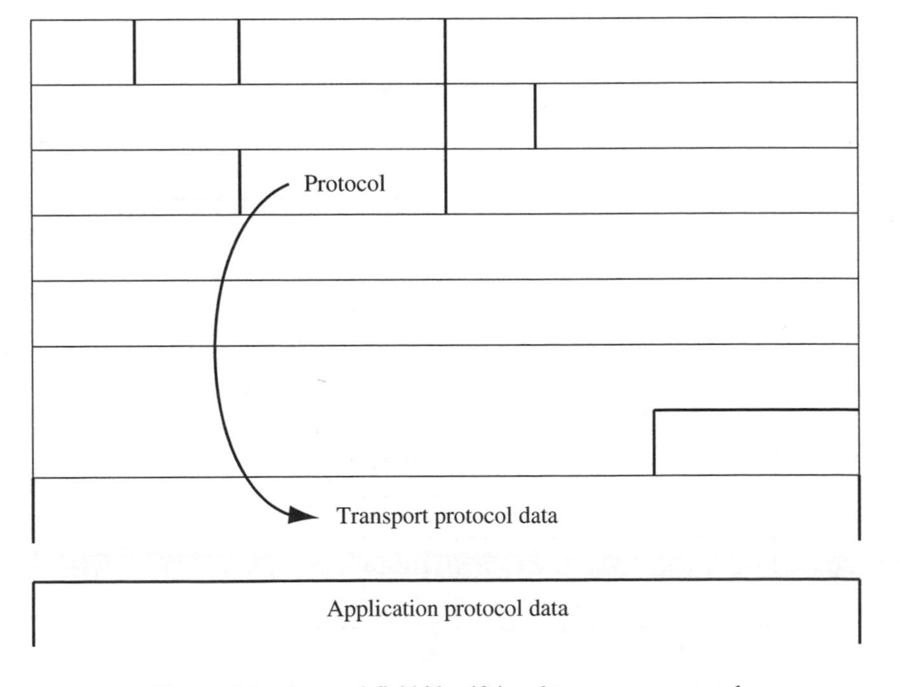

Figure 3.3 Protocol field identifying the transport protocol

3.3 IPv6 Header

The basic IPv6 header [RFC2460] is simpler than the IPv4 header: it has 8 fields instead of 12. The IP header length and checksum fields are removed, the flow label field is added and the fragmentation fields are moved in the extension headers. The size of the basic header is 40 octets, which is double the IPv4 header size. This is due to the 4 fold increase of each address field, from 32 bits to 128 bits each, or from 4 octets to 16 octets each. Figure 3.4 shows the IPv6 header.

Important to remember:

The IPv6 header size is doubled compared to the IPv4 header size, but the size of the addresses is four times.

The Next Header field is the equivalent of the IPv4 Protocol field and identifies what is in the payload after the basic header, as shown in Figure 3.5.

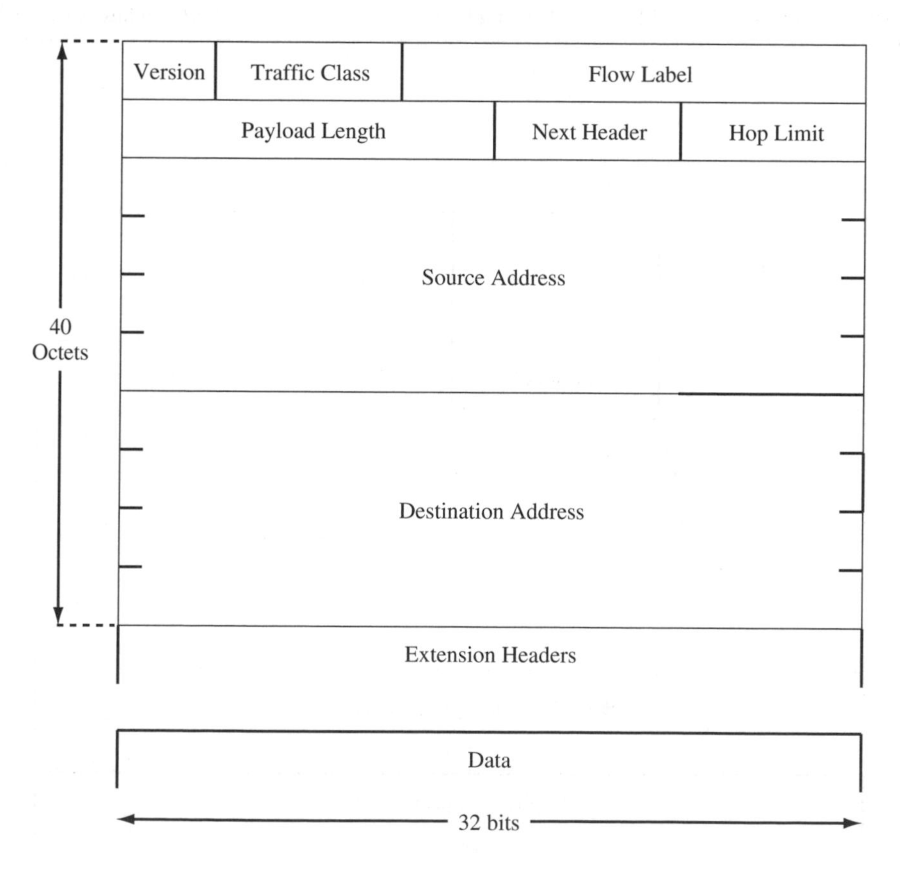

Figure 3.4 IPv6 header

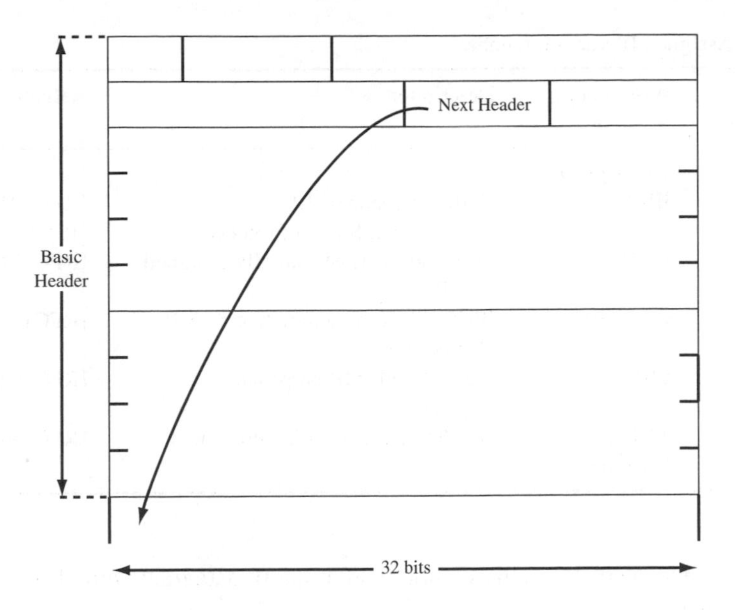

Figure 3.5 Use of the Next Header field

Extension headers are optional and placed after the basic header. They are similar to the options in the IPv4 header. After the extension headers, the transport header follows in the payload.

3.4 Header Fields

Table 3.2 list all the fields. Each field is described in the order it appears in the header, starting with the version field.

3.4.1 Version

The 'Version' field has 4 bits and identifies the version of the IP protocol . This enables the operating system receiving the datagram to relay it to the right stack. The current Internet

Table 3.2 IPv6 basic header fields

Name	Size	Description
Version	4 bits	= 6 for IPv6
Traffic Class	8 bits	Diffserv and Explicit Congestion Notification bits
Flow Label	20 bits	Per-flow identification
Payload Length	16 bits	Size of the inner datagram (after the basic header)
Next Header	8 bits	Identification of the inner datagram
Hop Limit	8 bits	Maximum number of hops
Source Address	128 bits	Source address of the datagram
Destination Address	128 bits	Destination address of the datagram

Table 3.3 Assigned IP version numbers

Value of the Version field	Assignment	Description	Reference
0–3	Unassigned		
4	IPv4	Current version of IP	[RFC791]
5	ST	Experimental Stream Protocol	[IEN119, RFC1190]
6	IPv6	Successor of IPv4; initially assigned to SIP	[RFC1710, RFC1752]
7	CATNIP	IPv7, TP/IX; an alternative to SIP. Deprecated.	[RFC1707, RFC1475]
8	PIP	Merged with SIP proposal. Deprecated.	[RFC1621, RFC1622]
9	TUBA	An alternative to SIP. Deprecated.	[RFC1347]
10–15	Unassigned		

Protocol has the value of '4' in the version field of the IP datagram while IPv6 has the value of '6', obviously.

Table 3.3 lists the current assignments of version numbers [IANAVERSIONNUMBERS]. Version 5 is an experimental streaming protocol, discussed briefly in Chapter 14. Version 6 was assigned to the SIP (and later SIPP) proposal that later was chosen as the basis of the new IP protocol: it was then renamed IPv6. Versions 7, 8 and 9 were assigned to competitive proposals of SIP and these assignments are now being deprecated because they are not used. Since it is a 4 bit field, only 16 assignments are possible.

3.4.2 *Traffic Class*

The 'Traffic class' field has 8 bits. It is defined as 'Type of Service' in IPv4 and the bits were initially assigned to serve different levels of service for the datagrams [RFC791]. The processing of these bit assignments in IPv4 stacks is more or less implemented. These assignments, shown in Figure 3.6, have been redefined [RFC2474] to differentiated services (diffserv) code points (DSCP). Diffserv enables quality of service in the network and is discussed in Section 14.2.

The last two unused bits are defined for explicit congestion notification (ECN) [RFC3168], discussed in Section 19.2.1.

These new assignments (Diffserv and ECN) of the bits are common to IPv4 and IPv6 and processing is identical. The only difference is the name of the field.

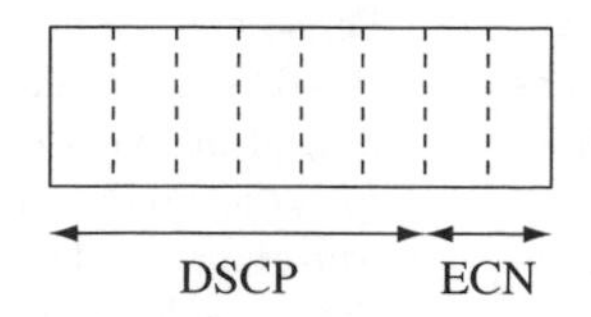

Figure 3.6 Diffserv Codepoints Assignments in the TOS/Traffic Class field

Important to remember:

Traffic Class bits assignment and processing are identical in IPv4 and IPv6.

3.4.3 Flow Label

In IPv4, the routers have to open the IP header and the transport header in order to identify flows and then to apply specific processing for quality of service. This incurs a high penalty on the router and introduces delays. Furthermore, it is very difficult to aggregate flows of protocols that allocate ports dynamically.

The IPv6 flow label field has 20 bits and is the only new field introduced in the header. It enables a specific per-flow processing of datagrams. This corresponds to the integrated services (intserv) approach for quality of service. The source tags the flows by putting a flow identifier in the flow label field. This enables the routers to process specifically and efficiently those tagged datagrams, since the routers have direct access to the flow label and don't have to guess or find the flow using the transport and application data.

3.4.4 Payload Length

In an IPv4 header, there are two length fields: header length defining the length of the header and total length, the length of the full datagram.

In an IPv6 header, only one length field is available: Payload Length, which is defined as the length of what follows the basic IP header. This means the transport and application data as well as extension headers if they are used. It has 16 bits, which means that the maximum payload length is 2^{16} or 65536 octets.

3.4.5 Hop Limit

In IPv4, the Time To Live (TTL) field was defined in units of seconds before the datagram dies, but since it is difficult to assume synchronized clocks over the Internet, it was implemented, as suggested in the specification [RFC791], by decreasing the value by one each time a router forwards the datagram. When a router decreases the value to zero, it rejects the datagram and sends a 'Time exceeded' ICMP error message back to the source. This behavior is integrated in IPv6 and the name of the field is renamed appropriately to 'hop limit'. It also has the same length of 8 bits.

3.4.6 Next Header

The 'Next Header' field has 8 bits and identifies the data inside the payload of the IP datagram. Typically, this is the transport protocol, like TCP or UDP, or in the case of encapsulation of IPsec, it is the ESP or AH IPsec headers. This is the exact same semantic as the Protocol field in IPv4 and it shares the same values for both IP protocols. The list of registered values is handled by IANA[IANAPROTOCOLNUMBERS]. Table 3.4 lists a few usual values.

The only difference in IPv6 regarding the 'Next Header' field is the processing of options. These are defined in extension headers with their own next header value.

Table 3.4 Next Header typical values

Value	Keyword	Description	Reference
6	TCP	Transmission Control Protocol	[RFC793]
17	UDP	User Datagram Protocol	[RFC768]

3.5 Extension Headers

The IP datagram header must be fixed to enable fast hardware-based processing. The basic IPv6 header is fixed: 40 octets with fixed-length fields. In order to process options and exceptions, extension headers are used. They are 64 bits aligned to enable fast hardware/register-based processing.

Each type of extension header is identified by a Next Header specific value, as listed in Table 3.5. This table contains only the specific IPv6 extension headers.

Extension headers are daisy-chained. The next header value of the IP header is pointing to the first extension header, the next header fields of the successive extension headers points to the next extension header until the last extension header where the next header points to the transport header.

Important to remember:

Extension headers are used for the optional info of the IPv6 datagram. They are daisy-chained up to the transport header.

Figure 3.7 shows two extension headers between the IP header and the transport header. The second extension header has its Next Header field, shown as 'NH', identifying the transport protocol following it.

Figure 3.8 shows an IPv6 datagram with a routing header (43 as listed in Table 3.5), an authentication header (51) and TCP transport (6).

Table 3.5 IPv6 Next Header values for extension headers

Value	Description	Book Section	Reference
0	Hop-by-Hop Option	3.6.4, 14.3, 15.2	[RFC2460]
43	Routing Header	3.5.2, 9.2	[RFC2460]
44	Fragment Header	3.6.3	[RFC2460]
50	Encapsulation Security Payload	3.5.5, 13.1	[RFC2406]
51	Authentication Header	3.5.5, 13.1	[RFC2402]
59	No Next Header	3.5.6	[RFC2460]
60	Destination Options	3.5.4,	[RFC2460]

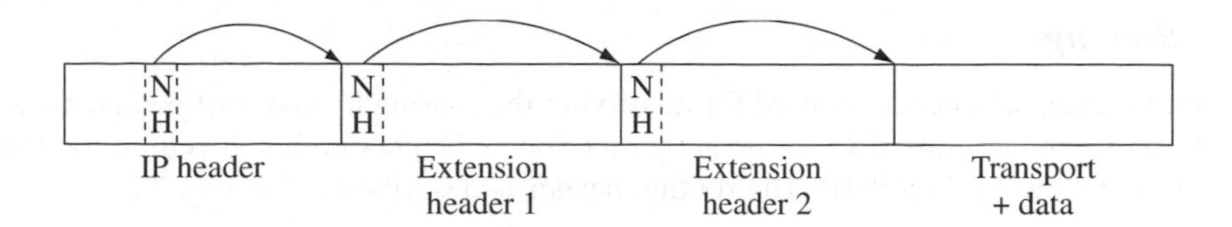

Figure 3.7 Extension headers are daisy-chained

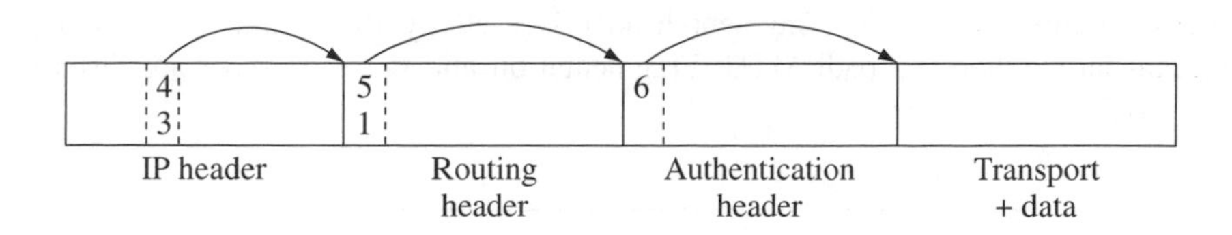

Figure 3.8 Example of 2 extension headers

3.5.1 Hop-by-Hop Option

This extension header has to be examined by every intermediate node (e.g. routers) along the path to the destination. There is currently two uses of the Hop-by-Hop options extension header: the router alert and the jumbogram.

3.5.1.1 Router Alert

This function implemented as a hop-by-hop option extension header is to alert the intermediate nodes in the path to process specifically the datagram [RFC2711]. Some protocols, like Resource Reservation Protocol (RSVP), need to send control messages in a timely fashion to all routers of a path by sending only one datagram. The usual way for routers to see these messages is to look in the transport data in all datagrams, in order to find those that are relevant. This is a very consuming process for a router. Router Alert in the header of an IPv6 datagram is an easier way to alert the routers of the path so that they will intercept only these tagged datagrams.

Inside the Router Alert header, only one field is used and it identifies the specific upper-layer protocol in the datagram. The current registered use of Router Alert [IANAROUTERALERT] is for:

- the Multicast Listener Discovery (MLD) protocol, discussed in Section 15.2;
- the Resource Reservation Protocol discussed in Section 14.3.1.

3.5.1.2 Jumbogram

Jumbograms are IPv6 datagrams larger than 64 K octets. They need special processing on each hop since they are oversized compared to the maximum of 16 bits of the Payload Length field. Jumbograms are discussed in Section 3.6.4.

3.5.2 *Routing*

The routing extension header is used for modifying the routing behavior of datagrams along the path. It is similar to the IPv4 Loose and Strict Source Route options. It is used for generic source routing and by MobileIP. The routing header is described in Section 9.2.

3.5.3 *Fragment*

Fragmentation in IPv6 is processed only by the end nodes. Intermediate nodes do not fragment or process fragmentation. The fragment header is used by the source node to fragment a datagram larger than the path MTU. Fragmentation and datagram size are discussed in Section 3.6.3.

> **Important to remember**:
>
> Fragmentation in IPv6 is done only by the end nodes, whereas in IPv4 it is handled by the routers.

3.5.4 *Destination Options*

This header is only processed by the destination node. MobileIP, discussed in Chapter 11, uses this header.

3.5.5 *Authentication and Encapsulating Security Payload*

These headers are the IPsec protocol headers. They are only processed by the destination node. In the case of tunneling, the destination address points to the end of the tunnel, while the inner datagram points to an end node inside the destination network. In this case, the destination node processing the IPsec header is the tunnel endpoint.

IPsec is identical in IPv4 and IPv6, while IPv6 enables its use more widely because NATs are no longer used and are harmful for IPv4, and because IPsec is mandatory in IPv6.

> **Important to remember**:
>
> IPsec is identical in IPv4 and IPv6, but it will be easier to use and deploy in IPv6.

IPsec headers are described in Section 13.1.

3.5.6 *No Next Header*

This header indicates no payload, which means that all relevant information of this datagram is in the header and extension headers. This happens when control protocol messages use extension headers and do not need to put extra information in the payload.

Table 3.6 Extension header order

Order	Header
1	Hop-by-Hop[1]
2	Routing[2]
3	Fragment
4	Authentication
5	Encapsulating Security Payload
6	Destination Options
7	Upper-layer

Notes:
[1] If IPv6-in-IPv6 tunneling is used, then the very first header should be the inner IPv6 header.
[2] If both routing and destination options headers are in the same datagram, then destination options should be before the routing header.

3.5.7 Order of the Extension Headers

Extension headers are daisy-chained and the order is important because of the following facts:

- Only one of them, the Hop-by-Hop Options, has to be processed by every intermediate node.
- The routing header has to be processed only by the intermediate nodes (e.g. routers) listed in the source route, as described in Section 9.2.
- At the destination, fragments should be processed before anything else.

The order of headers is specified to enable the most efficient processing for all intermediate nodes and the destination node, shown in Table 3.6.

Source nodes follow this order, but destination nodes are implemented to receive in any particular order. Intermediate nodes (e.g. routers) should only process hop-by-hop extension headers. This is much more efficient than IPv4 where the router must look at each variable-length option in the IPv4 header. This IPv4 processing is more difficult to implement in hardware. In IPv6, only a single lookup at a specific place in the IPv6 header determines if additional processing has to be done by the router.

3.6 Datagram Size

The maximum size of an IPv6 datagram depends on the Maximum Transmission Unit (MTU) and on the Payload Length field size. The Payload Length field is 16 bits wide, which means that a normal payload cannot be larger than 64 K octets, unless a specific feature named Jumbogram, discussed in Section 3.6.4, is used.

Table 3.7 MTU for IPv4 and IPv6

	IPv4	IPv6
Minimum MTU	68	1280
Most efficient MTU	576	1500

3.6.1 Maximum Transmission Unit

The MTU is the largest size a given link layer technology can support for datagrams. For any link, IPv4 mandates a minimum MTU of 68 octets, while the recommended MTU is 576 (see Table 3.7). This enables any IPv4 stack to send a 68 octet datagram with the assurance that it is guaranteed to be forwarded up to the destination. 68 octets is very small, since most current link layer technologies have a minimum MTU of 1500.

In IPv6, the minimum MTU is 1280 octets. However, the mandatory minimum fragment reassembly buffer size is 1500 octets, which tends towards a 1500 octet MTU as the most efficient MTU.

Important to remember:

The minimum MTU for IPv6 is 1280 octets and the most efficient MTU is 1500. The maximum datagram size is 64 K octets.

3.6.2 Path MTU Discovery

Since routers do not fragment IPv6 datagrams, if a datagram is larger than the size of one link in the path to the destination, the router connected to that link and receiving the datagram sends an ICMP error message to the source and the datagram is dropped. IPv6 nodes use Path MTU (PMTU) discovery to discover the right MTU to use for this destination.

Path MTU discovery is not new for IPv6 [RFC1981]. It is defined for IPv4 [RFC1191], but has been rarely used for IPv4. If an IPv6 stack does not use it, then it sends datagrams with the minimum MTU (1280 octets).

3.6.2.1 Path MTU Discovery Process

The discovery process of the Path MTU is illustrated in Figure 3.9. Source node A wants to send a datagram to destination node E.

Node A starts by sending a datagram to destination node E (step 1) using its own link-layer MTU (8122). This first datagram reaches router B which cannot forward the datagram because the MTU of the next link is smaller. B sends back to A an ICMP error message ‘Packet too big’ with the MTU of the next link of B (4352) in the ICMP error datagram. The source node A then uses this new received MTU (4352) and resends the datagram to E. The datagram is forwarded by B and C but stops at D because the MTU of the next link is

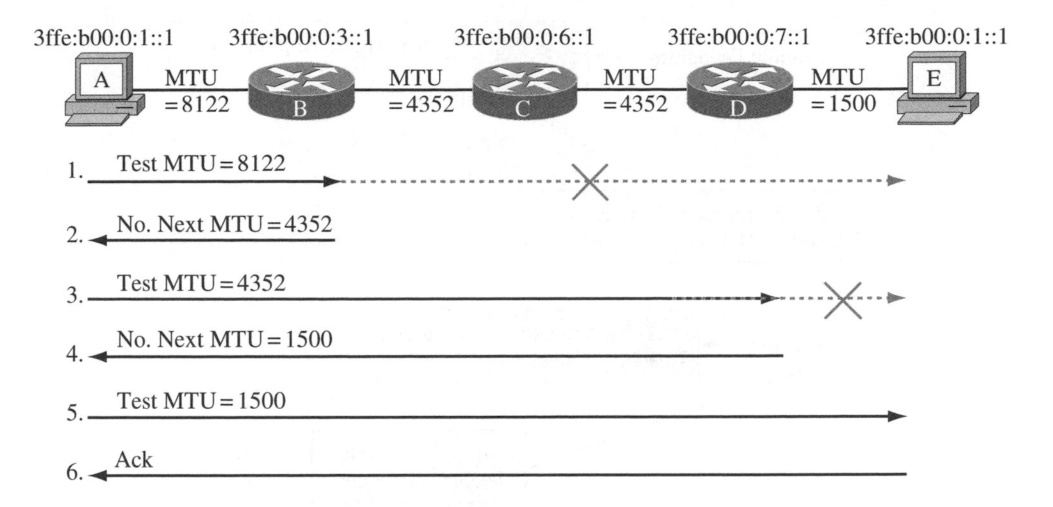

Figure 3.9 Path MTU discovery process

smaller. D sends back to A a similar ICMP error message with the MTU of the next link of D (1500) in the ICMP error datagram. Node A then uses this new MTU (1500) and resends the datagram to E which receives it correctly. Node A then selects the last used MTU (1500) as the Path MTU for all future communications with node E.

Nodes build a table of Path MTUs for each of their destinations.

3.6.2.2 Considerations

Since the network topology changes over time between the two end nodes, this means that the actual forwarding path changes, which implies that the Path MTU may change by increasing or decreasing. But the source node is not informed directly about those network changes. So the discovery process should be done periodically to ensure the best elected Path MTU.

Since the transport layers are responsible for packeting the data, they need to know the Path MTU at all time. This means that there is a close interaction between this PMTU discovery process and the transport layer process.

There is essentially no cost to PMTU considering that the first packet is used to test the PMTU.

3.6.3 Fragmentation

If the source node finds that the PMTU is smaller than the size of the datagram to be sent, then the source node has to fragment. This might not happen often, but it is a mandatory feature of any IPv6 stack. Fragmentation in IPv6 is done only by source nodes, not by routers as in IPv4. Destination nodes reassemble fragments.

The process and the type of information needed to support fragments are mostly identical to IPv4 fragmentation, but in IPv6 it is done using extension headers instead of options header fields in IPv4. A datagram is cut into many fragments. Figure 3.10 shows the original

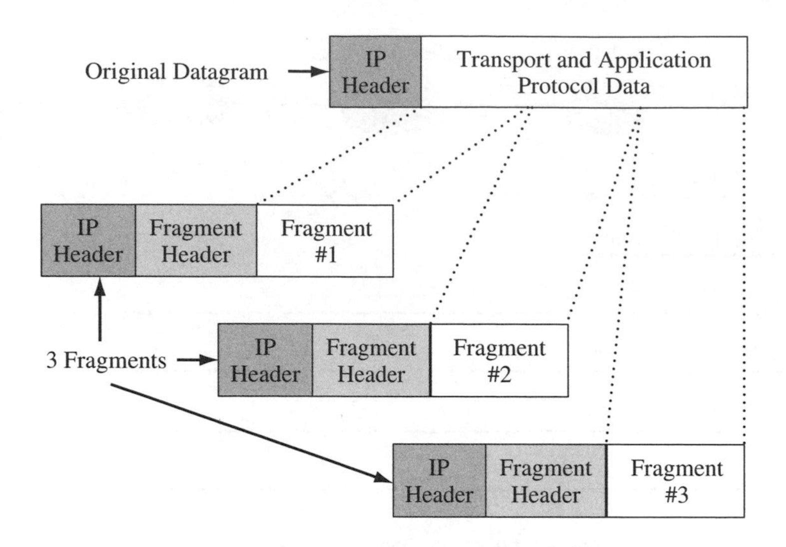

Figure 3.10 Original datagram and fragments

datagram and the resulting three fragments. Each fragment size is smaller or equal to the PMTU.

Fragments are IPv6 datagrams with the header of the original datagram and an additional fragment header, which contains the needed additional information as described in Table 3.8.

All fragments of the same datagram have the same identification, which is the 32 bit identification field in the fragment header. The order of fragments is specified by a 13 bit field called 'fragment offset'. A flag also identifies the last fragment for that datagram. Since the fragments can be received by the destination in a random order, the destination node uses the identification field to group them and the fragment offset to reassemble the original datagram in the right order.

3.6.4 Jumbogram

While some special link-layer technologies handle MTUs greater than 64 K octets, the upper limit of the IP datagram size is still limited to 64 K octets by the 16 bits of the Payload Length field in the header, as described in Section 3.4.4. This prohibits the use of datagrams larger than 64 K octets.

Table 3.8 Fields in the fragment header

Field name	Description	Length
Fragment Offset	Offset to order the fragments	13
M flag	Identify the last fragment	1
Identification	Group all fragments of the same datagram together with this ID	32

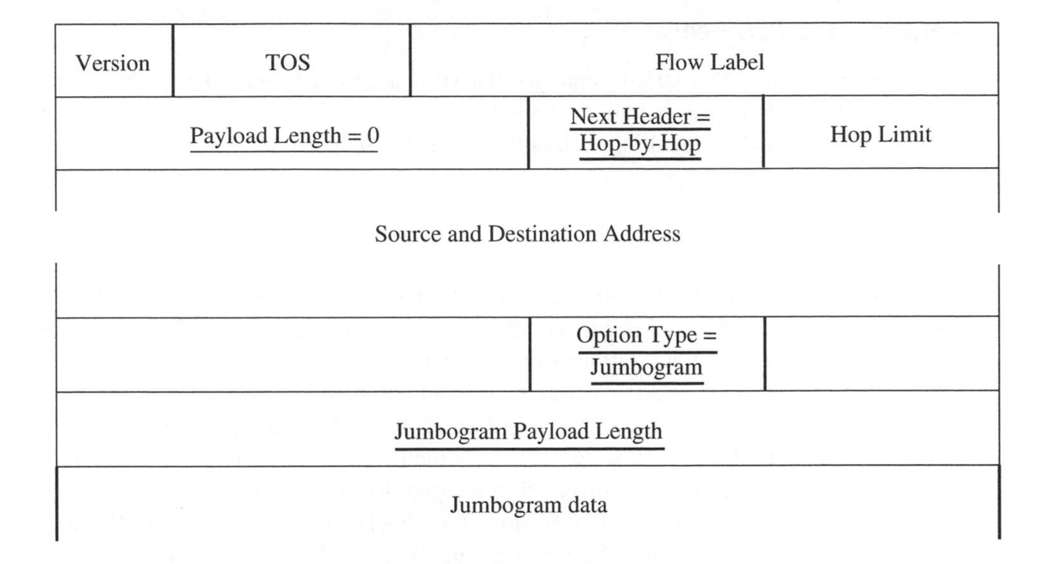

Figure 3.11 Jumbogram header

Jumbograms are datagrams larger than 64 K octets that need special processing by all intermediate nodes (i.e. routers). A jumbogram is identified by the following fields, shown in Figure 3.11:

- The Payload Length header field is set to zero.
- The Next Header is set to Hop-by-Hop.
- The Hop-by-Hop extension header identifies a jumbogram and contains a 32 bit value specifying the size of the datagram. Therefore, jumbograms have a maximum size of 4G octets (2^{32}).

When a jumbogram-aware stack sees these conditions, it can handle the jumbogram. Note that jumbograms don't have to be supported in every IPv6 stack. In fact, jumbograms are used essentially for fast and large links inside or between supercomputers.

Modifications of UDP, TCP, (both limited by 16 bit counters) and ICMP, within the context of jumbograms, are also defined [RFC2675].

3.6.5 Header Compression

The IPv6 header doubles in size to 40 octets compared to the IPv4 header. The IPv6 large header size is an important issue on limited bandwidth link layers, such as radio. It is also an important penalty for small datagrams. Multiple header compression schemes have been designed over time. The Robust Header Compression (RoHC) standard, the last IETF header compression standard, is discussed in Section 12.2. It is specifically used in some wireless technologies.

3.7 Upper-layer Protocols

As a new version of IP, IPv6 strictly changes the IP header but it should not change the upper layer protocols. However, small changes were necessary: i) the mandatory checksum in the transport layer, and ii) handling of literal IPv6 addresses.

3.7.1 Checksum

A checksum is applied to the IPv4 header. This enables the IPv4 stack to verify if the header changed during the trip of the packet between the source and the destination. TCP also uses a checksum for the transport and UDP uses an optional checksum.

No checksum is applied to the IPv6 header. Instead, IPv6 relies on both the link-layer and the transport layer checksums, as shown in Figure 3.12. The rationale is based on the fact that most link-layer already have checksums, which means that there is a very small probability that the datagram will be corrupted in the path to its destination. If this happens, then the transport layer checksum will find it, since this checksum also takes many IP header fields in its calculation. Not doing checksum at the IP layer leaves the routers free for calculating checksums on each forwarded datagram, which gains efficiency and scalability in the long run.

UDP checksums are optional in IPv4. UDP checksums are mandatory over IPv6. Other transport protocols must also use a checksum.

> **Important to remember**:
>
> IPv6 has no IP checksum, but mandates the UDP checksum.

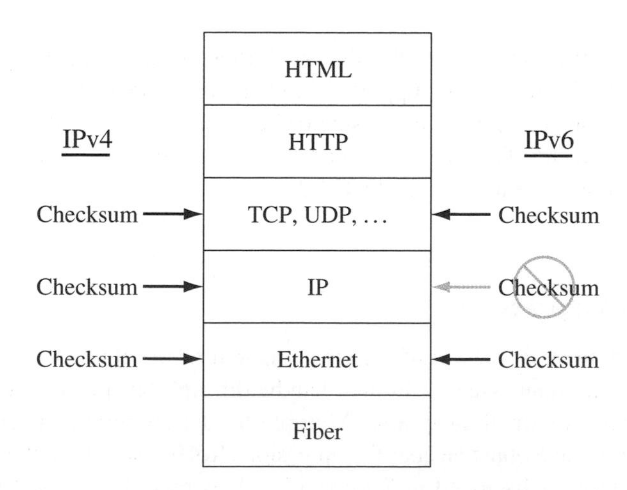

Figure 3.12 Checksums in IPv4 and IPv6

3.7.2 Implications in Application Protocols

While IPv6 does not mandate any change in the upper-layer protocols, the protocols or their implementation still need to be updated. Some upper-layer protocols have to handle IP addresses in their mechanism. If these protocols were recently designed or updated, then the handling of IPv6 addresses is taken care of. If not, then they would have to be modified accordingly.

For example, the FTP protocol [RFC959] exchanges IP addresses in its control connection. The protocol has been updated [RFC1639, RFC2428] to support IPv6 addresses.

Some protocols may be modified to use the new functions in IPv6. For example, MobileIPv6 uses the routing and the destination options headers to optimize the routing of the MobileIP datagrams. Mobile nodes also use anycast addresses to find agents. MobileIP is discussed in Chapter 11.

3.8 Summary

The IPv6 header is similar to the IPv4 header, but:

- the header size is 40 octets (instead of 20);
- the options are processed as daisy-chained extension headers;
- the fragmentation is an extension header, and is only done by source nodes;
- there is no checksum in the IPv6 header, but mandatory transport protocol checksums;
- the path MTU discovery is mandatory.

3.9 References

[IANAPROTOCOLNUMBERS] IANA, 'Protocol numbers', http://www.iana.org/assignments/protocol-numbers, October 2004.

[IANAROUTERALERT] IANA, 'IPv6 Router Alert Option Values', http://www.iana.org/assignments/ipv6-routeralert-values, October 2001.

[IANAIPVERSIONNUMBERS] IANA, 'IP Protocol Numbers', ftp://ftp.iana.org/assignments/version-numbers, November 2001.

[IEN119] Forgie, J., 'ST–A Proposed Internet Stream Protocol', IEN 119, September 1979.

[RFC768] Postel, J., 'User Datagram Protocol', STD 6, RFC 768, August 1980.

[RFC791] Postel, J., 'Internet Protocol', STD 5, RFC 791, September 1981.

[RFC793] Postel, J., 'Transmission Control Protocol', STD 7, RFC 793, September 1981.

[RFC959] Postel, J. and Reynolds, J., 'File Transfer Protocol', STD 9, RFC 959, October 1985.

[RFC1190] Casner, S., Lynn, C., Park, P., Schroder, K. and Topolcic, C., 'Experimental Internet Stream Protocol: Version 2 (ST-II)', RFC 1190, October 1990.

[RFC1191] Mogul, J. and Deering, S., 'Path MTU Discovery', RFC 1191, November 1990.

[RFC1347] Callon, R., 'TCP and UDP with Bigger Addresses (TUBA), A Simple Proposal for Internet Addressing and Routing', RFC 1347, June 1992.

[RFC1475] Ullmann, R., 'TP/IX: The Next Internet', RFC 1475, June 1993.

[RFC1621] Francis, P., 'Pip Near-term Architecture', RFC 1621, May 1994.

[RFC1622] Francis, P., 'Pip Header Processing', RFC 1622, May 1994.

[RFC1639] Piscitello, D., 'FTP Operation Over Big Address Records (FOOBAR)', RFC 1639, June 1994.

[RFC1707] McGovern, M. and Ullmann, R., 'CATNIP: Common Architecture for the Internet', RFC 1707, October 1994.

[RFC1710] Hinden, R., 'Simple Internet Protocol Plus White Paper', RFC 1710, October 1994.

[RFC1752] Bradner, S. and Mankin, A., 'The Recommendation for the IP Next Generation Protocol', RFC 1752, January 1995.

[RFC1981] McCann, J., Deering, S. and Mogul, J., 'Path MTU Discovery for IP version 6', RFC 1981, August 1996.
[RFC2402] Kent, S. and Atkinson, R., 'IP Authentication Header', RFC 2402, November 1998.
[RFC2406] Kent, S. and Atkinson, R., 'IP Encapsulating Security Payload (ESP)', RFC 2406, November 1998.
[RFC2428] Allman, M., 'FTP Extensions for IPv6 and NATs', RFC 2428, September 1998.
[RFC2460] Deering, S. and Hinden, R., 'Internet Protocol, Version 6 (IPv6) Specification', RFC 2460, December 1998.
[RFC2474] Nichols, K., Blake, S., Baker, F. and Black, D., 'Definition of the Differentiated Services Field (DS Field) in the IPv4 and IPv6 Headers', RFC 2474, December 1998.
[RFC2675] Borman, D., Deering, S. and Hinden, R., 'IPv6 Jumbograms', RFC 2675, August 1999.
[RFC2711] Partridge, C. and Jackson, A., 'IPv6 Router Alert Option', RFC 2711, October 1999.
[RFC3168] Ramakrishnan, K., Floyd, S. and D. Black, 'The Addition of Explicit Congestion Notification (ECN) to IP', RFC 3168, September 2001.

4

Addressing

As discussed in Chapter 1, a central rationale behind IPv6 is a much larger address space. This address space not only fulfills the exhaustion requirements, but also enables the use of the addressing structure for protocol means, which gives IPv6 a much more flexible protocol for the new applications requirements. Therefore, this chapter is very important for the understanding of IPv6.

The address space is first presented followed by the format of an IPv6 address. Each type of address is then discussed: unicast, multicast and anycast. Finally, the address architecture is detailed.

4.1 Address Space

IPv4 addresses are 32 bits wide, which expands to a maximum of 4,294,967,296 unique addresses. IPv6 addresses are 128 bits wide, which expands to a maximum of 340,282,366,920,938,463,463,374,607,431,768,211,456 unique addresses, or 3.4×10^{38}. This is 4 times the number of bits or 79,228,162,514,264,337,593,543,950,336 times the number of IPv4 unique addresses. Given the Earth's population of around 7 billion people, this is 48,611,766,702,991,209,066,196,372,490 (4.8×10^{28}) addresses per person on the planet (see Table 4.1). Assuming the Earth's surface is 511,263,971,197,990 square meters, then this is 665,570,793,348,866,943,898,599 (6.6×10^{23}) addresses per square meter of the Earth's surface.

As in IPv4 and other numbering systems like the telephone numbering system, we will not be able to use all of this address space efficiently. Even worst, we are using a significant part of the address space for protocol use as described in the next sections. Nevertheless, we still have a lot of address space (don't ask me when the IPv6 addresses will be exhausted ... you never know ...).

Migrating to IPv6: A Practical Guide to Implementing IPv6 in Mobile and Fixed Networks Marc Blanchet

Table 4.1 IPv6 address space compared with Earth metrics

Metric	Value	IPv6 address space
Earth population	~ 7 billion	4.8×10^{28} addresses per person on the earth
Earth surface	5.1×10^{14} square meters	6.6×10^{23} addresses per square meter of the earth

4.2 Format of an Address

IPv4 addresses are represented in text as 4 fields of decimal numbers, from 0 to 255, representing 8 bits each, such as 192.0.2.156. Using the same representation technique with IPv6 addresses will give 16 fields of decimal numbers, such as the following: 134.109.89.132.123.192.111.222.9.101.167.189.245.243.190, which is obviously pretty long and prone to errors when typing. Don't use this format, it is illegal in IPv6! IPv6 uses hexadecimal notation to better compress the representation of addresses. Rules are also introduced to compress the representation even more.

4.2.1 Text Representation of Addresses

IPv6 addresses are represented as 8 fields of hexadecimal numbers (0–F), each field representing 16 bits using 4 hexadecimal digits and fields are separated by a colon ':' [RFC3513]. For example, 2001:0000:1234:0000:0000:C1C0:ABCD:0876 is a valid address.

The following rules can be applied to address representations:

(a) Letters are case-insensitive. For example, 'AB09' equals 'ab09'.
(b) Leading zeros in a field are optional. For example, '00c1' equals 'c1'.
(c) Successive fields of '0' are represented as '::', but only once in an address.[1]

For example,

2001 : 0000 : 1234 : 0000 : 0000 : C1C0 : ABCD : 0876

can be represented, using rule a), by:

2001 : 0000 : 1234 : 0000 : 0000 : c1c0 : **abcd** : 0876

which can be compressed using rule b) to:

2001 : **0** : 1234 : **0** : **0** : c1c0 : abcd : **876**

which can be further compressed using rule c) to:

2001 : 0 : 1234::c1c0 : abcd : 876

[1] Why? A parser that sees a '::' in an address will do the following: expand the other parts of the address, count 128 bits – the number of bits in the other parts – and assign 0 to the remaining bits. A parser that reads an address which has multiple '::' cannot guess how many 0s were in the original non-compressed address.

Table 4.2 Examples of address representation

Full representation	Compact representation
2001:0000:1234:0000:0000:C1C0:ABCD:0876	2001:0:1234::c1c0:abcd:876
3ffe:0b00:0000:0000:0001:0000:0000:000a	3ffe:b00::1:0:0:a
FF02:0000:0000:0000:0000:0000:0000:0001	ff02::1
0000:0000:0000:0000:0000:0000:0000:0001	::1
0000:0000:0000:0000:0000:0000:0000:0000	::

which is a much more compact written form of the same address. Note that rule c) can only be applied once in an address. Any address containing more than one '::' is invalid.

Table 4.2 contains examples of applying the above rules to various addresses.

Note that the full representation, as well as any compressed forms, are all valid address representations.

4.2.2 Text Representation of Prefixes

A prefix represents a range of addresses. The format is <address>/<prefix length>. The prefix length identifies the separation of the address. There is no limitation nor boundaries on the prefix length, as in IPv4 with Classless Interdomain Routing [RFC1519].

In the context of separating the network part and the host part of an address, the prefix length represents the equivalent of the IPv4 subnet mask. For example, a host with 3ffe:b00:c18:1::1/64 identifies the first 64 bits (from the /64) as the network number (3ffe:b00:c18:1 or 3ffe:0b00:0c18:0001) and the remaining 64 bits ($128 - 64 = 64$) as the host part (0000:0000:0000:0001). If 3ffe:b00:c18:1::1/**124** is used, then the network part has 124 bits and the host part has 4 bits ($128 - 124 = 4$).

In the context of a site address space, such as 3ffe:0b00:c18::/48, the prefix represents the range of addresses available for numbering the networks and the hosts.

Table 4.3 shows examples of prefixes.

4.2.3 Addresses in URL

Uniform Resource Locators (URL) [RFC3896] usually contain domain names, such as http://www.example.com/example.html instead of IP addresses like http://192.0.2.1/example.html. If one uses a port number with an IPv4 address, it is written as: http://192.0.2.1:8080/example.html.

IPv6 addresses use the colon ':' as field separator and URLs use the colon ':' to separate the IPv4 address to the port number, which makes both incompatible with IPv6.

Table 4.3 Examples of Prefixes

Prefix	Left Part *Right Part*
3ffe:b00:c18:1::1/64	**3ffe:0b00:0c18:0001**:*0000:0000:0000:0001*
3ffe:b00:c18:1::1/124	**3ffe:0b00:0c18:0001:0000:0000:0000:000***1*
3ffe:b00:c18:1::1/40	**3ffe:0b00:0c***18:0001:0000:0000:0000:0001*

Table 4.4 Examples of URLs with IPv6 addresses

Address	URL
2001:0:1234::c1c0:abcd:876	http://[2001:0:1234::c1c0:abcd:876]/test.html
3ffe:b00::1:0:0:a	ftp://[3ffe:b00::1:0:0:a]/test.txt
::1	ftp://[::1]
3ffe:b00::1:0:0:a on port 8080 with http	http://[2001:0:1234::c1c0:abcd:876]:8080/test.html

An IPv6 address in a URL must be enclosed in square brackets: '[' and ']' [RFC3986]. Table 4.4 shows some examples of URLs with IPv6 addresses.

Obviously, users prefer to type domain names instead of IP addresses in URLs, even more with IPv6!

The important consequence of adding square brackets in URLs is that all applications parsing URLs must be modified to support IPv6 URLs.

Some command-line shells consider the brackets as special characters, so these IPv6 brackets have to be escaped when used as arguments of commands. The following is an example of the ftp command on FreeBSD within a csh shell.

```
% ftp http://[3ffe:b00:c18:1::10]:80/index.htm
ftp: No match.
% ftp 'http://[3ffe:b00:c18:1::10]:80/index.htm'
Requesting http://[3ffe:b00:c18:1::10]:80/index.htm
100% |*******************************| 11196        00:00 ETA
Successfully retrieved file.
%
```

The first try, without escaping the brackets, makes the shell try to expand the brackets to some local filename which obviously does not exist, so it responds with 'no match' (of a file). By escaping the whole URL inside apostrophes ('), the HTTP request succeeds.

4.3 Unicast Addresses

Three kinds of addresses exist in IPv6: unicast, multicast and anycast. Unicast addresses are used for communications between two nodes. A unicast address is a one-to-one address. Multicast addresses are used for communications between one node and many nodes and anycast addresses are used for communications between one node and the nearest node among a group of nodes.

Multicast addresses start with 'ff' as the leftmost octet. Any other value of the leftmost octet ('00' to 'fe') identifies a unicast address. Anycast addresses are formed using the unicast address space, so they cannot be distinguished from unicast addresses.

4.3.1 Global Unicast Addresses

Global addresses are used for communications between nodes on the Internet. They are the normal addresses that every node uses. These addresses, called global unicast addresses

[RFC 3513], are currently assigned as 001 as the 3 leftmost bits in the 128 bit address. This corresponds to addresses from 2000:: to 3fff:ffff:ffff:ffff:ffff:ffff:ffff:ffff, or 2000::/3.

This address space has been defined to use the leftmost 64 bits for the network prefix and the rightmost 64 bits for the host part. So, except in specific cases, all subnets in IPv6 have the same prefix length of 64 bits (/64). The host part, now precisely named as the interface identifier because hosts and interfaces may have multiple addresses, is based on the IEEE EUI-64 standard described in Section 5.2.1.

Important to remember:

All subnets in the global unicast address space 2000::/3 have the same prefix length of 64 bits (/64).

IPv6 enforces the use of provider-based addresses. This means that an organization cannot get an IP address block directly from a registry, but will get it from its upstream provider. The main benefit is the aggregation of prefixes done by the providers which results in a smaller global routing table. However, multiple peerings and multihoming may annihilate this aggregation, as discussed in Section 9.11.

As shown in Figure 4.1, the basic structure of a global unicast address contains 3 parts, from left to right:

(a) The leftmost part has a length of 48 bits and is the prefix allocated by the provider to a site. The details of allocation and assignments inside the 48 bits are discussed in Section 23.3.1.
(b) The middle part has a length of 16 bits and contains the subnet numbers inside a site. 2^{16} subnets are available inside any site.
(c) The rightmost part has a length of 64 bits and contains the host part, also called interface identifier. This identifies the host in the subnet. 2^{64} addresses are available for hosts in each subnet. This is a very large space spend, but it enables autoconfiguration by embedding the MAC address in this part of the IP address, as described in Section 5.2.1.

The important consequences of this structure are:

(a) /64 is the separation between the network and the host parts. Contrary to IPv4, IPv6 uses fixed length subnet masks.[2]

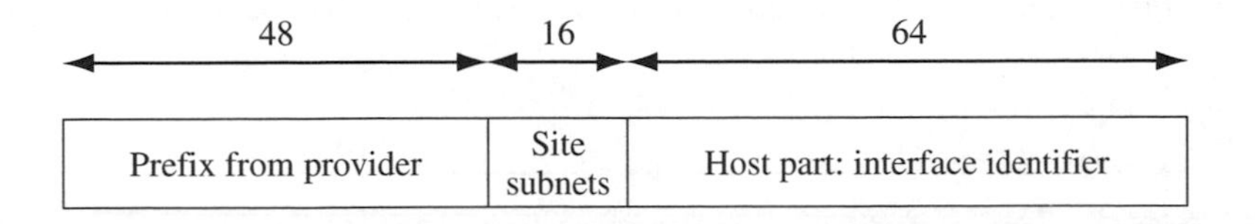

Figure 4.1 Basic structure of an address

[2] There are some exceptions to this, but the generic case is /64.

(b) /48 is the separation between the provider prefix and the site. Contrary to IPv4, IPv6 always uses 16 bits for subnet numbers in a site.
(c) These boundaries are fixed. All sites receive a /48 and all subnets are using a /64.

Important to remember:

Any site, of any size, receives a /48 prefix. All subnets have a fixed /64 prefix.

With IPv4, organizations have many constraints and complications with subnet masks and the lack of address space, so network managers have to use tricks such as variable-length subnet masks and secondary addresses on routers to support multiple subnets on the same link. These issues are all gone. In IPv6, an address plan for a site is very easy: one could start numbering each subnet using the 16 bit site space, which means a maximum of 2^{16} subnets in a site. Each subnet can contain up to 2^{64} nodes, which means there is no need to have multiple subnets on the same link for the purpose of adding nodes on the same link. The only concern one may have about subnet addresses is to locate appropriately the subnet numbers to do aggregation of routes in the internal routing protocol. An optimized algorithm for an addressing plan for providers, organizations or sites is discussed in Section 23.3.1.

4.3.2 Scoped Addresses

IPv6 introduces scopes in the address space: link and site scopes.

4.3.2.1 Link-local

Link-local addresses are scoped addresses restricted to the link: they can only be used on the specific link connected to the interface. In other words, they can only be used between two nodes on the same link and are never forwarded by a router.

The structure of a link-local address is 'fe80:0:0:0:<interface identifier>', as shown in Figure 4.2. Link-local addresses are automatically configured on each IPv6-enabled interface, using the IEEE EUI-64 format for the interface identifier, as described in Section 5.2.1.

By automatically configuring interfaces with link-local addresses, the communication between nodes is enabled on the same link, without any manual configuration of these interfaces. For example, one can setup an ad hoc LAN of IPv6 devices communicating

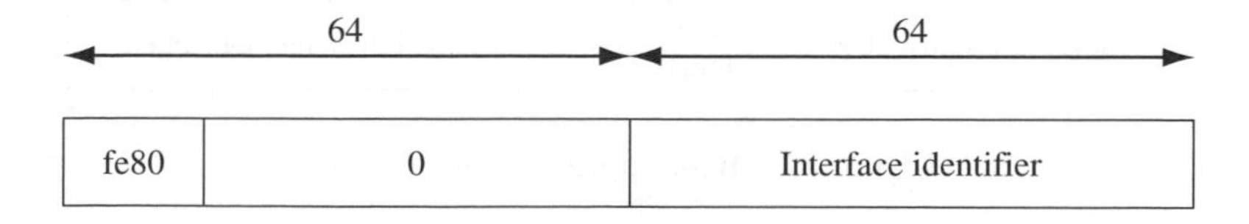

Figure 4.2 Link-local address structure

together without any DHCP server or any router and without any manual configuration. By having the link-local address as the source address, the other nodes can reply directly to that valid IP address.

> **Important to remember:**
>
> Link-local addresses have the scope of the link, are automatically configured on each interface and are defined as: fe80::<interface identifier>.

Two separate links use the same fe80:0:0:0::/64 link-local prefix, as shown in Figure 4.3.

Nodes A and B can communicate together, as well as C and D. But the router R does not forward a datagram containing link-local addresses, so A or B cannot communicate with C or D.

> **Hint on troubleshooting connectivity between two nodes on the same link:**
>
> To troubleshoot the connectivity between two IPv6 nodes on the same link, use the link-local addresses of their interfaces, since they do not rely on router advertisements or DHCP servers.

Care should be taken not to advertise link-local addresses in DNS entries or outside a link, since they are only valid within a link.

Link-local addresses are often used in protocols when there is a need for link-local scope exchanges between two nodes.

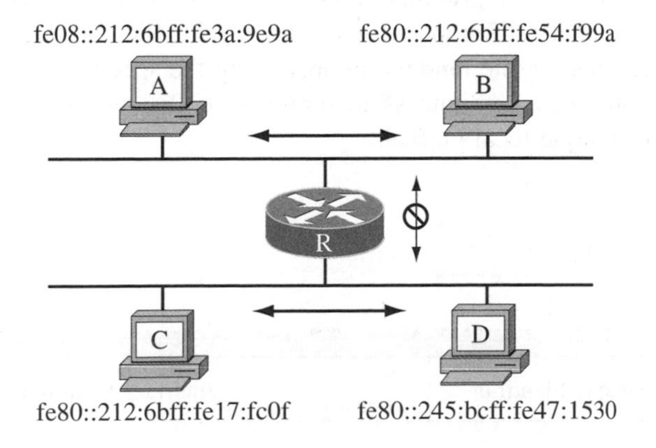

Figure 4.3 Link-local use on different links

4.3.2.2 Scope Ambiguity

In Figure 4.3, if router R wants to send a packet to a link-local address (fe80::), the interface to send the packet is ambiguous: the packet can be sent to any of the links, since the link-local prefix is identical on each link. To resolve the scope ambiguity, the user has to specify the interface. In many implementations, the user adds the '%' character after the address and then adds the interface name or number. For example, on FreeBSD, one would type: ping fe80::245:bcff:fe47:1530%fxp0 to send a packet on the fxp0 interface. The interface specification is generalized to the concept of zones [RFC4007].

4.3.2.3 IPv4 Link-local

Microsoft and Apple proposed for use and implemented the 169.254/16 range [RFC3927] in IPv4 for a similar use, but the current IP stacks allow no special processing with these addresses. There is no easy way to retrofit this functionality in IPv4: current routers will forward these addresses since they are not aware of this new special IPv4 range semantic. Moreover, because IPv4 interfaces usually have only one IP address, the stack and applications have difficulty handling the switch over to a normal IPv4 address, for example when a DHCP server appears to be available and giving addresses.

4.3.2.4 Unique Local Address Space

Unique local addresses [RFC4193] are unicast addresses, globally unique but used locally within a site. Any site can have /48 for its private use. Each /48 is globally unique, so no collision of identical address space is possible in the future when two organizations connect together their sites.

The structure of a unique local address is 'fdUU:UUUU:UUUU:<subnet number>:<interface identifier>' where U indicates bits of a unique identifier, as shown in Figure 4.4. The unique identifier is assigned to a site.

The <subnet number> is a 16 bit field that can be used to number the site subnets. This field has the same length and is located at the same place as in a global address. The unique local prefix (/48) is similar to a provider-based /48 prefix, but for private use.

To create the unique identifier part (bits 9 to 48) for the fd00::/8 address space, the organization should generate the random number using the specified algorithm [RFC4193], which guarantees sufficient randomness to minimize collisions. The book website has a program to generate unique local prefixes.

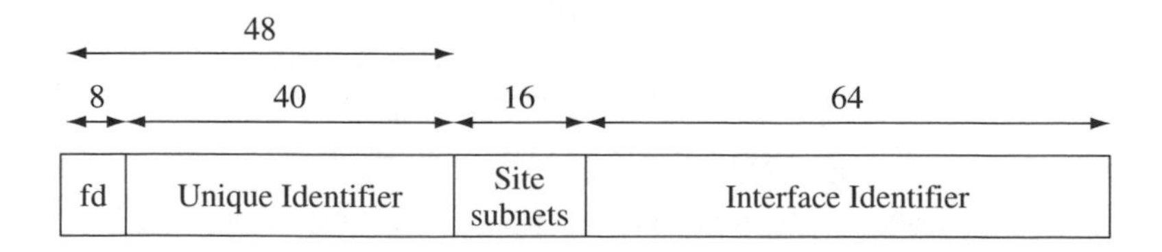

Figure 4.4 Unique local address structure

Consequences of unique local addresses are:

- Any site can use a /48 from fd00::/8 as its prefix for its site numbering.
- Local devices that do not need to be known outside, like printers, can be addressed using unique-local addresses.

Hint for site numbering:

One way to deploy IPv6 in a site is to start using the unique local numbering, by numbering the subnets with the 16 bit field. When the site connects to a provider and receives a prefix from the provider, the provider prefix is advertised inside the network. The same subnet numbers are used. The site can either use both address spaces or can withdraw the unique-local addresses.

4.3.2.5 IPv4 Private Address Space

IPv4 has a private address space [RFC1918], namely 10/8, 172.16/12 and 192.168/16, for a similar use. Owing to the address space exhaustion, the need for provider independence and the impression of security through hidden addresses, most organizations use IPv4 private addresses while connecting to the IPv4 Internet using network address translation at the border gateway.

However, IPv4 is designed for only one IP address per interface, which disables the node to use private address space for internal connections, and global addresses for global connections, as possible in IPv6.

4.3.3 Protocol Use Addresses

Some unicast addresses are defined to trigger specific behavior in the protocol. The 6to4 addresses are used for automatic tunnels and the IPv4-mapped addresses are used for representation of IPv4 addresses as IPv6 addresses.

4.3.3.1 6to4

6to4 is one mechanism to establish automatic IPv6 in IPv4 tunnels and to enable complete IPv6 sites to be connected through automatic tunnels.

A 6to4 address has the following structure: 2002:<ipv4 address>:<subnet>:<interface identifier>, as shown in Figure 4.5. Since it starts with 2002, it is a unicast address and conforms to the structure of the global addresses, as discussed in Section 4.3.1.

The <ipv4 address> field is 32 bits wide and contains the IPv4 address of the border router of the IPv6 site connected using 6to4. The <subnet> field is 16 bits wide and is used for numbering the subnets within the site. The rightmost 64 bits are for the interface identifier.

The 6to4 mechanism is an optional transition mechanism, so not all IPv6 stacks have implemented it. In the context of a complete site, only the border router needs to implement 6to4. This mechanism is further discussed in Section 16.2.5.

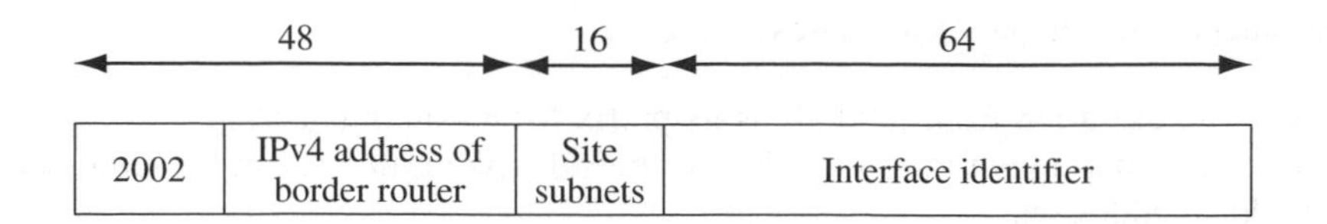

Figure 4.5 6to4 address structure

4.3.3.2 IPv4-mapped Address

IPv4-mapped addresses are used inside applications to represent an IPv4 address of a node using the IPv6 format. This enables the application to care only about the IPv6 address format, the IPv4 address being a special case of an IPv6 address.

An IPv4-mapped address has the following structure: 0:0:0:0:ffff:<ipv4 address>, as shown in Figure 4.6.

These IPv4-mapped addresses should not be used as a source or destination address of generic IPv6 datagrams: they are used internally within the kernel, applications and API on the nodes.

4.3.4 Unspecified Address

IPv4 stacks sometimes use 0.0.0.0 as a placeholder when no address is known or to specify the default route in routing tables. The address '0:0:0:0:0:0:0:0' or '::', as shown in Figure 4.7, is the equivalent in IPv6 and is called the 'unspecified' address.

It is used in preliminary communications such as in the duplicate address detection process, described in Section 6.3.

For example, the `netstat -A inet6 -r -n` command shows the IPv6 routing table of a Linux host.

```
% netstat -A inet6 -r -n | grep '::/0'
Kernel IPv6 routing table
Dest   Next Hop                  Flags Metric Ref    Use Iface
::/0   fe80::200:d1ff:feed:ac3   UGDA  1024   25     1   eth0
```

This extract of the routing table shows the unspecified address as the default route (::/0) for the destination, pointing to the default router fe80::200:d1ff:feed:ac3 as next hop through the eth0 interface.

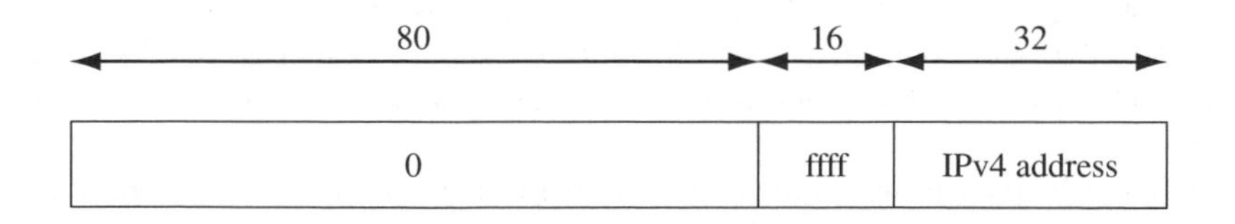

Figure 4.6 IPv4-mapped address structure

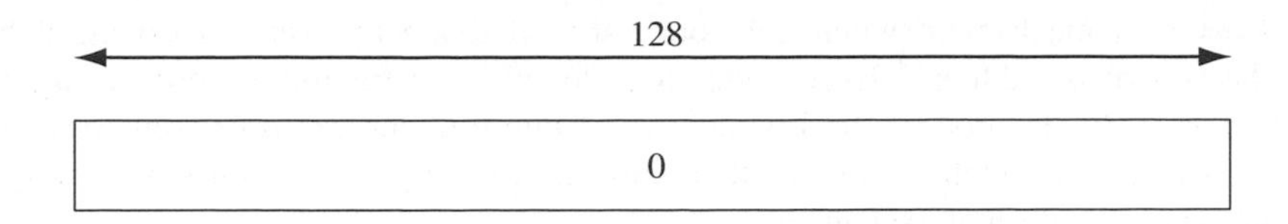

Figure 4.7 Unspecified address representation

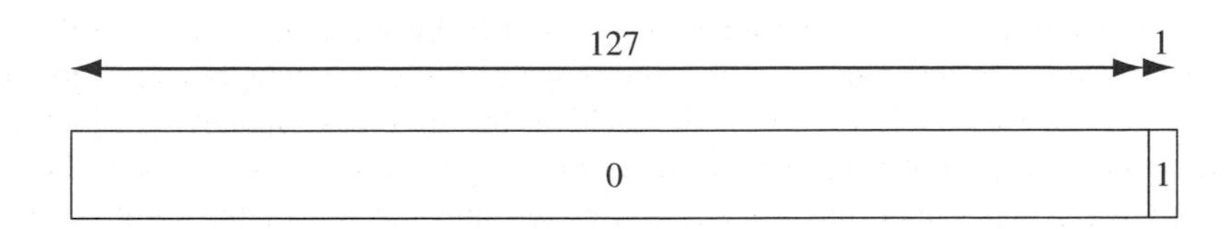

Figure 4.8 Loopback address representation

4.3.5 Loopback Address

IPv4 defines the loopback address as 127.0.0.1. This enables an IP stack to verify connectivity with itself, without going on a physical interface.

IPv6 defines the loopback address as '0:0:0:0:0:0:0:1', or '::1', as shown in Figure 4.8. The IPv4 semantic applies to IPv6: it is used to identify itself.

Hint for verifying an IPv6 stack:

To find if an IPv6 stack is functioning properly, ping the loopback address of the node: `ping6 ::1`.

4.4 Multicast Addressing

Multicast enables the efficient use of network bandwidth by sending the minimum number of datagrams to the maximum number of nodes. On a local link, one multicast datagram is sent to a possibly unlimited number of nodes. A special address prefix[3] identifies a multicast datagram, and a specific address inside that prefix identifies each group of nodes. For example, a videoconference would have one specific multicast destination address that each node, listening to the videoconference, processes these multicast datagrams. This multicast address is often called a multicast group. Multicast is a one-to-many addressing mechanism.

[3] In IPv4, the multicast prefix is 224.0.0.0/8.

Broadcast is a mechanism where all nodes should listen to every broadcast datagram sent to the broadcast address.[4] Even nodes that do not care for this specific datagram are interrupted in order to process the broadcasted datagram. For example, routers exchange routing information through broadcasts that cause an interrupt on all hosts, including those that do not care about route exchanges.

IPv4 uses broadcasts for many purposes but IPv6 does not, using instead multicast. This enables more fine grain targets of datagrams that have to be sent to many nodes on the network. For example, IPv6 routers exchange information using a specific multicast address identifying all routers. A host will not listen to that specific multicast address since it is not a router.

The scope of multicast groups is important in order to limit the forwarding of these datagrams to only the interested nodes. In IPv4, the scope of multicast datagrams is implemented using the Time-To-Live (TTL) IPv4 header field in the multicast datagram, which defines the scope based on the number of hops. In IPv6, the scope is implemented inside the address of the multicast group, which has the key advantage that the use of the right scope is not related to the number of hops, which may vary in real-time. Routers can easily enforce the scope of the multicast traffic by looking at the scope of the multicast address.

Multicast addresses are assigned for temporary use, such as a one-hour videoconference, as well as for permanent use like an address identifying all routers on a link. IPv6 includes a bit in the address to identify permanent or temporary addresses. No such bit exists in IPv4.

The structure of an IPv6 multicast address is shown in Figure 4.9. An IPv6 multicast address begins with 'FF' as the first octet.

The next 4 bits, shown as '*L*' in Figure 4.9, contains a one bit flag identifying the lifetime of the multicast group. If the multicast address is permanent as a permanent well-defined group, then the bit is zero (0), otherwise it is temporary with a value of one (1). All other 3 leftmost bits are currently undefined and equal to 0. Table 4.5 shows the values of the '*L*' 4 bit field.

The next 4 bits, shown as '*S*' in Figure 4.9, identifies the scope of the address. The bits are assigned as listed in Table 4.6.

A scope value of 1 means interface-local scope, which is for multicasting inside a node, like for inter-process communications. This address scope might not be used often. Link-local scope (2) and site scope (5) have the same meanings as unicast scopes, described in Section 4.3.2. A link-local scope multicast datagram is only valid on the local link and should not be forwarded by routers. A site scope multicast datagram is only valid on the site and should not be forwarded to the Internet or to another site by the site border router.

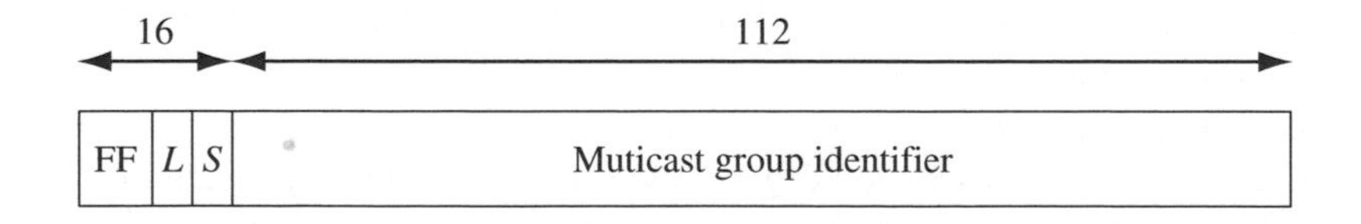

Figure 4.9 Multicast address structure

[4] In IPv4, the broadcast address is all 1s in the host part of the address for a subnet. If the subnet is 192.0.2.0/24, then the broadcast address on that subnet is 192.0.2.255, where 255 is all 1s in the host part.

Table 4.5 Permanent group bit in a multicast address

Value of L (4 bits binary)	Value of L (4 bits hex)	Description
0000	0	Permanent group
0001	1	Temporary group

Table 4.6 Scope bits in a multicast address

Value of *S* (4 bits binary)	Value of *S* (4 bits hex)	Scope description
0001	1	Interface
0010	2	Link
0100	4	Admin
0101	5	Site
1000	8	Organization
1110	E	Global
All others		Unassigned or reserved

Organization-local scope (8) is for an organization that has multiple sites. Global scope is for the whole Internet. All other possible values are undefined or reserved.

Important to remember:

Multicast addresses have bits in the address to specify the lifetime (*L*) and the scope (*S*) of the multicast group and are defined as: ff<L><S>:<multicast group identifier>.

Permanent multicast addresses are assigned by IANA[IANAMULTICASTADDRESS], but many of them have been predefined in the specifications [RFC2375, RFC3513]. Temporary addresses are used on demand without any prior registration or assignment.

Table 4.7 lists some of the permanent assignments. Section 15.7 lists most current assignments.

Hint for discovering the nodes of a link:

To discover all the nodes on a link, ping the multicast address of the 'all-nodes' on the link: `ping6 ff02::1`.

Multicast is discussed in Chapter 15.

Table 4.7 Permanently assigned multicast addresses

Multicast address	Scope	Group	Description
FF01::1	1 = interface	::1 = All nodes	All nodes on the interface
FF02::1	2 = link	::1 = All nodes	All nodes on the link
FF01::2	1 = interface	::2 = All routers	All routers on the interface
FF02::2	2 = link	::2 = All routers	All routers on the link
FF05::2	5 = site	::2 = All routers	All routers in the site
FF02:0:0:0:0:1:FFXX:XXXX	2 = link	Solicited-node. See Section 6.1	

4.5 Anycast

Anycast enables the sending of a datagram to one node belonging to a group of nodes that are either on the same subnet, or topologically located on different links on a network. The routing identifies one or many nodes in the group that will receive the datagrams. It could also be used to provide a global address but routed to the nearest server for that anycast address, thus enabling redundancy and optimum routing for distributed servers.

The anycast mechanism is used for discovery of services on a network or to provide a redundancy and closest service. Two anycast groups are currently defined:

- Routers must join the all-routers-in-the-subnet anycast group so that nodes can find a router on the subnet.
- MobileIPv6 uses an anycast group to locate a home-agent in the home network, as explained in Chapter 11.

An anycast address starts with the prefix of the targeted network followed by the host part that identifies the anycast group. An anycast address looks just like a regular unicast address. Anycast addresses can only be used as destination addresses. Section 15.8 describes the anycast mechanism.

4.6 Addressing Architecture

Table 4.8 shows a summary of most types of addresses and their basic structure.

Table 4.9 shows a summary of the current assignments of the address space for all types of addresses described previously.

One has to remember that in IPv4, the subnet broadcast is all 1s in the host part. There is no pre-defined meaning of the value 0 or 1 in any bit of an IPv6 address.

Figure 4.10 shows a pie-chart representation of the address space. Note that 86% of the space is unassigned and undefined.

13% of the address space is currently assigned to the use of global unicast addresses. This does not mean they are currently used. It only says that this space is reserved to be allocated for Internet use. The Internet might take a while to consume all this reserved address space. To get a sense of the size, the reserved space for allocation is 13% of the IPv6 address space, but corresponds to $2^{(128-3-32)} = 9,903,520,314,283,042,199,192,993,792$ (9.9×10^{27}) times the full IPv4 address space. This 13% also corresponds

Table 4.8 Types of addresses

<table>
<tr><th>Type</th><th colspan="7">Structure (16 bit boundaries)</th></tr>
<tr><td>Unicast global</td><td colspan="2"><provider prefix></td><td>subnet</td><td colspan="4"><interface identifier></td></tr>
<tr><td>Link-local</td><td>fe80</td><td colspan="2">0</td><td colspan="4"><interface identifier></td></tr>
<tr><td>Unique-local</td><td>fc00</td><td>0</td><td>subnet</td><td colspan="4"><interface identifier></td></tr>
<tr><td>Unique-local</td><td>fd00</td><td>0</td><td>subnet</td><td colspan="4"><interface identifier></td></tr>
<tr><td>IPv4-mapped</td><td colspan="4">0</td><td>ffff</td><td colspan="2"><ipv4 addr></td></tr>
<tr><td>6to4</td><td>2002</td><td><ipv4 addr></td><td>subnet</td><td colspan="4"><interface identifier></td></tr>
<tr><td>Unspecified</td><td colspan="7">0</td></tr>
<tr><td>Loopback</td><td colspan="6">0</td><td>0001</td></tr>
<tr><td>Multicast</td><td>ff<ls></td><td colspan="6"><multicast group></td></tr>
</table>

Table 4.9 Current assignments of the address space

Prefix (binary)	Start (hex)	End (hex)	Usage	Space used (%)
0000 0000	0000::	00ff:ffff:ffff:ffff:ffff:ffff:ffff:ffff	Unspecified, localhost.	0.3
0000 0001	0100::	01ff:ffff:ffff:ffff:ffff:ffff:ffff:ffff	Unassigned	0.3
0000 001	0200::	03ff:ffff:ffff:ffff:ffff:ffff:ffff:ffff	Unassigned	0.6
0000 010	0400::	05ff:ffff:ffff:ffff:ffff:ffff:ffff:ffff	Unassigned	0.6
0000 011	0600::	07ff:ffff:ffff:ffff:ffff:ffff:ffff:ffff	Unassigned	0.6
0000 1	0800::	0fff:ffff:ffff:ffff:ffff:ffff:ffff:ffff	Unassigned	3
0001	1000::	1fff:ffff:ffff:ffff:ffff:ffff:ffff:ffff	Unassigned	6
001	2000::	3fff:ffff:ffff:ffff:ffff:ffff:ffff:ffff	Unicast global, 6to4, Anycast	13
010-110	4000::	dfff:ffff:ffff:ffff:ffff:ffff:ffff:ffff	Unassigned	60
1110	e000::	efff:ffff:ffff:ffff:ffff:ffff:ffff:ffff	Unassigned	6
1111 0	f000::	f7ff:ffff:ffff:ffff:ffff:ffff:ffff:ffff	Unassigned	3
1111 10	f800::	fbff:ffff:ffff:ffff:ffff:ffff:ffff:ffff	Unassigned	1
1111 110	fc00::	fdff:ffff:ffff:ffff:ffff:ffff:ffff:ffff	Unique-local	0.6
1111 1110 0	fe00::	fe7f:ffff:ffff:ffff:ffff:ffff:ffff:ffff	Unassigned	0.2
1111 1110 10	fe80::	febf:ffff:ffff:ffff:ffff:ffff:ffff:ffff	Link-local	0.1
1111 1111	ff00::	ffff:ffff:ffff:ffff:ffff:ffff:ffff:ffff	Multicast	0.3

to $2^{(64-3)} = 2,305,843,009,213,693,952$ (2.3×10^{18}) subnets on the Internet or $2^{(48-3)} = 35,184,372,088,832$ (3.5×10^{13}) sites. If the Internet consumes even all this 13%, there is still 6 times more space unassigned!

The unassigned space can be defined in the future with different requirements. For example, the current unicast space defines a fixed boundary of 64 bits for the network part and 64 bits for the host part of an address. In the future, the IETF might decide to define a new address structure and identify another part of the address space to use that new address structure. This gives us room for finding new ways of using the address space or changing our mind

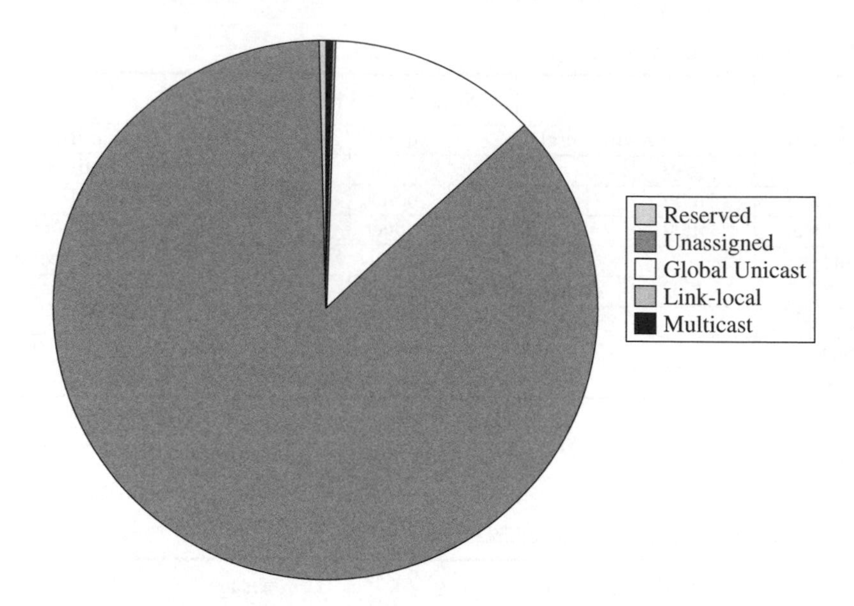

Figure 4.10 Address space utilization

if the current one has issues that we may discover in the future. This also means that the current IPv6 implementations in hosts and routers must not guess the address structure for the unassigned space.

4.7 Summary

IPv6 addresses are 128 bits wide, which expands to a maximum of 340,282,366,920,938,463, 463,374,607,431,768,211,456 unique addresses. IPv6 addresses are represented as 8 fields of hexadecimal numbers (0–F), each field representing 16 bits using 4 hexadecimal digits and fields are separated by a colon ':'. For example, 2001:0000:1234:0000:0000: C1C0:ABCD:0876 is a valid address.

Unicast global addresses are defined within 2000::/3, where /64 is a link and /48 is a site. Link-local addresses are defined as fe80::<interface identifier>, are of link scope and autoconfigured on each interface. Unique-local addresses are defined fd00::<subnet number>:<interface identifier>, are of site scope and not autoconfigured.

Multicast addresses are defined with ff<L><S>::/16, where the L bit is the lifetime of the multicast group and the S bit defines the scope of the multicast group.

Unicast global address assignment uses 13% of the address space. The remaining 86% is unassigned and undefined.

4.8 References

Hinden, R. and Haberman, B., 'Unique Local IPv6 Unicast Addresses', draft-ietf-ipv6-unique-local-addr-09 (work in progress), January 2005.

[IANAMULTICASTADDRESS] IANA, 'Internet Protocol Version 6 Multicast Addresses', http://www.iana.org/assignments/ipv6-multicast-addresses, June 2005.

[RFC1519] Fuller, V., Li, T., Yu, J. and Varadhan, K., 'Classless Inter-Domain Routing (CIDR): an Address Assignment and Aggregation Strategy', RFC 1519, September 1993.
[RFC1918] Rekhter, Y., Moskowitz, R., Karrenberg, D., Groot, G. and Lear, E., 'Address Allocation for Private Internets', BCP 5, RFC 1918, February 1996.
[RFC2375] Hinden, R. and Deering, S., 'IPv6 Multicast Address Assignments', RFC 2375, July 1998.
[RFC3513] Hinden, R. and Deering, S., 'Internet Protocol Version 6 (IPv6) Addressing Architecture', RFC 3513, April 2003.
[RFC3927] Cheshire, S., Aboba, B. and Guttman, E., 'Dynamic Configuration of IPv4 Link-Local Addresses', RFC 3927, March 2005.
[RFC3986] Berners-Lee, T., Fielding, R. and L. Masinter, 'Uniform Resource Identifier (URI): Generic Syntax', STD 66, RFC 3986, January 2005.
[RFC4007] Deering, S., Haberman, B., Jinmei, T., Nordmark, E. and Zill, B., 'IPv6 Scoped Address Architecture', RFC 4007, March 2005.

4.9 Further Reading

[RFC3587] Hinden, R., Deering, S. and Nordmark, E., 'IPv6 Global Unicast Address Format', RFC 3587, August 2003.
[RFC3879] Huitema, C. and Carpenter, B., 'Deprecating Site Local Addresses', RFC 3879, September 2004.
[RFC4048] Carpenter, B., 'RFC 1888 Is Obsolete', RFC 4048, April 2005.

5

Configuring Node Addresses

While IPv6 introduces new features, it inherits most of the attributes of IPv4 and does not prohibit the features available in IPv4. For example, an IPv6 node can be autoconfigured by the new IPv6 mechanism, but can still be configured manually or by a DHCP server, as in IPv4.

This chapter presents static, server-based and autoconfiguration of addresses on interfaces. It describes the router advertisements and solicitations that are part of the autoconfiguration process. DHCP is also described as another method for providing configuration data to hosts. The end of the chapter shows examples of major implementations.

5.1 Static Address Configuration

Static configuration means assigning the information, in a static file or on a command line, needed for the host to configure itself. At a minimum, this includes the IP address, prefix length (or subnet mask) and DNS servers. IPv6 nodes can be configured manually as IPv4 nodes.

The following example shows the FreeBSD command `ifconfig` to configure a static IPv6 address (`3ffe:b00:0:1::a`) and the /64 prefix length on an interface (`fxp0`) using the `inet6` address family.

```
# ifconfig fxp0 inet6 3ffe:b00:0:1::a/64
```

5.2 Address Auto-Configuration

A key new feature of IPv6 is autoconfiguration [RFC2462], where nodes configure themselves without prior static configuration as soon as they are connected on a network. Their first task is to find their address by concatenating the prefix of the network with the host part of the address.

Migrating to IPv6: A Practical Guide to Implementing IPv6 in Mobile and Fixed Networks Marc Blanchet

IPv6 uses the uniqueness of link-layer addresses by inserting the link-layer address of the interface inside the host part of the address. The layer 2 address (e.g. Ethernet MAC address) is effectively embedded in the layer 3 (e.g. IP) address. This mechanism, called interface identifier, should avoid simultaneous use of the same address chosen by two nodes on the same link.

5.2.1 Interface Identifier

Figure 5.1 shows the process of composing an interface identifier. The IEEE EUI-64 [IEEEEUI64] procedure is used with a slight change.

1. The link-layer address is extracted from the interface. The example shows a 48 bit Ethernet address: 00:12:6b:3a:9e:9a.
2. The 16 bit 'fffe' string is inserted in the middle of the 48 bits, which results in a string of 64 bits. 'fffe' is reserved by IEEE specifically to convert a 48 bit MAC address into a 64 bit address.
3. The second bit of the leftmost octet is used to identify the uniqueness of the MAC address. It is set to 1 if the MAC address is unique. A MAC address may be not unique if the administrator of the node changes the MAC address of the interface, which is not useful to do in most cases.

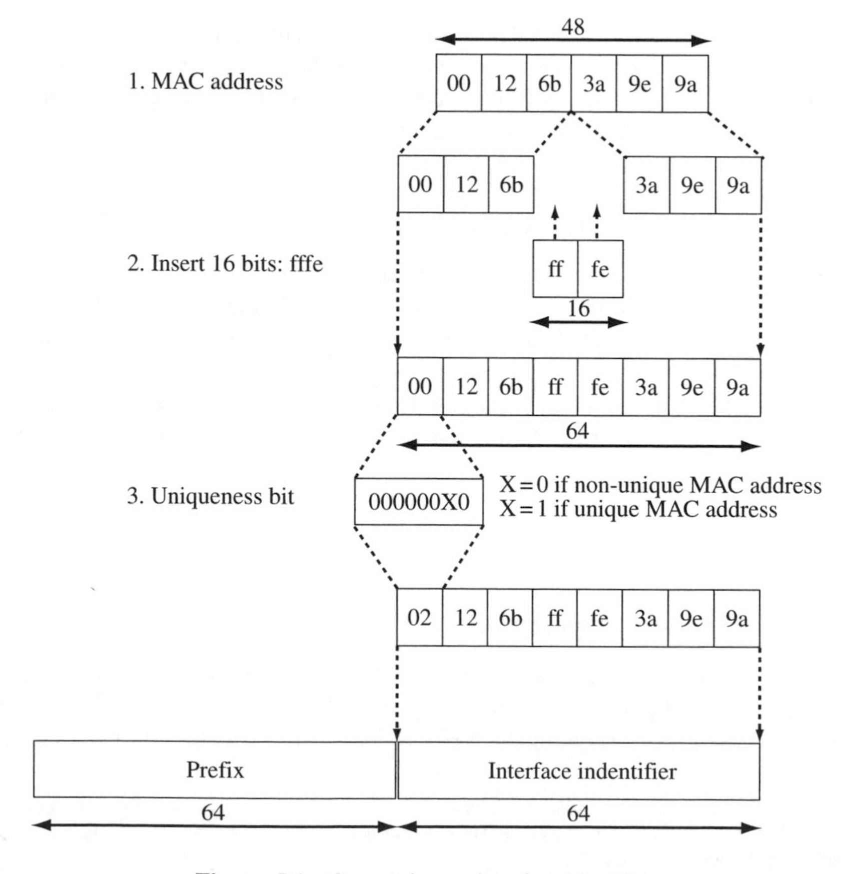

Figure 5.1 Composing an interface identifier

This new 64 bit address is called the interface identifier and is used as the host part of an IPv6 address. The leftmost 64 bits are the prefix of the network.

Important to remember:

For typical hardware interfaces, the interface identifier is formed by inserting fffe in the middle of a 48 bit MAC addresses and placing a '1' in the 7th leftmost bit. This generates a 64 bit interface identifier to be used in the IPv6 address.

IEEE manages the assignment of the MAC addresses. The first 24 bits (of the 48 bits) are assigned by the IEEE registration authority to interface manufacturing companies. The other 24 bits are assigned by the company. IEEE reserved FFFF and FFFE for inserting 48 bit addresses in 64 bit addresses. IEEE intends to define 64 bit MAC addresses in future protocols, since the 48 bit space is being consumed rapidly. By reserving 64 bits in the address, IPv6 autoconfiguration is ready for the IEEE 64 bit MAC addresses.

Many addresses are composed using the EUI-64 interface identifier. All autoconfigured addresses on the same interface will have the rightmost 64 bits identical, but the leftmost 64 bits will be different.

5.2.2 Router Advertisements and Solicitations

Node autoconfiguration works by concatenating the prefix of the network to the interface identifier described in the previous section. The prefix of the network is sent by the routers in router advertisements. These router messages are part of a protocol named Neighbor Discovery (ND) which takes care of most interactions on the local link between hosts and routers and between hosts.

Neighbor discovery uses ICMP for its transport, so neighbor discovery messages are effectively ICMP messages. Neighbor discovery is described in Section 7.2.

5.2.2.1 Router Advertisement

Router advertisement (RA) messages are sent by the routers to inform the hosts about the link information so the hosts can autoconfigure themselves. The key information included is the prefix(es) of the link and the default router(s).

These messages are sent periodically every 5 minutes by the routers to all nodes on the link.[1] The source address of these router advertisement (RA) messages is the link-local address of the interface of the router and the destination address is FF02::1, the all-nodes on the link multicast address. All nodes process that datagram. Unless specifically prohibited by configuration, routers send advertisements on all their interfaces.

[1] The periodicity is randomized in order to avoid simultaneous advertisements sent by multiple routers. The default range of periodicity is from ~200 to 600 seconds.

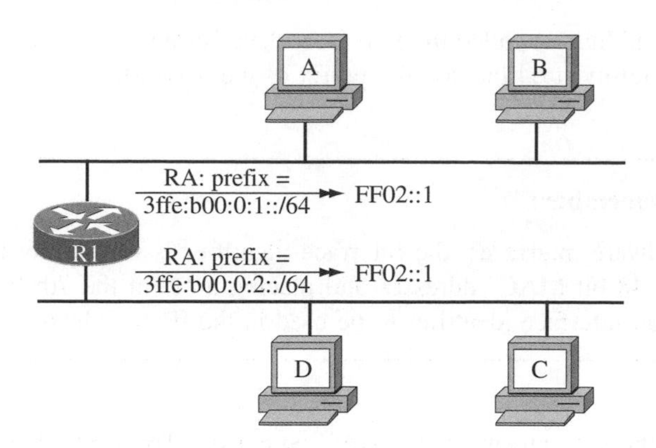

Figure 5.2 Router advertisements

Figure 5.2 shows the R1 router advertising the 3ffe:b00:0:1::/64 prefix to all hosts on LAN1, which includes hosts A and B. R1 also advertises 3ffe:b00:0:2::/64 to the other LAN.

If host A has a MAC address of 00:12:6b:3a:9e:9a, then its interface identifier is 0212:6bff:fe3a:9e9a as shown in Figure 5.1. After receiving the RA of the 3ffe:b00:0:1::/64 prefix from R1, the host A configures the assembled 3ffe:b00:0:1:212:6bff:fe3a:9e9a IPv6 address on its network interface.

All nodes configured to use autoconfiguration always listen to the router advertisements. Each time an advertisement is received, the nodes parse the information and configure themselves accordingly. This enables easy renumbering of nodes, as described in Section 5.3.

Important to remember:

All nodes configured to use autoconfiguration always listen to the router advertisements and process them immediately when received.

Table 5.1 shows the typical fields found in the router advertisement. The detailed field list is described in Section 5.9.1.

A router can send multiple prefixes in the same advertisement. The nodes will then autoconfigure their interface with all prefixes sent by the RA.

Table 5.1 Router Advertisement Datagram Typical Fields

Field	Description
Source address	Link-local address of the router interface sending the router advertisement.
Destination address	All-nodes on the link multicast address (FF02::1)
ICMP type	Value = 134. Identifies a router advertisement.
Prefix	One or more prefixes with associated lifetimes, onlink, managed bits and other info.
Router lifetime	Identifies if the router is a default router.

Important to remember:

An IPv6 node usually has multiple addresses per interface, many of them can be made from the router advertisements.

Two lifetimes, preferred and valid, are associated with each prefix. This is to help renumbering as discussed in Section 5.3.

A router can also identify itself as a default router, using the router lifetime field: a value of 20 minutes (expressed in seconds in the specification as well as in most implementations) means that the router can be used by the hosts as a default router for the next 20 mins. Since this value is resent with every router advertisement, the value will then be refreshed by the hosts so the router will always be a valid default router. The advertised value should be large enough to keep the router as default router even if the router advertisements are not received for a few cycles. If the value is zero, the router sending the advertisement should not be used as a default router.

5.2.2.2 Multiple Routers Sending Router Advertisements

If a node receives RAs from multiple routers on the link, it makes the union of these advertisements. Router R1 sending one prefix and R2 sending another prefix makes the node autoconfigured with the two prefixes.

If both routers advertise themselves as default router (router lifetime in RA not equal to zero), then the node either decides to pick one and always use the same one or do some round robin cycling between the default routers. This is the choice of the implementation, but most implementations use the same router until it comes unreachable, where it switches to a new default router.

5.2.2.3 On-link in Router Advertisements

The on-link parameter is sent inside router advertisements. It specifies if the prefix is valid on the link. Nodes will send a direct neighbor solicitation (see Section 6.2) to on-link destinations, but will send packets to the default router for off-link destinations. If the on-link parameter is not set for a specific prefix, then nodes on the link sending to a destination with this off-link prefix, send the packet to the default router and do not perform neighbor solicitation. This is used to force the forwarding to some prefixes to go through the router. A good example of its use is multilink subnets, described in Section 6.8, and mobileIP, described in Section 11.12.

5.2.2.4 Router Solicitation

As soon as a router advertisement is sent, all nodes on the link process it and configure themselves accordingly. These advertisements are sent every five minutes. When a node is booting, it might not be lucky if the advertisement might has been sent just a few seconds before. It would have to wait for the next advertisement, in four minutes, to autoconfigure itself. In order to receive a router advertisement immediately during its booting sequence, the

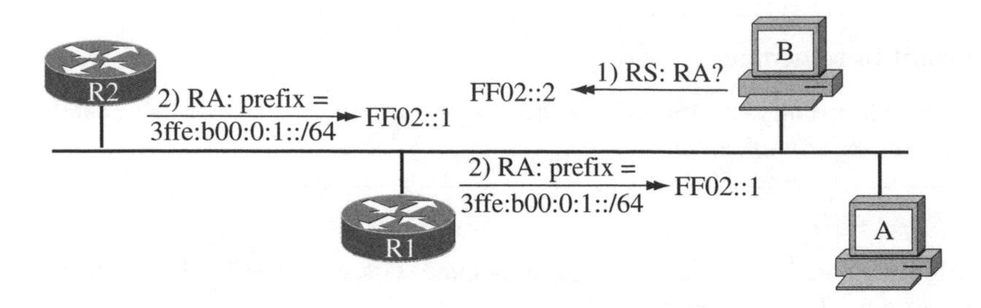

Figure 5.3 Router solicitation

Table 5.2 Router Solicitation Datagram Typical Fields

Field	Description
Source address	Link-local address of the node if available or :: (unspecified) if link-local not available.
Destination address	All-routers on the link multicast address (FF02::2)
ICMP type	Value = 133; identifies a router solicitation.

node sends a router solicitation (RS) message to all routers on the link, requesting them to send router advertisements straight away so the node can autoconfigure itself immediately.

Figure 5.3 shows node B sending a router solicitation to all routers on the link (FF02::2). Then routers R1 and R2 send their router advertisement to all nodes on the link (FF02::1). All nodes on the link listen to these router advertisements and configure their interface accordingly. The booting node B autoconfigures itself right away instead of waiting for the next scheduled router advertisement.

Table 5.2 describes the typical fields in the router solicitation message. The detailed field list is described in Section 5.9.1.

The source address of the node could be either the unspecified address (::) or the link-local address. Since nodes can send traffic with a link-local source address and link-local addresses are used by protocols, implementations usually start by using the link-local address and then ask for router solicitation for the other addresses. So the source address of the router solicitation is usually the link-local address, unless it cannot be used.

5.3 Lifetime of Advertised Prefixes

In IP networks, as well as in telephone networks, there is always a trade-off between the ownership of an address space and the routing efficiency of a global network.

If the address space of an organization is independent of the provider, often called portable addresses, then the provider has to announce that address space to the Internet. If all organizations own their address space, then the number of route entries in the global routing table is equal to the number of organizations, which may be a very large number in the long run. This is generally considered as non-scalable or at least a difficult engineering challenge.

The way to decrease this global routing problem is to keep the ownership of the organization's address space in the hands of the provider. In other words, an organization receives its address space from its provider. If the organization changes provider, then it receives a new address space from the new provider and the former provider reassigns the old address space to other customers. There are at least two implications: the minimization of the global routing table since the number of route entries would be somewhat proportional to the number of providers, and so orders of magnitude smaller than if all organizations owned their address space; the trouble and inconvenience of renumbering the network of the organization when it changes provider.

Until the mid-1990s, the IPv4 addresses were allocated to organizations so they owned their address space. The global routing table explosion meant that the policy changed to allocate addresses only to providers. Organizations now do not own their IPv4 address space and have to renumber when they change provider.

In IPv4, there is no provision or mechanism for handling renumbering, so renumbering an IPv4 network is a real pain. Today, organizations try to avoid renumbering in IPv4 using the following possibilities:

- For those organizations that have portable addresses, they keep using it.
- For those organizations that do not, they use private addresses [RFC1918] with network address translation. Only the computers directly reachable from the Internet which are assigned public addresses from the provider, like Web servers and firewalls, have to be renumbered if they change providers.

Good news! IPv6 includes mechanisms for renumbering hosts and routers, which helps organizations when they have to renumber.

In autoconfiguration mode as well as in DHCP, the 64 bit network prefix received by the node in the router advertisements has two associated lifetimes: the preferred and the valid lifetimes.

The valid lifetime is the lifetime of the prefix. After the lifetime is expired, the address, composed of the prefix and the interface identifier, must not be used. The preferred lifetime is the period where the address is a preferred address. When the preferred lifetime is expired but the valid lifetime is not expired, the address can still be used but is not preferred: it is deprecated. The address should not be used for new communications initiated by the node, but could still be used for communications that already occurred during the preferred lifetime period. If a communication is started by another node to an address which is deprecated, the node will answer and use the deprecated address. This enables servers to continue serving clients with its old address before the renumbering event is completely over. If the network manager carefully adjusts the lifetimes during a renumbering event, it enables the smooth handover without breaking the already opened connections.

Important to remember:

When an address is deprecated (preferred lifetime is zero, but valid lifetime is not zero), already established connections remain and use the deprecated address, new communications use the preferred addresses (preferred lifetime is not zero) and new communications started by other nodes can still use the deprecated address.

Let's explain the lifetimes with a simple renumbering example. An organization uses 3ffe:b00:1::/48 prefix received from its current provider and is going to use 3ffe:b00:2::/48 prefix received from its new provider. To ensure a smooth handover, both providers will be connected to the organization network for an overlap period of 2 days (t0–t2), so both prefixes will be valid during that overlap period, as illustrated by Figure 5.4.

Two days before the start of the overlap period (at $t = -2$), the valid lifetime of the current prefix (3ffe:b00:1::/48) is set to 4 days, because it is $2+2$ days before the current provider link is shutdown (at $t = +2$). At the same time ($t = -2$), the preferred lifetime of the current prefix is set to 2 days, because at $t = 0$, the new provider link will be up and we want nodes to switch over to the new provider. Those two lifetimes will be decreased over the two days ($t = -2$ to $t = 0$). For example, 1 day before the new provider link is up ($t = -1$), the valid lifetime of the current prefix (3ffe:b00:1::/48) is 3 days and the preferred lifetime is 1 day, as shown in Table 5.3.

When the new provider link is up, the new prefix (3ffe:b00:2::/48) is announced with greater lifetimes (for example, valid lifetime = 2 weeks and preferred lifetime = 1 week) since this provider link will be up for many months. The old prefix is now deprecated by setting the preferred lifetime to zero: any new connections will now use the new prefix but already started connections using the old prefix still work until the valid lifetime comes also to zero, which happens when the old provider link is shutdown (at $t = +2$).

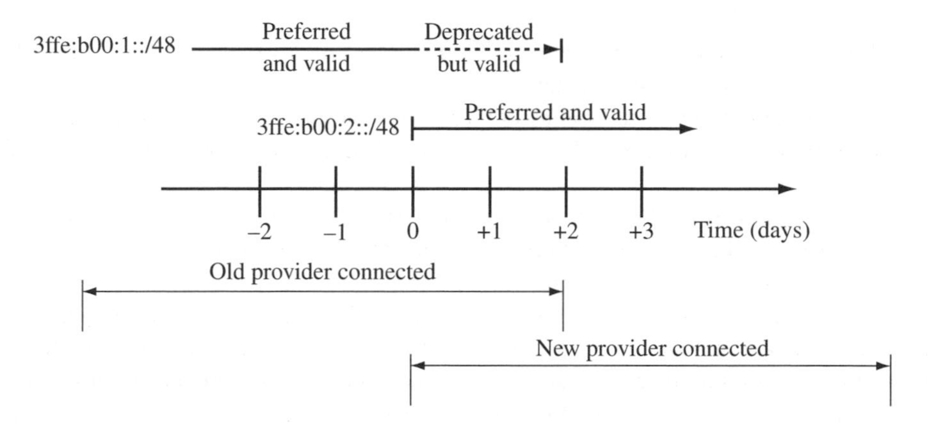

Figure 5.4 Timeline of lifetime for typical renumbering

Table 5.3 Time line of lifetimes for typical renumbering

Lifetimes Time line	−2	−1	0	1	2	3+
3ffe:b00:1::/48						
Preferred lifetime	2	1	0	0	0	N/A
Valid lifetime	4	3	2	1	0	N/A
3ffe:b00:2::/48						
Preferred lifetime	N/A	N/A	7	7	7	7
Valid lifetime	N/A	N/A	14	14	14	14

These lifetimes are tied to the prefixes as part of the router advertisements. The network manager has to change the lifetimes advertised by the routers. Since the hosts always listen to the router advertisements, they will pick up the new lifetimes.

To implement this renumbering scheme as smoothly as possible, the routers have to decrease linearly the lifetimes for each advertisement. If the previous advertisement was 5 minutes ago, then the lifetimes of the next advertisement should be 5 minutes less than the previous lifetimes.

Lifetimes are represented in seconds in the specification [RFC2461] and most implementations also use seconds in their configuration files and commands.

During t = +2 and t = +3, the old prefix is still announced with lifetimes equal to zero to make sure that nodes completely erase the old prefix in their configuration. This is just a useful trick, it is not a protocol requirement. One can also decide to disable the announcements of the old prefix as soon as the old provider link is shutdown.

Hint for clearing the old prefix when renumbering:

After a prefix become invalid, you should continue to announce it for a little while with zero values for both preferred and valid lifetimes. This insures the nodes will unconfigure immediately their old prefix.

5.4 Node Booting Process

Table 5.4 summarizes the autoconfiguration process, starting with the boot sequence of a node. For simplicity, the node is assumed to have only one physical interface.

Table 5.4 Node Booting Process

	Source	Destination	Description
1	Before booting, node is configured for autoconfiguration. Alternatively, its default configuration does autoconfiguration, instead of static address configuration.		
2	After kernel boots, a link-local address is created based on the MAC address of the interface. In this example, the link-local address is fe80::212:6bff:fe3a:9e9a, based on the Ethernet address 00:12:6b:3a:9e:9a.		
3	::	ff02::1:ff3a:9e9a	Node verifies that this link-local address is OK to use on the local link by doing a duplicate address detection, described in Section 6.3. Destination address is the solicited-node multicast address (see Section 6.1) of the tentative link-local address of the node.

(*continued overleaf*)

Table 5.4 (*continued*)

Source	Destination	Description
4 If another node responds by claiming the same link-local address, then the autoconfiguration process stops and an error is sent to the user.		
5 Link-local address of the node (fe80::212:6bff:fe3a:9e9a).	ff02::2	Assuming no errors in step 4, node sends a router solicitation to all routers on the link.
6 Link-local address of a router.	Link-local address of the node (fe80::212:6bff:fe3a:9e9a).	Router advertisement sent by a router.
7 For each prefix in the router advertisement, concatenate the received prefix with the EUI-64 of the MAC address and configure the interface with this additional address.		
8 ::	ff02::1:ff3a:9e9a	For each interface, verify the usability of the address by doing duplicate address detection.

The minimal information needed for a node autoconfiguration is the address, prefix length, default router and DNS server address. The router advertisements do not contain any DNS server information. DNS server discovery is discussed in Section 8.4.

If no router advertisement is received, the node resends router solicitation messages a few more times. If no router advertisement is received, then the node might try DHCP requests. If DHCP is not available, then the node can only use its link-local address.

5.5 DHCPv6

The auto configuration mechanism creates an address where the network part is provided by the router advertisements and the host part is created by the host, based on its link-layer address. DHCP, instead, is a centralized way of handling node configuration where the DHCP server gives all information, both the network and the host part, to the node. It is also capable of sending additional information, such as DNS servers, time servers, printers, font server, etc., not available in the router advertisements.

5.5.1 Basic Behavior

DHCP in IPv6 (DHCPv6) [RFC3315] does the same function as DHCP in IPv4. The DHCP server sends IP addresses, DNS server addresses [RFC3646] and other possible data to the DHCP client and the client configures itself accordingly. As opposed to stateless autoconfiguration described in Section 5.2.2.1, DHCP gives to the network manager more control over the configuration of the client and is more appropriate where tight central network management is required.

The server also sends a lease time of the address and time to recontact the server to renew the address. The client has then to resend a request to renew the address.

DHCP servers do not have to be on the same link as their clients. A DHCP relay, usually implemented in routers, listens to client requests on the link and forward all requests to the known DHCP servers. If dynamic DNS is used, then the DHCP and DNS can be glued together to register a DNS dynamic name for the DHCP client.

The DHCPv6 server sends IPv6 addresses to the DHCP client using IPv6 packets. Typically, it would also send the IPv6 addresses of the DNS servers. A DHCP server could also send IPv6 addresses of printers, time servers, etc.

However, there are many differences with DHCP in IPv4, as summarized in Table 5.8. The following sections describe the DHCPv6 process.

5.5.2 Initial Exchange

The process for a DHCPv6 booting client is illustrated in Figure 5.5, when the DHCP server is on the link.

Table 5.5 describes the initial communications when the DHCP client is booting. The first step for the client is to find routers by sending a router solicitation message to all routers on

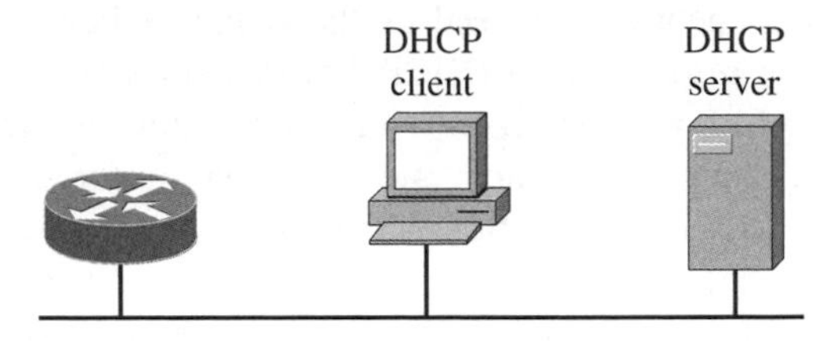

Figure 5.5 DHCP Client and Server

Table 5.5 Booting DHCP client to a Server on the Same Link

From → To	Type of packet	Source and Destination Addresses	Description and Content
1 Client → Routers	ND RS	Src: client link-local Dst: all-routers on the local link multicast: FF02::2	A router solicitation is sent at boot time to get router advertisements.
2 Routers → Client	ND RA	Src : router link-local Dst: client link-local	Routers send router advertisements.
3 Client	Process the received RA to find if the 'Managed Configuration Flag' is set. If the flag is set or if no RA is received after multiple RS, then continue with next step, otherwise do not use DHCP and try autoconfiguration.		
4 Client → Server	DHCP Solicit	Src: client link-local Dst: all-DHCP-agents on local link multicast: FF02::1:2	Client tries to find DHCP servers.
5 Server → Client	DHCP Advertise	Src: server link-local Dst: client link-local	Server responds to client.

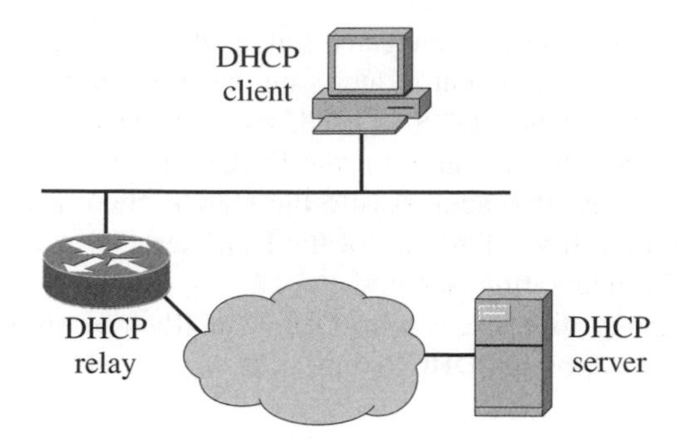

Figure 5.6 DHCP Client and Server with a Relay

the link. If one or more routers respond with a router advertisement, then the client parses the message to see if the 'Managed Configuration Flag' is set. If it is set, as described in Section 5.9.1, then the client can proceed by the DHCP Solicit message to find DHCP servers. The DHCP server responds by a DHCP Advertise message.

If a DHCP relay is present and the server is on another link, as illustrated in Figure 5.6, then the communications between the client and the server are forwarded by the relay.

The communications between the client and the server will be the same, except that the relay will intercept the request and forward it by sending a DHCP relay-forward message to the server which includes the original client message together with the relay address and prefix length. By including the relay address, the server knows where the request came from so it can reply with the appropriate and valid address for the client. These communications are summarized in Table 5.6. The new messages showing the presence of the relay are labeled 4.5 and 5.5 in the table.

This relay behavior is similar to the IPv4 case, although the IPv6 relay may forward the request to the 'All-DHCP-Servers on the site' multicast address (FF05::1::3), if the relay does not know the IPv6 addresses of the servers in the site. IPv6 DHCP relays do not need the list of site DHCP servers if the multicast address is used.

The 'Managed Configuration Flag' in the router advertisements is a new concept in IPv6 where routers are now policing what nodes could do for their configuration. For managers, it is a way to control the configuration of IPv6 nodes, by policy. None of this is possible in IPv4.

5.5.3 Data Exchange

After the initial steps described in the preceding table, the client and the server exchange the configuration data. This is done by the client sending the DHCP Request message asking for data to the server and the server responding with the data in the DHCP Reply message. Then the client confirms the data by sending a DHCP Confirm message to the server. This is illustrated in Table 5.7. Note that the relay behavior is not shown, since it is illustrated in Table 5.6.

Table 5.6 Booting DHCP Client to a Server with a Relay

	From → To	Type of datagram	Source and Destination Addresses	Description and Content
1	Client → Routers	ND RS	Src: client link-local Dst: all-routers on the local link multicast: FF02::2	A router solicitation is sent at boot time to get router advertisements.
2	Routers → Client	ND RA	Src: router link-local Dst: client link-local	Routers send router advertisements.
3	Client	Process the RA to find if the 'Managed Configuration Flag' is set. If the flag is set or there is no RA received after multiple RS, then continue with next step, otherwise do not use DHCP and try autoconfiguration.		
4	Client → Server	DHCP Solicit	Src: client link-local Dst: all-DHCP-agents on local link multicast: FF02::1:2	Client tries to find DHCP servers.
4.5	Relay → Server	DHCP relay-forward	Src: relay address Dst: DHCP server or all-DHCP-servers in the site multicast (FF05::1::3)	Relay forwards the request by including it in a relay-forward message to the server.
5	Server → Relay	DHCP relay-reply	Src: server address Dst: relay address	Server responds to the client by sending the response to the relay, which will forward it to the client.
5.5	Relay → Client	DHCP Advertise	Src: relay address Dst: client link-local	Response from the server sent by the relay.

Table 5.7 Data Exchange between the Client and the Server

From → To	Type of datagram	Source and Destination Addresses	Description and Content
6 Client → Server	DHCP Request	Src: Client link-local Dst: Server address	Client requests data to the server.
7 Server → Client	DHCP Reply	Src : Server address Dst: Client Link-local	Server sends data to the client.
8 Client → Server	DHCP Confirm	Src: Client link-local Dst: Server address	Client confirms the data to the server.

IPv6 prefixes received by the clients have lifetimes as in stateless autoconfiguration. Lifetimes are bound to IPv6 addresses, but DHCP lease times are bound to the renewal of all DHCP data. Lifetimes are used to help renumbering and are discussed in Section 5.3.

5.5.4 DHCPv6 Prefix Delegation

DHCPv6 configuration is extended to provide an IPv6 prefix [RFC3633] to the requesting DHCPv6 client, which is in this case a router. This mechanism is known as DHCPv6 prefix delegation (DHCPv6-PD). A typical scenario [RFC3769] is the home gateway which receives a prefix for the home network by the service provider's DHCP service. The prefix is then advertised on the home links using router advertisements. The nodes on the home network are using RA as described in Section 5.2.2.1, not DHCPv6.

5.5.5 Differences Between DHCPv4 and DHCPv6

With DHCPv4, a network manager has no easy way to signal new configuration data immediately to the clients when a change has to be done: for example, when the IP address of the DNS servers has to be changed. With DHCPv6, a Reconfig-init message can be sent by the server to its clients. This will signal the clients to come back to the server immediately for additional data. In the event of renumbering a network as well as for adding new services or changing addresses of DNS servers, this feature in DHCPv6 is very useful.

DHCPv6 is a generalized version of the DHCP protocol, since it supports multiple servers sending information to the same clients, and clients can receive multiple addresses for the same interface, for example a unique-local and a global address. The DHCPv4 client-server association based on an IP address cannot be used anymore for IPv6. To implement these new features, the DHCPv6 protocol defines the concept of an identity association to bind all requests and data together in this generalized context.

DHCPv6 is also used with one transition mechanism where the DHCPv6 server will send temporary IPv4 addresses to IPv6 clients that need an IPv4 address for some connection. This mechanism, called Dual Stack Transition Mechanism (DSTM), is explained in Section 17.1.3.

Table 5.8 summaries the key differences between DHCPv4 and DHCPv6. As one can see, DHCPv6 is a definitive improvement over DHCPv4.

Table 5.8 Key differences between DHCPv4 and DHCPv6

Feature	DHCPv4	DHCPv6	Benefit
Managed Configuration Flag	Not available	The router using router advertisements flags controls whether the nodes can use DHCP or not.	Configuration of nodes can be managed by a network policy.
Destination address of the initial request.	Broadcast	Multicast to All-DHCP-Agents	More specific signaling. More efficient use of link.
Source address of the initial request.	0.0.0.0	Link-local address of the client	More specific signaling. More efficient use of link.
Relay forwarding	Needs a static list of DHCP servers on the site	May use the All-DHCP-Servers on the site multicast address.	Easier to manage on a large site. More fault-tolerant and redundant.

Reconfiguration message	Not available	The server can ask the clients to update their configuration info.	Easier to trigger any site-wide reconfiguration.
Identity Association	Not available	Clients can handle multiple DHCP servers and receive multiple addresses.	More scalable use of DHCP.
DSTM	Not applicable	DHCPv6 used to give temporary IPv4 addresses in the DSTM transition mechanism (see Section 17.1.3).	Efficient use of remaining IPv4 address space.

5.5.6 Dual Stack DHCP Clients

For nodes that have both IPv4 and IPv6 stacks configured as DHCP clients, the IPv4 and IPv6 stacks send separate DHCP requests and receive separate DHCP answers from different servers, using the respective IP protocol datagrams. It may happen that the two servers are located on the same computer, running a combined software, for easy management, but the requests will still be sent separately.

5.5.7 Renumbering with DHCP

With DHCP, preferred and valid lifetimes of prefixes are only sent in the DHCP exchanges. However, there is no interaction of routers with hosts. In the event of renumbering, DHCP clients are not informed as with the router advertisements. To inform the nodes about renumbering, the network manager must:

1. Configure the DHCP servers with the new prefix data.
2. Send the Reconfig-init message to all its clients to signal the clients to go back to servers for additional info.

The clients then automatically re-request the servers for updated data, get the new prefix and configure themselves accordingly, which is changing their IPv6 addresses.

5.6 Node Addresses

A key difference in IPv6 is the fact that a node has many addresses defined. Table 5.9 shows all addresses that a node may have, after successful configuration.

5.7 Configuring Interfaces and Router Advertisements on Hosts and Routers

This section describes how to enable IPv6, and configure interfaces and router advertisements on the various implementations. Skip the sections in which you are not interested.

Table 5.9 Required node addresses

Quantity	Address	Requirement	Context
One	link-local	must be defined	on each interface
One	loopback	must be defined	on each node
Zero to many	unicasts	may be defined	on each interface
Any	unique-local	may be defined	on each interface
One	the all-nodes multicast	must be joined	on each interface
One	the solicited-node multicast (See Section 6.1)	must be joined	for each unicast and anycast addresses defined
Any	multicast group	may be joined	on each interface

5.7.1 Network Example

Figure 5.7 shows the network used in the following configuration examples.

5.7.2 FreeBSD

The following sections describe enabling IPv6, configuring an interface and configuring router advertisements for the FreeBSD 5.2 implementation.

Most configuration is done in the `/etc/rc.conf` file. The `/etc/defaults/rc.conf` file provides all the default values for each variable and a basic documentation on each variable that could be used in the `/etc/rc.conf`.

5.7.2.1 Enabling IPv6

In FreeBSD, IPv6 is enabled by setting the `ipv6_enable` variable to `yes` in the `/etc/rc.conf` configuration file.

```
# cat /etc/rc.conf
ipv6_enable=yes
```

This configuration file is read at boot time by the boot scripts, namely `/etc/rc` and `/etc/rc.network6`.

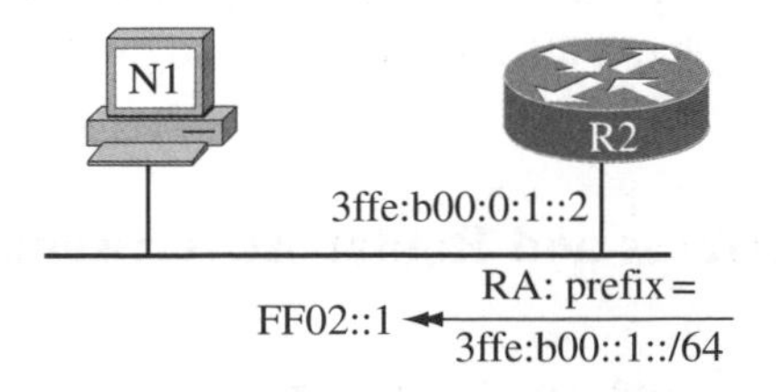

Figure 5.7 Network example for interface and RA configurations

5.7.2.2 Configuring an Interface

By default, all interfaces are autoconfigured if IPv6 is enabled as above. Autoconfiguration is done by sending a router solicitation on each interface and configuring interfaces based on received router advertisements.

The `ifconfig` command is used to configure manually an IPv6 address. The arguments include '`inet6`' for the address family, the IPv6 address and the prefix length. The example below configures the `fxp0` interface with the `3ffe:b00:0:1::a` address and a `/64` prefix length.

```
# ifconfig fxp0 inet6 3ffe:b00:0:1::a/64
```

The address configured by the ifconfig command is not saved for the next reboot. If the static address needs to be permanently assigned, then the `ipv6_ifconfig_fxp0` variable is set to the address in the `/etc/rc.conf` file, as shown below.

```
# cat /etc/rc.conf
ipv6_ifconfig_fxp0=''3ffe:b00:0:1::a/64''
```

For the purpose of the FreeBSD examples throughout this book, `fxp0` is shown where an interface name is used.

5.7.2.3 Configuring Router Advertisements

When behaving as a router, FreeBSD is configured with IPv6 forwarding enabled (see section 10.2.1) and to advertise prefixes on links by setting the `rtadvd_enable` variable in `/etc/rc.conf`.

```
# cat /etc/rc.conf
rtadvd_enable=''yes''
```

By default, router advertisements are sent on all IPv6 enabled interfaces, with default values. The advertised prefixes are taken from the routing table.

To restrict the advertisements to a specified set of interfaces, the `rtadvd_interfaces` variable is set to the list of interfaces.

```
# cat /etc/rc.conf
rtadvd_interfaces=''fxp0''
```

This variable is particularly useful when the device is a gateway to some external network (like the Internet or a direct provider) and router advertisements must be only sent to the internal network.

Router advertisements are configured using the '`/etc/rtadvd.conf`' file. This file uses the termcap format where the index is the interface name. Variables and values are separated by the # character. Since the termcap separator is ':', any prefix or address must be written within double quotes (`""`). The following file enables router advertisements of the `3ffe:b00:0:1::/64` prefix on the `fxp0` interface. The '`addrs`' variable is set to the number of prefixes, in example 1. The '`addr`' variable contains the advertised prefix. The '`prefixlen`' variable contains the prefix length (64).

```
# cat /etc/rtadvd.conf
fxp0:addrs#1:addr="3ffe:b00:0:1::":prefixlen#64:
```

5.7.2.4 Troubleshooting

To verify the IPv6 addresses on an interface, use `ifconfig <interface_name>`.

```
% ifconfig fxp0
```

Pinging to self is a good way to see that basic IPv6 is working.

```
# ping6 ::1
```

IPv6 kernel variables starts with '`net.inet6`' and can be managed with the `sysctl` command.

```
# sysctl -a | grep inet6
```

The variables are set by the operating system scripts and modify the kernel behavior. The `net.inet6.ip6.accept_rtadv` variable is the one to look at when troubleshooting, as shown in Table 5.10.

Sometimes an interface is temporarily disconnected. When the interface is back, an immediate router solicitation can be forced by the '`rtsol`' command. The following example triggers a router solicitation on the '`fxp0`' interface.

```
# rtsol fxp0
```

5.7.3 Linux

The following sections describe enabling IPv6, configuring an interface and configuring router advertisements for the Linux RedHat 8.0 implementation.

Table 5.10 Important FreeBSD IPv6 kernel variables

Variable	Value	Description
`net.inet6.ip6.accept_rtadv`	0	Do not accept/listen to router advertisements. All interfaces must not be autoconfigured. Interfaces should be configured statically or by other means.
`net.inet6.ip6.accept_rtadv`	1	Accept router advertisements, to enable autoconfiguration on interfaces. This is the normal value for a host.

5.7.3.1 Enabling IPv6

In Linux RedHat, IPv6 is enabled by setting the `NETWORKING_IPV6` variable to `yes` in the `/etc/sysconfig/network` configuration file.

```
# cat /etc/sysconfig/network
NETWORKING_IPV6=yes
```

The file is read by the scripts at boot time. However, at anytime, one can restart network services, enabling IPv6 at the same time, with the '`service`' command.

```
# /sbin/service network restart
```

5.7.3.2 Configuring an Interface

Linux is identical to FreeBSD for the `ifconfig` command used to configure manually an address on an interface.

```
# ifconfig eth0 inet6 3ffe:b00:0:1::a/64
```

For the purpose of the Linux examples throughout this book, `eth0` is shown where an interface name is used.

5.7.3.3 Configuring Router Advertisements

When behaving as a router, Linux is configured with IPv6 forwarding enabled (see Section 10.3.1). Redhat 8.0 distribution does not include the router advertisement daemon. The radvd rpm (this book uses radvd-0.7.1-3.i386.rpm) must be installed for the Linux host to do router advertisements.

The radvd daemon is configured by the `/etc/radvd.conf` file. The following file enables router advertisements of the `3ffe:b00:0:1::/64` prefix on the `eth0` interface.

```
# cat /etc/radvd.conf
interface eth0 {
 advSendAdvert on;
 prefix 3ffe:b00:0:1::/64 {
  AdvOnLink on;
 };
};
```

5.7.3.4 Troubleshooting

The first step is to verify if your kernel supports IPv6, by looking at the existence of the `/proc/net/if_inet6` file. If it exists, then the kernel supports IPv6.

```
# more /proc/net/if_inet6
```

The IPv6 module should also be loaded. The `lsmod` command should show the IPv6 module.

```
# lsmod | grep ipv6
```

To verify the IPv6 addresses on an interface, use `ifconfig <interface_name>`.

```
% ifconfig eth0
```

Pinging to self is a good way to see that basic IPv6 is working.

```
% ping6 ::1
```

Router advertisements received on an interface can be viewed with the `radvdump` command.

```
# radvdump
```

5.7.4 Solaris

The following sections describe enabling IPv6, configuring an interface and configuring router advertisements for the Solaris 9 implementation.

5.7.4.1 Enabling IPv6

In Solaris, IPv6 is enabled as soon as one interface is configured for IPv6.

5.7.4.2 Configuring an Interface

To enable IPv6 on an interface for autoconfiguration, a `/etc/hostname6.<interface_name>` file is created with no content. The following enables IPv6 for autoconfiguration on the `le0` interface.

```
# touch /etc/hostname6.le0
```

The `ifconfig` command is used to configure manually an IPv6 address. The arguments include '`inet6`' for the address family, the IPv6 address and the prefix length. The example below configures the `le0` interface with the 3ffe:b00:0:1::a address and a /64 prefix length.

```
# ifconfig le0 inet6 addif 3ffe:b00:0:1::a/64
```

The address configured by the `ifconfig` command is not saved for the next reboot. If the static address needs to be permanently assigned, then the arguments of the `ifconfig` command are put in the corresponding /etc/hostname6.<interface_name>, as shown below.

```
# cat /etc/hostname6.le0
addif 3ffe:b00:0:1::a/64
```

For the purpose of the Solaris examples throughout this book, le0 is shown where an interface name is used.

5.7.4.3 Configuring Router Advertisements

Router advertisements are managed by the ndpd (neighbor discovery protocol daemon) daemon, which is configured by the /etc/inet/ndpd.conf file. The file contains global variables, variables per prefix and variables per interfaces. The following file enables router advertisements of the 3ffe:b00:0:1::/64 prefix on the le0 interface.

```
# cat /etc/inet/ndpd.conf
ifdefault AdvSendAdvertisements on
prefix 3ffe:b00:0:1::/64 le0
```

5.7.4.4 Troubleshooting

To verify the IPv6 addresses on an interface, use ifconfig <interface_name>.

```
% ifconfig le0
```

Pinging to self is a good way to see that basic IPv6 is working. For IPv6, use the –A inet6 to specify the IPv6 address family.

```
% ping –A inet6 ::1
```

5.7.5 Windows

The following sections describe enabling IPv6, configuring an interface and configuring router advertisements for the Windows XP implementation.

5.7.5.1 Enabling IPv6

In Windows XP, IPv6 modules are loaded on the disk as part of the installation, but are not enabled by default. To enable IPv6, the following command needs to be done once by an administrator.

```
C> netsh interface ipv6 install
```

Enabling IPv6 is also possible by installing the 'Microsoft TCP/IP version 6' protocol for a LAN connection in 'Network Connections'.

5.7.5.2 Configuring an Interface

The 'add address' subcommand is used to configure manually an address on an interface. The example below configures the 3ffe:b00:0:1::a address on the 'Local Area Connection' interface.

```
C> netsh interface ipv6 add address "Local Area Connection"
3ffe:b00:0:1::a
```

Different from most implementations, the prefix length is not specified with the address, which forces a /64 as the prefix length.

For the purpose of the Windows examples throughout this book, 'Local Area Connection' is shown where an interface name is used.

5.7.5.3 Configuring Router Advertisements

Windows can be configured as a router that forwards traffic between interfaces and advertise prefixes on links. However, only static routes can be installed because no routing protocol is available.

The 'set interface' subcommand is used to enable forwarding and router advertisements on an interface. The example below configures, on the 'Local Area Connection' interface, the forwarding with the 'forwarding=enable' option and router advertisements with the 'advertise=enabled' option.

```
C> netsh interface ipv6 set interface "Local Area Connection"
forwarding=enabled advertise=enabled
```

Enabling router advertisements on the interface is not enough to advertise prefixes. The add route subcommand is used to specify the prefix to advertise with the 'publish=yes' option. In fact, only the routes specifically published and present in the routing tables are advertised. The following example will generate router advertisements of the 3ffe:b00:0:1::/64 on the 'Local Area Connection' interface.

```
C> netsh interface ipv6 add route 3ffe:b00:0:1::/64 "Local Area
Connection" publish=yes
```

By default, the prefixes are advertised with an infinite lifetime, which is not ideal for most cases. The 'validlifetime' and 'preferredlifetime' options, expressed in minutes, are used to set the valid and preferred lifetimes. The following example advertises the 3ffe:b00:0:1::/64 prefix with a preferred lifetime of 1440 minutes (1 day) and a valid lifetime of 10080 minutes (1 week).

```
C> netsh interface ipv6 add route 3ffe:b00:0:1::/64 "Local Area
Connection" publish=yes validlifetime=10080
preferredlifetime=1440
```

5.7.5.4 Troubleshooting

The '`netsh interface ipv6 show`' command is used for troubleshooting addresses and interfaces.

The '`show address`' subcommand displays the IPv6 addresses and the interface indexes for all interfaces, as shown below.

```
C> netsh interface ipv6 show address
```

The '`ipconfig`' command shows the IPv4 and IPv6 addresses, DNS configuration and default router for all interfaces.

```
C> ipconfig
```

The '`show interface`' subcommand displays the interface information, as shown below.

```
C> netsh interface ipv6 show interface
```

The '`level=verbose`' option displays additional information about the interfaces.

```
C> netsh interface ipv6 show interface level=verbose
```

The '`ping`' command can be used to verify the reachability of another node. Ping accepts both a hostname or an IP address as argument. If the hostname has both an IPv4 and an IPv6 address in the DNS, then the IPv6 address is preferred.

```
C> ping 3ffe:b00:0:1::2
```

To force the use of IPv6, the '`–6`' option can be used.

```
C> ping –6 www.example.org
```

In the service pack 2 of XP [Microsoft, 2005], a host firewall for IPv6 is enabled by default. Some problems that you may experience with applications not connecting might be related to some restrictions imposed by the firewall default rules. Use the `netsh firewall show` command to see the status of the firewall.

```
C> netsh firewall show
```

5.7.6 Cisco

The following sections describe enabling IPv6, configuring an interface and configuring router advertisements for the Cisco IOS 12.2(13)T implementation.

5.7.6.1 Enabling IPv6

In Cisco IOS, IPv6 is enabled as a host (non-routing) as soon as one interface is configured for IPv6. This enables the Cisco device to act as a host. To enable IPv6 forwarding, the `ipv6 unicast-routing` statement is used in the configuration mode.

```
configure terminal
ipv6 unicast-routing
```

5.7.6.2 Configuring an Interface

The '`ipv6 address`' statement under the interface level is used to configure the address of an interface. The following example configures the `Ethernet0` interface with the `3ffe:b00:0:1::1` prefix and `/64` as the prefix length.

```
configure terminal
interface Ethernet0
 ipv6 address 3ffe:b00:0:1::1/64
```

IOS can also use the MAC address to make the host part of the address, by specifying the '`eui-64`' argument to the `ipv6 address` statement.

```
configure terminal
interface Ethernet0
 ipv6 address 3ffe:b00:0:1::/64 eui-64
```

Additional link-local addresses can be configured by adding the '`link-local`' argument and by specifying a link-local address. The following example adds the `fe80::1/64` link-local address on the `Ethernet0` interface.

```
configure terminal
interface Ethernet0
 ipv6 address fe80::1/64 link-local
```

Since IPv6 supports multiple addresses on the same interface, IOS does not need the '`secondary`' argument used in IPv4 addresses.

As in IPv4, the address of an interface can be set to the one on another interface, using the '`ipv6 unnumbered`' statement. The following example configures the address of the Ethernet0 based on the Loopback interface address.

```
configure terminal
interface Loopback0
 ipv6 address 3ffe:b00:0:1::1/64
interface Ethernet0
 ipv6 unnumbered Loopback 0
```

For the purpose of the Cisco examples throughout this book, `Ethernet0` is shown where an interface name is used.

5.7.6.3 Configuring Router Advertisements

Router advertisements in IOS are set per interface using the `ipv6 nd prefix-advertisement` statement.

```
configure terminal
interface Ethernet0
 ipv6 nd prefix-advertisement 3ffe:b00:0:1::/64
```

Note that by default, the prefix is advertised on-link.

The prefix advertisement can be further specified with the following options on the `ipv6 nd prefix-advertisement` statement:

- The valid lifetime of the prefix, in seconds, is set as the first argument after the prefix. The default is 2592000 seconds (30 days).
- The preferred lifetime of the prefix is set as the second argument after the prefix. The default is 604800 seconds (7 days).

5.7.6.4 Troubleshooting

To verify the IPv6 addresses on an interface, use the `show ipv6 interface <interface_name>` command.

```
enable
show ipv6 interface ethernet0
```

Pinging to self is a good way to see that basic IPv6 is working.

```
> ping ipv6 ::1
```

To show a summary of the IPv6 traffic, use the '`show ipv6 traffic`' command.

```
enable
show ipv6 traffic
```

To debug IPv6 packets, use the '`debug ipv6 packet`' command.

```
enable
debug ipv6 packet
```

To debug IPv6 neighbor discovery packets and processing, use the '`debug ipv6 nd`' command.

```
enable
debug ipv6 nd
```

5.7.7 Hexago

The following sections describe enabling IPv6, configuring an interface and configuring router advertisements for the Hexago HexOS implementation.

5.7.7.1 Enabling IPv6

In HexOS, IPv6 is enabled as soon as one interface is configured for IPv6. IPv6 forwarding is also enabled by default.

5.7.7.2 Configuring an Interface

The '`ipv6 address`' statement under the interface level is used to configure the address of an interface. The following example configures the `fastethernet0` interface with the `3ffe:b00:0:1::1` prefix and `/64` as the prefix length.

```
configure terminal
interface fastethernet0
 ipv6 address 3ffe:b00:0:1::1/64
```

For the purpose of the HexOS examples throughout this book, `fastethernet0` is shown where an interface name is used.

5.7.7.3 Configuring Router Advertisements

Router advertisements in HexOS are set per interface using the `ipv6 nd prefix` statement.

```
configure terminal
interface fastethernet0
 ipv6 nd prefix 3ffe:b00:0:1::/64
```

By default, the preferred lifetime is 604800 seconds (7 days), the valid lifetime is 2592000 seconds (30 days), and the onlink and autoconfig flags are set. These default values can be changed as options to the `ipv6 nd prefix` statement.

5.7.7.4 Troubleshooting

To verify the IPv6 addresses on an interface, use the `show ipv6 interface <interface_name>` command.

```
enable
show ipv6 interface FastEthernet0
```

Pinging to self is a good way to see that basic IPv6 is working.

```
> ping ipv6 ::1
```

To debug IPv6 packets, use the 'debug ipv6 packet' command.

```
enable
debug ipv6 packet
```

5.7.8 Juniper

The following sections describe enabling IPv6, configuring an interface and configuring router advertisements for the Juniper JunOS 5.5 implementation.

5.7.8.1 Enabling IPv6

In JunOS, IPv6 is enabled as soon as one interface is configured for IPv6.

5.7.8.2 Configuring an Interface

JunOS presents a hierarchical view of the configuration statements. At the 'unit' sub-level of 'interfaces', an interface is configured for IPv6 by using 'family inet6' instead of using 'family inet' for IPv4.

The following example configures the Fast Ethernet interface with a static and fully qualified IPv6 address of 3ffe:b00:0:1::2 and a /64 prefix length on the link.

```
interfaces fe-0/0/1 {
 unit 0 {
  family inet6 {
    address 3ffe:b00:0:1::1/64;
  }
 }
}
```

To configure the host part with the interface identifier based on EUI-64, the address object is specified only by the first 64 bits of the address (3ffe:b00:0:1::/64) and the 'eui-64' attribute is added in the address sub-level, as shown below.

```
interfaces fe-0/0/1 {
 unit 0 {
  family inet6 {
   address 3ffe:b00:0:1::/64 {
    eui-64;
   }
  }
 }
}
```

For the purpose of the Juniper examples throughout this book, fe-0/0/1 unit 0 is shown where an interface name is used.

5.7.8.3 Configuring Router Advertisements

To configure RAs on JunOS, the '`router-advertisement`' hierarchy is used. For each interface enabled for RA with the interface statement, the prefix sub-statement is used to specify the prefixes to be announced on the link of the interface.

5.7.8.4 Basic RA Configuration

The following statement configures the router advertisement of the 3ffe:b00:0:1::/64 prefix on the LAN1 link.

```
protocols {
 router-advertisement {
  interface fe-0/0/1 {
   prefix 3ffe:b00:0:1::/64;
  }
 }
}
```

By default, the prefix is advertised on-link.

The prefix advertisement can be further specified with the following options:

- The preferred lifetime of the prefix is set with the '`preferred-lifetime`' statement. The default is 604800 seconds (7 days).
- The valid lifetime of the prefix is set with the '`valid-lifetime`' statement. The default is 2592000 seconds (30 days).

5.7.8.5 Options

At the '`interface`' sub-level, the following options can be added (this list is not exhaustive and represents the typical options most networks will use; please refer to the manufacturer manuals):

- The hop limit to be configured by the nodes is set with the '`current-hop-limit`' statement. The default is 64.
- The default router lifetime is set with the '`default-lifetime`' statement. The default is three times the advertisement interval.
- The stateful autoconfiguration flags are set with the '`managed-configuration`' and '`other-stateful-configuration`' statements. The default is no stateful autoconfiguration.

The statements below show the same router advertisement with all the above options set: a 3ffe:b00:0:1::/64 is advertised on the fe-0/0/1 interface, with a preferred lifetime of 3 days (259200 seconds), a valid lifetime of 7 days (604800 seconds), a hop limit of 128, a default router lifetime of 10 minutes (600 seconds) and the stateful autoconfiguration flags set.

```
protocols {
 router-advertisement {
   interface fe-0/0/1 {
     prefix 3ffe:b00:0:1::/64 {
       preferred-lifetime 259200;
       valid-lifetime 604800;
   }
   current-hop-limit 128;
   default-lifetime 600;
   managed-configuration;
   other-stateful-configuration;
  }
 }
}
```

5.7.8.6 Troubleshooting

To clear all the router advertisements, use the following command.

```
clear ipv6 router-advertisement
```

To show the router advertisements, use the following command.

```
show ipv6 router-advertisement
```

The ‘conflicts’ option of this command is very useful to find out if other routers are sending different and conflicting router advertisements on the link, when nodes seem to have some instability in their configuration.

5.7.9 Debugging Autoconfiguration and Router Advertisements

The default periodicity of the router advertisements is between 200 and 600 seconds. The preferred and valid lifetime should be greater than the router advertisement periodicity, by a few times. If the node do not get one or two advertisements (because of a busy link, etc.), it does not unconfigure its address. Always make sure that lifetimes are greater than 30 minutes (3×600 seconds).

Some router advertisement implementations define the valid and preferred lifetimes as infinite by default. If any mobile device, like a laptop, is receiving the infinite lifetime, it will theoretically keep the address until a new reboot, even if it moves to a new network. This keeps an invalid address in the kernel that might cause timeouts. A good practice is to define the preferred and valid lifetime as non-infinite. Any value should be based on some assumptions about the router advertisement ‘service’, but valid and preferred lifetimes of hours or days are good practice.

5.8 Summary

Addresses of IPv6 nodes can be configured statically, with autoconfiguration or with DHCP. For nodes to use autoconfiguration, a router solicitation message is sent at boot time to all routers on the link. Routers on the link respond with router advertisements which include one or many prefixes of the link, the default router, and some additional information.

DHCP enables autoconfiguration of the address as well as any options such as DNS server, printers, etc.

Renumbering of a site is easier in IPv6 using either DHCP or address autoconfiguration.

5.9 Appendix

This appendix lists each parameter and options of the protocol messages described in this chapter.

5.9.1 Router Advertisement and Solicitation Message Formats

Table 5.11 lists all fields in the router advertisement packet.

The router advertisement includes information about the prefix, listed in Table 5.12.

Table 5.11 Router Advertisement Packet

IP Header Fields	Length (bits)	Value	Description
Source address	128	Link-local address of the router	Link-local address of the router interface sending the router advertisement.
Destination address	128	FF02::1	All-nodes on the link multicast address.
Hop limit	8	255	See Section 7.3 on the use of 255.
ICMP Header Fields	**Length (bits)**	**Value**	**Description**
Type	8	134	134 is the ICMP type for a neighbor discovery router advertisement message.
Code	8	0	
Router Advertisement Fields	**Length (bits)**	**Value**	**Description**
Managed address configuration flag	1		Set if DHCP is allowed to be used on the link.
Other stateful configuration flag	1		Set if other information such as DNS servers can be obtained from the DHCP server.
Router lifetime	16		If non-zero, identifies the router as a default router for the specified period.

Reachable time	32		Timer to be used by nodes to determine the reachability of another node on the link, discussed in Section 6.4.
Retransmission time	32		Delay, in seconds, between neighbor solicitation requests, discussed in Section 6.2.
Prefix		See Table 5.12	Prefix to be used by the node to autoconfigure.
Link-layer address of the source (router)	128		The link-layer address of the source router.
MTU	32		The MTU of the link.

Table 5.12 Prefix fields in router advertisements

Name	Length (bits)	Value	Description
Type	8	3	Identifies the type of option; prefix information in this case.
Length	8	4	Total length of the option fields.
Prefix length	8	0–128, usually 64	Length of the prefix below, indicating the number of useful bits in the prefix for node autoconfiguration.
L	1	0,1, usually 1	Set when the prefix is used on the link. Usually set.
A	1	0,1, usually 1.	Set to signal the node to use the prefix for autoconfiguration. Usually set.
Valid lifetime	32		Duration in seconds of the prefix validity. Value of 0xffffffff defines infinity.
Preferred lifetime	32		Duration in seconds of the preferred status of the address, which means the client can use this address as a source address when preferred is different to zero.
Prefix	128		Prefix available to the node for address autoconfiguration.

5.9.2 DHCP Variables, Addresses and Ports

DHCP has a number of variables and fields. Table 5.13 lists the port numbers used by the DHCP protocol.

Table 5.14 lists all the protocol addresses for DHCP.

Table 5.15 lists the DHCP message types used between clients, relays and servers.

DHCPv6 uses a concept of identity association to manage the identity of a client, because the uniqueness of an address is no more available as an identifier of the client. Table 5.16 shows the fields of the identity association option.

Table 5.13 DHCP port numbers

Name	Number	Transport	Description
dhcpv6-client	546	UDP	Destination port used by servers and relays for messages sent to clients.
dhcpv6-server	547	UDP	Destination port used by clients for messages sent to agents.

Table 5.14 DHCP addresses

Name	Address	Description
All DHCP Agents	FF02::1:2	Destination address used by clients to find the agents on the link.
All DHCP Servers	FF05::1:3	Destination address used by clients (or relays) to find the servers on the site.

Table 5.15 DHCP message types

Name	Value	From	To	Description
SOLICIT	1	client	all agents	Sent by clients to find servers.
ADVERTISE	2	servers	client	Sent by servers to show their presence.
REQUEST	3	client	server	Sent by the client asking for configuration information.
CONFIRM	4	client	server	Sent by the client to confirm the received configuration information.
RENEW	5	client	server	Sent by the client to renew the previously received configuration information.
REBIND	6	client	servers	Sent by the client requesting any server to renew the previously received configuration information.
REPLY	7	servers	client	Sent by the server containing the configuration information for the client.
RELEASE	8	client	server	Sent by the client to release the configuration information to the server. Happens when the client is about to shutdown and does not need the address anymore.
DECLINE	9	client	server	Sent by the client to tell the server that the received configuration information is not usable.
RECONFIG-INIT	10	server	client	Sent by the server to tell the client that configuration information has been updated and client should re-request configuration information.
RELAY-FORW	11	relay	server	Sent by relays to forward client messages to server.
RELAY-REPL	12	server	relay	Sent by servers to send replies to client through relays.

Table 5.16 Identity association fields

Name	Length (bits)	Value	Description
OPTION IA	16	1	Identifier of this option.
option-len	16		Length of this message.
IAID	32		Unique identifier of this identity association, defined by the client.
T1	32		Time to re-contact the server to renew the address.
T2	32		Time to re-contact any server to renew the address.
IA status	8		Status of the identity association.
num addrs	8		Number of addresses included.
For each included address			
T	1		Set to 1 to indicate a temporary/privacy address (see Section 13.4)
addr status	7		Status of the address.
prefix length	8		Length of the prefix.
IPv6 address	128		IPv6 address sent by the server to the client.
preferred lifetime	32		Preferred lifetime of the address.
valid lifetime	32		Valid lifetime of the address.

5.10 References

Microsoft, 'Internet Protocol version 6 (IPv6) Internet Connection Firewall', April 2005, http://www.microsoft.com/technet/itsolutions/network/security/ipv6fw/ipv6fwov.mspx.

[IEEEEUI64] Guidelines for 64-bit Global Identifier (EUI-64™) Registration Authority, *IEEE*, http://standards.ieee.org/regauth/oui/tutorials/EUI64.html

[RFC1918] Rekhter, Y., Moskowitz, R., Karrenberg, D., Groot, G. and E. Lear, 'Address Allocation for Private Internets', BCP 5, RFC 1918, February 1996.

[RFC2461] Narten, T., Nordmark, E. and Simpson, W., 'Neighbor Discovery for IP Version 6 (IPv6)', RFC 2461, December 1998.

[RFC2462] Thomson, S. and Narten, T., 'IPv6 Stateless Address Autoconfiguration', RFC 2462, December 1998.

[RFC3315] Droms, R., Bound, J., Volz, B., Lemon, T., Perkins, C. and Carney, M., 'Dynamic Host Configuration Protocol for IPv6 (DHCPv6)', RFC 3315, July 2003.

[RFC3633] Troan, O. and Droms, R., 'IPv6 Prefix Options for Dynamic Host Configuration Protocol (DHCP) version 6', RFC 3633, December 2003.

[RFC3646] Droms, R., 'DNS Configuration Options for Dynamic Host Configuration Protocol for IPv6 (DHCPv6)', RFC 3646, December 2003.

[RFC3769] Miyakawa, S. and Droms, R., 'Requirements for IPv6 Prefix Delegation', RFC 3769, June 2004.

6

Link-layer Integration

IPv6 packets are encapsulated in layer 2 frames. Interaction with layer 2, such as IPv6 address to MAC address mapping, is key to any communication. IPv6 brings more efficiency than IPv4 in this interaction with layer 2. This chapter uses Ethernet as the layer 2 example.

Solicited-node multicast address is first described. This special form of address enables efficient discovery of a neighbor's layer 2 address. Similar to IPv4 Address Resolution Protocol (ARP), neighbor solicitation and advertisement is then described. Duplicate address detection is shown based on neighbor solicitation. IPv6 over Ethernet is presented with the specifics of frame identifier and multicast mapping. The chapter ends with some configuration examples of major implementations.

6.1 Solicited-Node Multicast Address

IPv6 uses multicast instead of broadcast for network efficiency purposes, such as with the Neighbor Discovery (ND) protocol. In IPv4 ARP, a node has to broadcast to everyone on the link to query the MAC address corresponding to a known IP address. Since ARP is the mechanism used before the real IP exchange happens, there is no easy way to avoid broadcasts in IPv4. IPv6 is well designed to make the discovery of the MAC addresses a near to unicast query!

Neighbor discovery uses a special form of a multicast address, called the solicited-node multicast address. This address contains part of the targeted IPv6 address of the neighbor. The format is based on a fixed leftmost prefix of 104 bits defined as: ff02:0:0:0:0:1:ff00::/104. The rightmost 24 bits are taken from the rightmost 24 bits of the target IPv6 address. Figure 6.1 describes the structure.

Figure 6.2 shows the construction of the solicited-node multicast address from the 3ffe:b00:0:1:212:6bff:fe3a:9e9a address.

Table 6.1 gives some examples of IPv6 addresses and their corresponding solicited-node multicast address.

Migrating to IPv6: A Practical Guide to Implementing IPv6 in Mobile and Fixed Networks Marc Blanchet

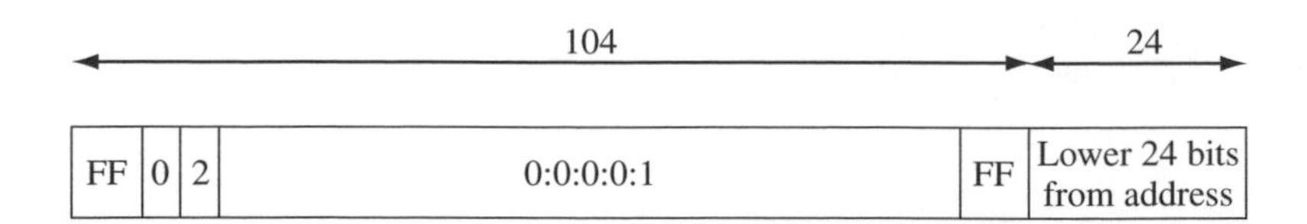

Figure 6.1 Solicited-node multicast structure

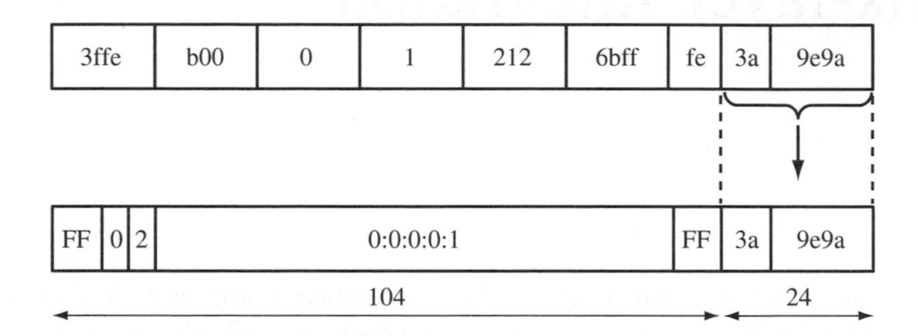

Figure 6.2 Solicited-node multicast address example

Table 6.1 Examples of Solicited-node Multicast Addresses

IPv6 address	Corresponding Solicited-node Multicast Address
3ffe:b00:0:1**::1**	ff02::1:ff**00:1**
3ffe:b00:0:1:212:6bff:fe**3a:9e9a**	ff02::1:ff**3a:9e9a**
fe80::212:6bff:fe**54:f99a**	ff02::1:ff**54:f99a**
fc00:0:0:1::aa**aa:a1**	ff02::1:ff**aa:a1**

If a node is configured with 3ffe:b00:0:1::1, then it must listen to the corresponding solicited-node multicast address: ff02::1:ff00:1. For all of its IPv6 addresses, a node listens to all the corresponding solicited-node multicast addresses on the respective interfaces. For example, if a node has three addresses (3ffe:b00:0:1::1, 3ffe:b00:0:1:212:6bff:fe54:f99a, fe80::212:6bff:fe54:f99a) assigned to an interface, then the node must listen to the corresponding solicited-node multicast addresses (ff02::1:ff00:1, ff02::1:ff54:f99a). Note that there are only two multicast addresses for three unicast addresses: because the last two unicast addresses share the rightmost 24 bits, the same multicast address is used. This is often the case since the same interface identifier is used by the autoconfiguration mechanism for many addresses on the same interface.

> **Important point:**
>
> A node listens to all solicited-node multicast addresses corresponding to all of its unicast and anycast addresses.

Solicited-node multicast addresses are used by the neighbor solicitation process, described in Section 6.2.

6.2 Neighbor Solicitation and Advertisement

The Neighbor Discovery (ND) protocol is used for different purposes. Router solicitation and advertisements are used by the routers to send network information to hosts, as described in Section 5.2.2. Neighbor solicitation and advertisements are also part of the neighbor discovery protocol.

Neighbor solicitation is the mechanism for a node to request the link-layer address of a target node known by its IPv6 address, to enable communications between the two parties on the same link. The target node responds with a neighbor advertisement. The ARP is the corresponding protocol in IPv4.

6.2.1 IPv4 Address Resolution Protocol

When an IPv4 node wants to send a datagram to another IPv4 node, the source node uses ARP to find the link-layer address of the target node. The ARP request is sent by the source node to all nodes on the local link, using the link-layer broadcast address. The request contains the IP address of the target. All nodes look at the request and only the one that has the target IP address responds to the source node with an ARP reply containing the target link-layer address. Table 6.2 shows the process for Ethernet links.

6.2.2 IPv6 Differences

IPv6 does the same process but optimizes it using multicast and uses a different transport mechanism.

Section 6.1 describes how any node listens to the solicited-node multicast address for each of its IPv6 addresses. The solicited-node multicast addresses contain the last 24 bits of its IPv6 addresses. When an IPv6 node does the equivalent of the IPv4 ARP request, it sends

Table 6.2 IPv4 ARP request on Ethernet

From → To	Type of packet	Source and Destination Addresses	Description and Content
1 Src host (A) → All hosts	ARP request	Ethernet Src: MAC address of A Ethernet Dst: link broadcast Ethernet Type: ARP	MAC address of source (A) IP address of source (A) and target (B) Opcode = Request
2 Target host (B) → Src host (A)	ARP reply	Ethernet Src: MAC address of B Ethernet Dst: MAC address of A	MAC address of target (B) and source (A) IP address of source (A) and target (B) Opcode = Reply

the datagram to the solicited-node multicast address of the target. This means that only the nodes that have the same rightmost 24 bits in their IPv6 address listen to that request. Recall that in IPv4, the ARP request is broadcast. But as in IPv4, only the one IPv6 node concerned with the request (by looking at the target IP address in the request) answers.

From the layer 2 perspective, IPv4 ARP is a layer 3 protocol just as IP, so ARP must be redefined for each layer 2 technology. Instead, the IPv6 ND protocol uses ICMP as transport. So neighbor solicitation and neighbor advertisements messages are specific ICMP messages and are completely independent of the layer 2.

6.2.3 Neighbor Solicitation Process

When a node wants to send a datagram to a neighbor on the same link, it sends a neighbor solicitation message to the solicited-node multicast address corresponding to the IPv6 address of the neighbor. All nodes listening to that solicited-node multicast address look at the request but only the real target answers, since the request message contains the full IPv6 address of the target.

Table 6.3 shows the process. The neighbor solicitation message sent by source A is an ICMP message of type 135. The IPv6 source address is the link-local address of source A. The destination address is the solicited-node multicast address of target B.

A neighbor advertisement is sent as a response to a neighbor solicitation. Node A sends a neighbor solicitation to the solicited-node multicast address of the IPv6 address of node B. Node B responds with a neighbor advertisement sent directly to node A and includes node B's layer 2 address. The neighbor advertisement contains the link layer address of B.

The delay before resending a neighbor solicitation when no neighbor advertisement is received can be configured. As part of the router advertisements, routers can specify

Table 6.3 Neighbor solicitation and advertisement

From → To	Type of packet	Source and Destination Addresses	Description and Content
1 Src host (A) → All hosts who have the same rightmost 24 bits of the targeted IPv6 address	ICMPv6 Type = 135 (ND NS)	Link Src: link-layer address of source (A) Link Dst: link-layer multicast address of the solicited-node multicast address of target (B) IP Src: link-local address of source (A) IP Dst: solicited-node multicast address of B address	Neighbor solicitation: constains the IPv6 address of B.
2 Target host (B) → Src host (A)	ICMPv6 Type = 136 (ND NA)	Link Src: link-layer address of target (B) Link Dst: link-layer address of source (A) IP Src: link-local address of target (B) IP Dst: link-local address of source (A)	Neighbor advertisement: contains the link-layer address of B.

this delay, called 'retransmission time', to nodes. The retransmission time is described in Table 5.11.

6.3 Duplicate Address Detection

A new mechanism is defined in IPv6 to verify the usability of every address a node intends to use. This is to ensure that nobody else on the link has the same address, which would lead to unexpected behavior on the link. In IPv4, many recent implementations carry out an ARP query to their own address to verify if someone is using that address. This process is not mandatory in IPv4. However, in IPv6, this process, called duplicate address detection (DAD), is mandatory and implemented using a more efficient technique by using the solicited-node multicast address.

A node starts using a new address after receiving a router advertisement or at boot time when trying to use its link-local address or when the address is manually configured. Every time a new address is going to be used, the duplicate address detection takes place.

If node A wants to use its link-local address fe80::212:6bff:fe54:f99a, it checks if this address is already in use on the link by sending a neighbor solicitation to the solicited-node multicast address (ff02::1:ff54:f99a) corresponding to its to-be-used address. If a node respond by a neighbor advertisement, then this address cannot be used. This process is detailed in Table 6.4.

If a node responds to the neighbor solicitation with a neighbor advertisement, then a duplicate address is detected, the new address is not used and an error is generated to the user.

Since the default prefix length is /64 for a link and hosts are autoconfigured using their unique MAC address, duplicate addresses should not occur very often.

6.4 Neighbor Cache

As in IPv4, a node maintains a list of his neighbors. Each destination address on the local link appears in the neighbor cache. The possible states of a cache entry are listed in Table 6.5.

In normal operations, most of the entries in the neighbor cache will show the 'stale' status. The nodes that were recently contacted will have the 'reachable' status. A node that is kept in incomplete state means there is a problem reaching it.

Table 6.4 Duplicate address detection

From → To	Type of packet	Source and Destination Addresses	Description and Content
1 Src host (A) → All hosts who have the same rightmost 24 bits of the targeted IPv6 address (A)	ICMPv6 Type = 135 (ND NS)	Link Src: link-layer address of source (A) Link Dst: link-layer multicast address of the solicited-node multicast address of target (A) IP Src: link-local address of source (A) IP Dst: Solicited-node multicast address of A address	Neighbor solicitation: contains the IPv6 address of A.

Table 6.5 States of the neighbor cache

State	Description
incomplete	The node has sent a neighbor solicitation to this neighbor but has not yet received a neighbor advertisement.
reachable	Neighbor advertisement was received 'recently': the expiration time for this entry has not been reached yet.
stale	Neighbor advertisement was received for this entry but the expiration time has arrived. Any packet that needs to be sent to this neighbor should restart the neighbor solicitation process. This state is the most common state because the expiration time of the reachable state is very short.
delay	Like the stale state, but is an intermediate step towards resending a neighbor solicitation through the probe state. This state is an optimization for upper-layer protocols such as TCP.
probe	Neighbor solicitation is being sent to this node, after being in the delay state.

The expiration time from the reachable to the stale state is called the 'reachable time'. It can be specified in router advertisements, so routers can force the nodes to use this value. Reachable time is described in Table 5.11.

6.4.1 Neighbor Unreachability Detection

Neighbor Unreachability Detection (NUD)[RFC2461] is a process where nodes are active in keeping the reachability state of their neighbors. This is very useful for nodes that have multiple default routers and when one of them becomes unavailable. By active detection, the node can switch to other default routers. The process is to send unicast neighbor solicitations messages to the neighbors when the node has to send traffic to the neighbor and to keep the states in the neighbor cache.

6.5 EUI-64 and Neighbor Discovery

When sending a packet from a node A to another node B and given that the host part of an address embeds the MAC address, a node (A) might just extract the MAC address from the IPv6 address of the destination (B) and then use the MAC address as the destination layer 2 frame address. This would avoid the need for the neighbor solicitation and advertisement process, the duplicate address detection and it would be faster. However, for the following reasons this is not the case:

- The fact that a node has an address that looks like an EUI-64 does not necessarily mean that the MAC address is there.
- Some link-layers do not have unique MAC addresses.
- There is not necessarily a one-to-one relationship between the MAC address and the IPv6 address.
- Nodes are also using manually assigned addresses or temporary addresses, which have no EUI-64 host part. These addresses must use neighbor discovery.
- It is safer to implement this mechanism for all addresses to avoid duplicate addresses on the same link.

- The cost of an initial neighbor solicitation/advertisement exchange is low compared to the safety guard it provides.
- If multiple addresses on the same interface use the EUI-64 from the same MAC addresses, implementations can choose to make the neighbor solicitation only on the first one and skip for the others, increasing the efficiency of the neighbor solicitation and advertisement process.

For all these reasons, neighbor solicitation is always used even when the target address is EUI-64 formatted.

6.6 IPv6 over Ethernet

IPv6 is defined to work over most of the link-layer technologies, such as, but not limited to:

- Ethernet [RFC2464]
- PPP [RFC2472]
- FDDI [RFC2467]
- Token-Ring [RFC2470]
- Non-Broadcast Multiple Access (NBMA) [RFC2491]
- ATM [RFC2492]
- Frame-Relay [RFC2590]
- IEEE 1394 (Firewire) [RFC3146]
- ARCNET [RFC2497]
- MAPOS (Multiple Access Protocol Over SONET/SDH) [RFC3572]

However, for simplicity, this chapter presents IPv6 over Ethernet and PPP, the most commonly used link-layer technologies.

When a node is autoconfigured using the interface identifier, this identifier is based on the MAC Ethernet address of the interface, following the EUI-64 specification [IEEEEU164] described in Section 5.2.1.

In this section, Ethernet addresses are represented as six fields written as octets in hexadecimal with a '-' as field separator (e.g. 00-11-22-33-44-55).

6.6.1 Frame Identifier

Ethernet has a protocol field in the frame format to identify the type of protocol in the frame payload. IPv4 datagrams are identified with 0x0800 in the Ethernet protocol field. IPv4 ARP and RARP are identified with 0x0806 and 0x8035 respectively.

IPv6 datagrams are identified with 0x86DD in the Ethernet protocol field, as shown in Figure 6.3.

6.6.2 Multicast

IP Multicast is a way to send one datagram to many selected nodes at the same time. To remain efficient, this mechanism is also implemented on the link layer. If not, then IP

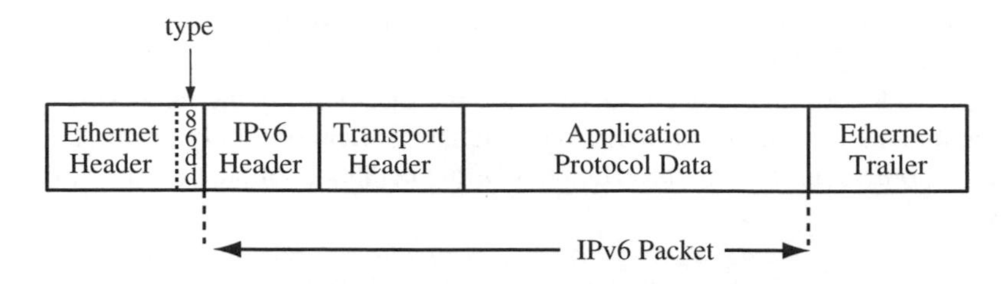

Figure 6.3 Ethernet frame ID for IPv6

multicast traffic is typically broadcasted which defeats the purpose of multicast on local links.

On Ethernet, multicast traffic is identified by the value of 1 in the 8th most significant bit of the Ethernet address, as shown in Figure 6.4. In other words, if the second hex digit of the Ethernet destination address is odd, then it is a multicast Ethernet frame.

Hint for identifying Ethernet multicast traffic:

If the second hex digit of an Ethernet destination address is odd, then the frame contains multicast traffic.

An Ethernet multicast address range is defined for the mapping between an IP multicast address and the corresponding Ethernet multicast address.

IPv4 multicast addresses are defined with 1110 as the leftmost four bits, which leaves 28 bits for the multicast group. The multicast range is 224.0.0.0 to 239.255.255.255. The corresponding Ethernet reserved address range for IPv4 multicast is 01-00-5E-xx-xx-xx. The rightmost 23 bits of an IPv4 multicast address are extracted and put as the rightmost 23 bits in the Ethernet multicast address, as shown in Figure 6.5.

For example, the Ethernet address corresponding to 224.**0.0.1** is 01-00-5E-**00-00-01**.

IPv6 multicast addresses are defined with 'ff' as the first octet. The second octet has fields for scope and permanency, which leaves 112 bits (128 – 8 – 8) for the multicast group. The Ethernet reserved address range for IPv6 multicast is 33-33-xx-xx-xx-xx. The rightmost 32 bits of an IPv6 multicast address are extracted and put as the rightmost 32 bits in the Ethernet multicast address, as shown in Figure 6.6.

For example, the Ethernet multicast address corresponding to ff02::1:**ff54:f99a** is 33-33-**ff-54-f9-9a**.

Link-layer devices such as interface cards in nodes and switches carry out filtering of multicast frames in order to process only the relevant multicast traffic. They compare

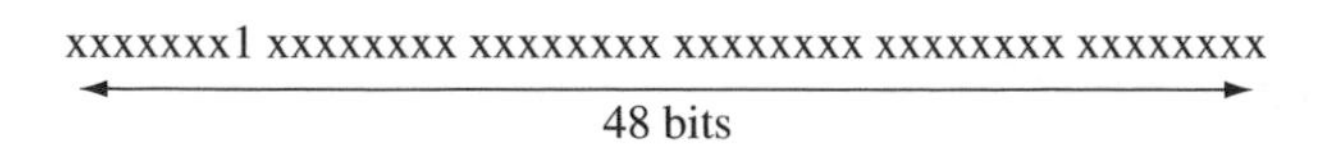

Figure 6.4 Multicast bit in an Ethernet address

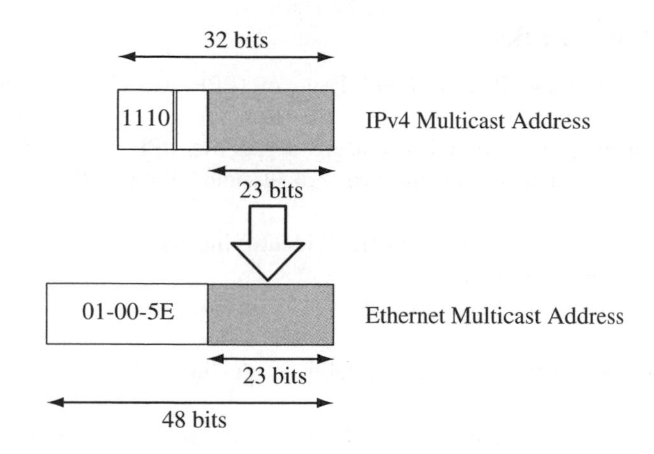

Figure 6.5 IPv4 to Ethernet multicast address mapping

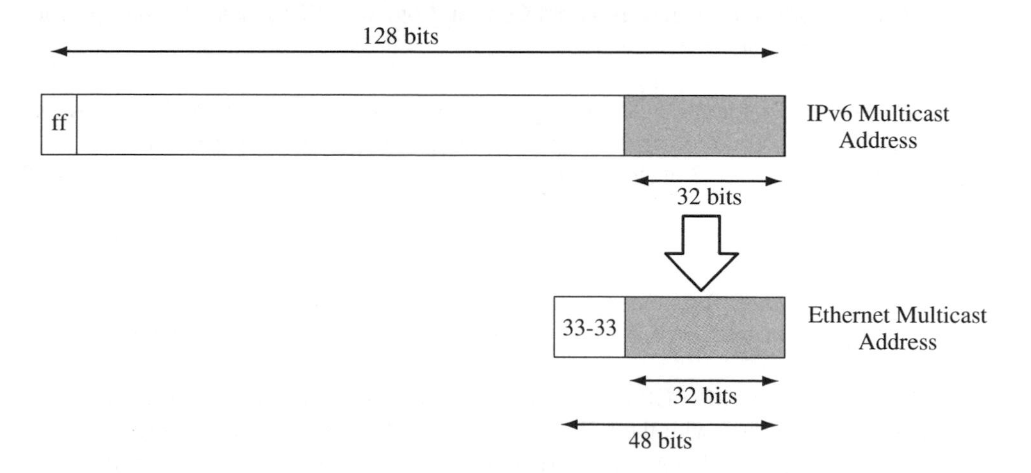

Figure 6.6 IPv6 to Ethernet multicast address mapping

the destination address of a multicast frame with their own list of registered multicast groups.

A multicast-aware switch usually snoops the Internet Group Management Protocol (IGMP) messages to identify which hosts join which group. This enables the switch to forward carefully the multicast datagrams to only the relevant hosts. If a switch is not aware of multicast, then it has to send the datagram on all its ports. Some issues have been discovered and documented [Christensen et al., 2004] with bad switch implementation of IGMP snooping. Experience has shown that switches are not a problem for IPv6 traffic.

6.6.3 Ethernet MTU

The Maximum Transmission Unit (MTU) of Ethernet is 1500. Larger MTU, such as 9 K are possible on gigabit Ethernet. Interfaces are usually configured with the right MTU of the link. A specific link MTU can be advertised to the nodes, as discussed in Section 6.9.

6.7 Point-to-Point Links

Point-to-point links use the Point-to-Point Protocol (PPP) [RFC1661] suite to negotiate the link-layer encapsulation, the control, the authentication and the layer 3 protocols using the link. Each function is accomplished using a specific PPP sub-protocol, which uses a specific PPP packet identified by the protocol id field in the PPP frame, as shown in Figure 6.7.

A PPP frame starts with a Protocol id field, identifying the type of payload. Padding is added after the payload if necessary.

There are three classes of PPP protocols:

- The Link-Control Protocol (LCP) which establishes the link, controls the basic link features and maintains the link state.
- Corollary protocols like authentication and compression protocols, for example, the Password Authentication Protocol (PAP) and the Challenge-Handshake Authentication Protocol (CHAP).
- Layer 3 control protocols, named Network Control Protocols (NCP), deal with the specifics of the layer 3 protocols. For example:
  - the IP Control Protocol (IPCP) manages the IPv4 behavior on the PPP link. As an example, IPCP is used to negotiate the IPv4 address of the incoming PPP node.
  - the Appletalk Control Protocol (ATCP) manages the Appletalk behavior on the PPP link.
  - the IPv6 Control Protocol (IPv6CP) manages the IPv6 behavior on the PPP link.

Table 6.6 lists some PPP protocols with their id and description. IPv4 is handled by the IPCP control protocol and IPv4 packets are put in the payload of PPP frames with the protocol id 0x0021.

IPv6 is handled by the IPv6CP control protocol [RFC2472] and IPv6 packets are put in the payload of PPP frames with the PPP protocol id 0x0057, as illustrated in Figure 6.8.

Multiple layer 3 protocols can be transported over the same PPP link since they have a specific control protocol and a specific PPP protocol identification to differentiate the packets. However, both ends of the PPP link have to support and agree on the layer 3 protocols.

Important point:

IPv4 and IPv6 can be transported over the same PPP link, if both ends support both protocols.

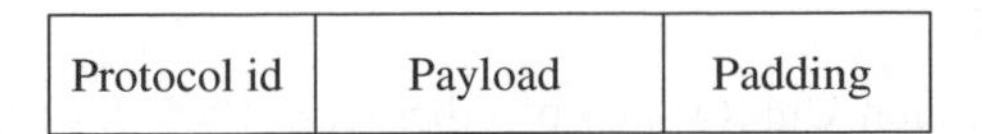

Figure 6.7 PPP frame

Table 6.6 PPP protocols

PPP protocol	Protocol id	Description
LCP	C021	Link-Control Protocol: the main protocol that establishes and controls the link.
PAP	C023	Password Authentication Protocol: one of the authentication protocols.
CHAP	C223	Challenge/Handshake Authentication Protocol: another authentication protocol.
IPCP	8021	The Network Control Protocol for IPv4
IPv4	0021	IPv4 packet in the PPP payload.
IPV6CP	8057	The Network Control Protocol for IPv6
IPv6	0057	IPv6 packet in the PPP payload.

PPP prot id	PPP payload	
0057	IPv6 header	IPv6 payload

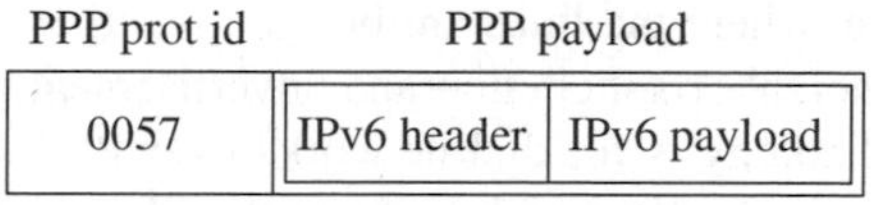

Figure 6.8 IPv6 packet in a PPP frame

The process for establishing a PPP connection can be simply described as:

1. Start with LCP frames to establish the link, negotiate characteristics of the link, authentication, compression, and layer 3 protocols.
2. Authentication is then performed using the negotiated authentication protocol.
3. Each network control protocol starts to negotiate its layer 3 specific parameters for the establishment of layer 3 connectivity.
4. Layer 3 packets are sent on the wire. LCP still continues to monitor the link by sending management frames.

IPv6 over PPP uses link-local addresses on both ends of the PPP link. However, point-to-point links usually do not have interface identifiers, so they have to be made and be obviously different between the two PPP peers. If a node already has some unique identifier like an EUI-64 from another interface (e.g. Ethernet card), then it uses it as the interface identifier on the PPP link. If no unique identifier is available on the node, then it generates one randomly. However, peers exchange their interface identifier using the IPV6CP, so they can verify the uniqueness on both ends. One peer could also ask the other peer for a unique interface identifier. Since this verification is done at the control layer, there is no need to do it again with the Duplicate Address Detection (DAD) process, so DAD is usually disabled on PPP links.

Nowadays, PPP links connect a network behind the connecting PPP device. In IPv4, an IPv4 NAT is used on the gateway, because the PPP link only gives one IPv4 addresses and there is no way to negotiate a range of addresses and to autoconfigure the nodes inside the network in IPv4. IPv6 has many ways to accomplish the goal without any address translation. One way is to use a prefix delegation method to the router, such as DHCPv6-PD (see Section 5.5.4) or TSP (see Section 16.2.9.11). Another way is the multi-link subnet, where a /64 is shared between the PPP link and the inside network, as discussed next.

6.8 Multi-link Subnets

With autoconfiguration, IPv6 helps the emergence of small autoconfigured networks, such as home, personal area and sensor networks. These networks usually have one gateway to the world, but they may have multiple links inside the small network. In a house, a link could be dedicated to home entertainment, another for Internet traffic and another for control such as alarm systems, each having a different set of policies. Multiple links require the configuration of the gateway router, with manual intervention and substantial networking skills, both not possible in small unmanaged networks. Moreover, a /64 prefix is sufficient as address space for the network.

Multi-link subnets are defined [Huitema and Thaler, 2002] to support multiple links using the same /64 address space, where the links are bridged together by the head router using a modified behavior of the neighbor solicitation and advertisement process on the router.

Another use of this technology is for dial networks over PPP. A PPP connection started by a home gateway gets only one IPv4 address from the PPP server. However, in IPv6, the PPP link itself uses a /64 prefix. Sharing the /64 prefix of the PPP link with the network behind the gateway enables the nodes to get reachable IPv6 addresses. This enables sharing of the PPP connection [Yamamoto and Hagino, 2001] with reachable addresses for all nodes behind the PPP gateway which avoids the need for address translation as done in IPv4 for the same application.

Figure 6.9 shows an example, where nodes N1 and N2 are behind router R1 connected to the IPv6 network through router R2 using PPP. The PPP link as well as LAN1 link share the same 3ffe:b00:0:1::/64 prefix even if they are two distinct links.

N1, N2, the two interfaces of R1 and the PPP interface of R2 share the prefix.

A more complex scenario, shown in Figure 6.10, includes multiple links, such as LAN1, LAN2 and LAN3 behind R1. However, the same /64 prefix is used for all the links behind R1.

Two methods are defined to enable multilink subnets: off-link and on-link.

The off-link method requires the gateway router (R1 in the examples) to send router advertisements of the prefix (3ffe:b00:0:1::/64 in the examples) with the off-link bit, which

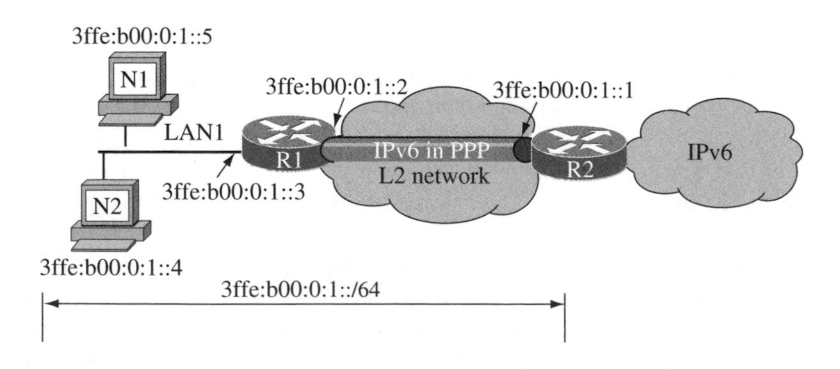

Figure 6.9 Multilink subnet example with PPP sharing

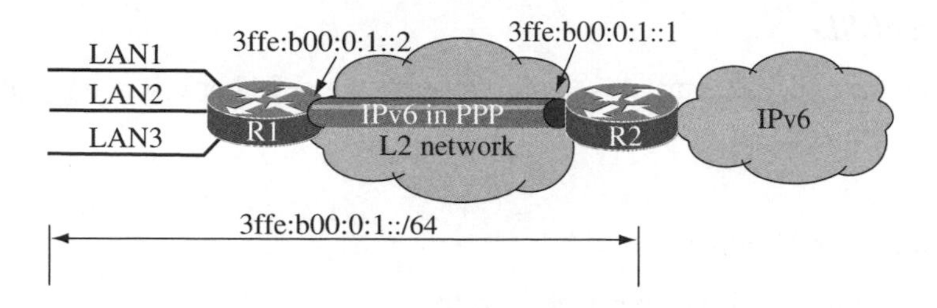

Figure 6.10 Multilink subnet example with multiple links

tells the nodes to configure themselves using the prefix but to consider any other node with the same prefix as off-link. This results in the nodes sending traffic to the default gateway when trying to reach other nodes on the same link. In Figure 6.10, when N1 wants to talk to N2, the first packet is sent to R1, instead of directly to N2, because of the off-link property in the prefix advertisement. R1 forwards the packet to the same physical link (LAN1) to N2 and sends a ICMP redirect message to N1 for the future packets.

The on-link method requires the gateway router to send router advertisements of the prefix with the on-link bit, which tells the nodes to configure themselves using the prefix and to consider any other node with the same prefix to be on the link, which is the normal expected behavior. The nodes do neighbor solicitations on nodes that appear to be on the same link because of the same prefix, but are not on the same link. Those off-link nodes are not responding to neighbor solicitations since, by default, they are not receiving them. In Figure 6.9, when N1 wants to reach 3ffe:b00:0:1::1, the PPP interface of R2, N1 sends a neighbor solicitation to R2 on LAN1 but R2 never receives it. To make multilink subnets work, R1 is a neighbor solicitation proxy for all nodes on the subnet, responding to neighbor solicitation from any node to any node on all the links of the subnet.

These methods do not change any protocol but affect the implementation of the routers involved in multilink subnets.

6.9 Router Advertisements of the Link MTU

Under some circumstances, the MTU of the link is not obvious. For example, when different layer 2 technologies are bridged together, there might be two different MTUs on the same logical link. Router advertisements can include the link MTU so that all nodes configure their interface using the received MTU. The MTU parameter in router advertisements is described in Table 5.11.

6.10 Managing Neighbors on Hosts and Routers

This section lists the implementation commands to manage neighbors and link specific configurations. Commands are shown to modify the neighbor cache. However, under normal operation, you do not need them.

6.10.1 FreeBSD

On FreeBSD, the 'ndp' command is used to manage the neighbor discovery protocol, such as managing neighbors.

6.10.1.1 Managing Neighbors

The 'ndp -a' command lists the neighbor cache.

```
# ndp -a
```

To add a static entry in the neighbor cache, use the '-s' argument. The following example installs the fe80::212:6bff:fe3a:9e9a neighbor with a link layer address of 00:12:6b:3a:9e:9a.

```
# ndp -s fe80::212:6bff:fe3a:9e9a 00:12:6b:3a:9e:9a
```

The following deletes the fe80::212:6bff:fe3a:9e9a neighbor in the cache.

```
# ndp -d fe80::212:6bff:fe3a:9e9a
```

The 'ndp -c' command flushes the neighbor cache.

```
# ndp -c
```

6.10.1.2 MTU in Router Advertisements

Setting the link MTU in router advertisements is done in the '/etc/rtadvd.conf' file. The syntax of the file is described in Section 5.7.2.3. The following example sets the link MTU to 1480 on the fxp0 interface.

```
# cat /etc/rtadvd.conf
fxp0:mtu#1480:
```

6.10.2 Linux

On Linux, the 'ip' command is used to manage many features of IPv6, such as managing neighbors.

6.10.2.1 Managing Neighbors

The 'ip -6 neighbour' command is used to manage the neighbor cache. The 'show' sub-command lists the neighbor cache.

```
# ip -6 neighbour show
```

Each entry in the cache is shown with the following information:

- IPv6 address;
- interface name;
- link-layer addresses;
- state of neighbor unreachability detection (same keywords as in Table 6.5);
- designation of the neighbor as a router.

To add a static entry in the neighbor cache, use the 'add' sub-command. The following example installs the fe80::212:6bff:fe3a:9e9a neighbor with a link layer address of 00:12:6b:3a:9e:9a on eth0 interface.

```
# ip -6 neighbour add fe80::212:6bff:fe3a:9e9a lladdr
00:12:6b:3a:9e:9a dev eth0 nud reachable
```

The following deletes the fe80::212:6bff:fe3a:9e9a neighbor in the cache.

```
# ip -6 neighbour delete fe80::212:6bff:fe3a:9e9a dev eth0
```

To flush the neighbor cache, use the 'flush' subcommand.

```
# ip -6 neighbour flush
```

6.10.2.2 MTU in Router Advertisements

The router advertisement daemon sends the link MTU with the 'advLinkMTU' statement in its /etc/radvd.conf configuration file. The following example sets the link MTU to 1480 on the eth0 interface.

```
# cat /etc/radvd.conf
if eth0 {
 advLinkMTU 1480;
};
```

6.10.3 Solaris

On Solaris, the router advertisement daemon sends the link MTU with the 'AdvLinkMTU' statement in its /etc/inet/ndpd.conf configuration file. The following example sets the link MTU to 1480 on the le0 interface.

```
# cat /etc/inet/ndpd.conf
if le0 advLinkMTU 1480
```

6.10.4 Windows

On Windows, to show the neighbor cache, use the '`show neighbors`' subcommand.

```
C> netsh interface show neighbors
```

Windows enables the user to see the full destination cache, which means information on destinations, including on-link and off-link.

```
C> netsh interface show destinationcache
```

To set the MTU on a interface, use the '`mtu`' option of the '`set interface`' subcommand, as shown below.

```
C> netsh interface ipv6 set interface interface="Local Area
Connection" mtu=1480
```

6.10.5 Cisco

On IOS, the neighbor cache is shown with the '`show ipv6 neighbors`' command.

```
show ipv6 neighbors
```

An interface name or address can be specified after the '`neighbors`' keyword to restrict the display.

To flush the neighbor cache, use the '`clear ipv6 neighbors`' command.

```
clear ipv6 neighbors
```

To add a neighbor, use the '`ipv6 neighbor`' command with the following information:

- IPv6 address of the neighbor;
- interface name on which the neighbor is reachable;
- link layer address of the neighbor.

The following example installs the `fe80::212:6bff:fe3a:9e9a` neighbor with a link layer address of `00:12:6b:3a:9e:9a` on `Ethernet0` interface.

```
configure terminal
ipv6 neighbor fe80::212:6bff:fe3a:9e9a Ethernet0
0:12:6b:3a:9e:9a
```

Modifying the default link MTU on an interface triggers the advertisement of that MTU in the router advertisements. The following example results in router advertisements of the 1480 MTU on the Ethernet0 link.

```
configure terminal
interface Ethernet0
  ipv6 mtu 1480
```

6.10.6 Hexago

On HexOS, the neighbor cache is shown with the '`show ipv6 neighbors`' command.

```
show ipv6 neighbors
```

To flush the neighbor cache, use the '`clear ipv6 neighbors`' command.

```
clear ipv6 neighbors
```

6.10.7 Juniper

On JunOS, the neighbor cache is shown with the '`show ipv6 neighbors`' command.

```
show ipv6 neighbors
```

To flush the neighbor cache, use the '`clear ipv6 neighbors`' command.

```
clear ipv6 neighbors
```

6.11 Summary

The neighbor discovery protocol is used to manage interactions on the link. Neighbor solicitation is used to query the link layer address of the neighbor. It uses the solicited-node multicast address of the neighbor as the destination address. The neighbor responds with a neighbor advertisement containing its link layer address.

Duplicate address detection is used to verify the usability of an address by sending a neighbor solicitation to its own address. If a neighbor advertisement is received, the address is not used.

IPv6 supports most link-layer technologies. On Ethernet, the frame protocol id is 0x86dd to identify IPv6 packets. Multicast IPv6 addresses are mapped to the 33-33-xx-xx-xx-xx Ethernet multicast address range. Ethernet MAC addresses are used for the EUI-64 format of an autoconfigured address.

Point-to-Point link are supported by the PPP protocol, which defines a new IPv6 control protocol (IPv6CP). IPv4 and IPv6 packets can share the same PPP link if both ends are configured appropriately.

MTU can be advertised through router advertisements.

IPv6 handling of link-layer interaction is more efficient than IPv4.

6.12 References

Christensen, M., Kimball, K. and Solensky, F., 'Considerations for IGMP and MLD Snooping Switches', Internet-Draft draft-ietf-magma-snoop-10, May 2004.

Huitema, C. and Thaler, D., 'Multi-link Subnet Support in IPv6', draft-ietf-ipv6-multilink-subnets-00 (work in progress), July 2002.

[IEEEEUI64] Guidelines for 64-bit Global Identifier (EUI-64™) Registration Authority, IEEE, http://standards.ieee.org/regauth/oui/tutorials/EUI64.html

[RFC1661] Simpson, W., 'The Point-to-Point Protocol (PPP)', STD ST, RFC 1661, July 1994.
[RFC2461] Narten, T., Nordmark, E. and Simpson, W., 'Neighbor Discovery for IP Version 6 (IPv6)', RFC 2461, December 1998.
[RFC2464] Crawford, M., 'Transmission of IPv6 Packets over Ethernet Networks', RFC 2464, December 1998.
[RFC2467] Crawford, M., 'Transmission of IPv6 Packets over FDDI Networks', RFC 2467, December 1998.
[RFC2470] Crawford, M., Narten, T. and Thomas, S., 'Transmission of IPv6 Packets over Token Ring Networks', RFC 2470, December 1998.
[RFC2472] Haskin, D. and Allen, E., 'IP Version 6 over PPP', RFC 2472, December 1998.
[RFC2491] Armitage, G., Schulter, P., Jork, M. and Harter, G., 'IPv6 over Non-Broadcast Multiple Access (NBMA) Networks', RFC 2491, January 1999.
[RFC2492] Armitage, G., Schulter, P. and Jork, M., 'IPv6 over ATM Networks', RFC 2492, January 1999.
[RFC2497] Souvatzis, I., 'Transmission of IPv6 Packets over ARCnet Networks', RFC 2497, January 1999.
[RFC2590] Conta, A., Malis, A. and Mueller, M., 'Transmission of IPv6 Packets over Frame Relay Networks Specification', RFC 2590, May 1999.
[RFC3146] Fujisawa, K. and Onoe, A., 'Transmission of IPv6 Packets over IEEE 1394 Networks', RFC 3146, October 2001.
[RFC3572] Ogura, T., Maruyama, M. and Yoshida, T., 'Internet Protocol Version 6 over MAPOS (Multiple Access Protocol Over SONET/SDH)', RFC 3572, July 2003.
Yamamoto, K. and Hagino, J., 'Requirements for IPv6 Dialup PPP Operation', draft-itojun-ipv6-dialup-requirement-02 (work in progress), November 2001.

6.13 Further Reading

[RFC3122] Conta, A., 'Extensions to IPv6 Neighbor Discovery for Inverse Discovery Specification', RFC 3122, June 2001.

7

Internet Control Message Protocol

The Internet Control Message Protocol (ICMP) is used to send information and errors to source nodes. This chapter describes the functionality of ICMP. The neighbor discovery protocol uses ICMP as a transport protocol. The chapter ends with configuration examples of major implementations.

7.1 ICMP

ICMP in IPv4 and IPv6 is used by intermediate or destination nodes to inform source nodes about problems and issues related to the delivery of datagrams. ICMP is also used for diagnostic purposes.

For example, in Figure 7.1, node A sends a datagram to B (step 1), but the router R1 cannot forward the datagram because B is unreachable. R1 sends an ICMP error message back to A (step 2), about the unreachability of B.

IPv6 ICMP [RFC2463] is similar to IPv4, but has a few enhancements:

- ICMP is carried in an IPv6 datagram identified by a specific Next Header field.
- A checksum is computed, since any layer above IPv6 should compute a checksum as discussed in Section 3.7.1.
- New messages are defined for the IPv6 specifics.
- In an ICMP error message, the original datagram is put within the error packet for easier recovery by the source.

An ICMP message is included in an IPv6 datagram by setting the Next Header field to 58, as shown in Figure 7.2. The ICMP header contains a type field which identifies the type of message. The code field further qualifies the specifics of the message. The next sections show the various type and code values and meanings. A checksum is computed (using a

Migrating to IPv6: A Practical Guide to Implementing IPv6 in Mobile and Fixed Networks Marc Blanchet

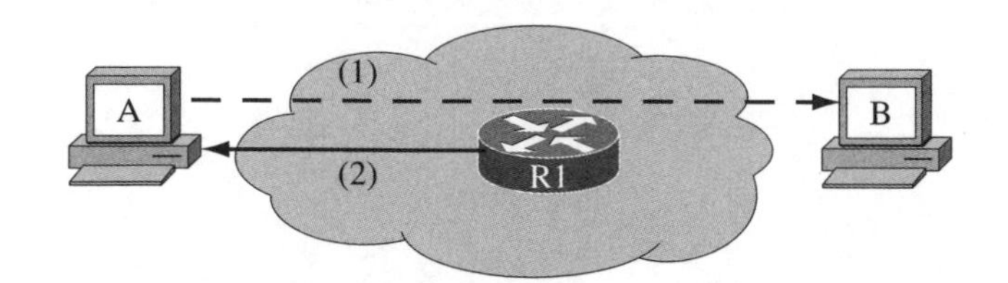

Figure 7.1 ICMP message sent by an intermediate node

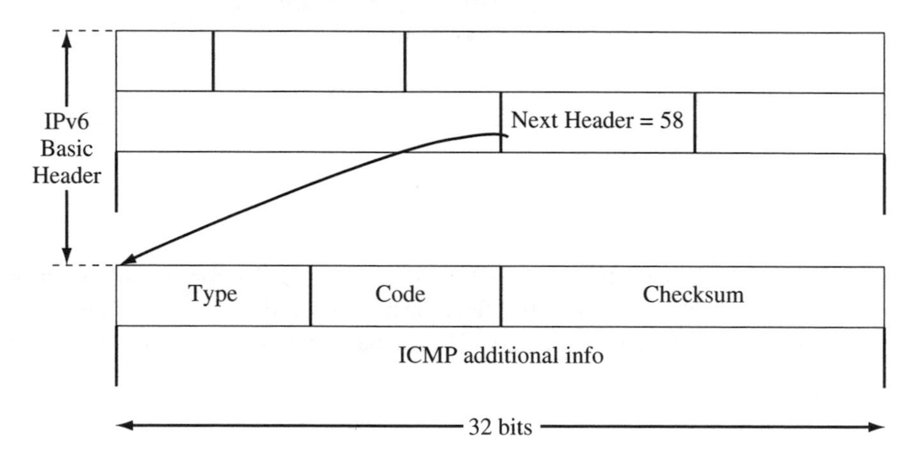

Figure 7.2 ICMP Header

pseudo-header, as described in Section 19.1) because ICMP is a transport protocol, relative to IPv6.

When an ICMP message is sent by an intermediate node because it cannot forward a datagram, the intermediate node sends the ICMP message back to the source of the original datagram. Any IPv6 node listens and processes ICMP messages.

Two classes of ICMP messages are defined: error messages that must be processed by nodes, and informational messages that may be ignored by nodes.

7.1.1 Error Messages

Error messages are identified by a type field value between 0 and 127.[1] Nodes sending an ICMP error message always include the original datagram in the payload of the ICMP message as shown in Figure 7.3. However, the resulting size of the ICMP datagram is less or equal to the minimum IPv6 MTU (1280 octets), to guarantee that under any circumstances the error message reaches the source.

Inserting the original IPv6 datagram in the ICMP error message enables the source node receiving the error message to find the context of the original datagram (source and destination addresses and ports), to process the error correctly and push it to the right upper-layer for further processing at the application level.

[1] In other words, an error message is when the most significant bit of the type field is set to '0'.

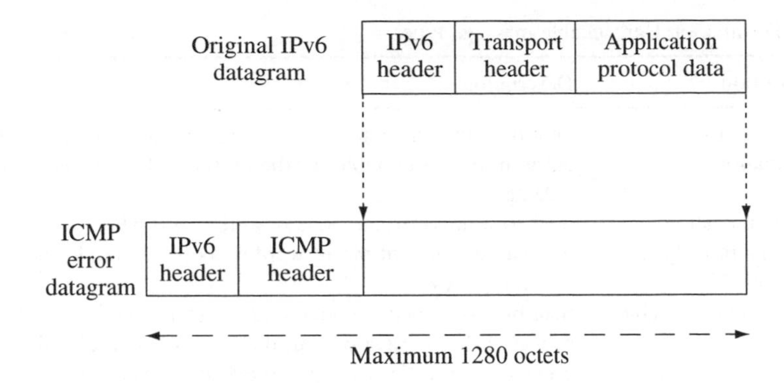

Figure 7.3 Original datagram in the ICMP error datagram payload

There are four types of error messages:

- Destination Unreachable;
- Packet Too Big;
- Time Exceeded;
- Parameter Problem.

The four types are distinguished by the type field value in the ICMP header.

7.1.1.1 Destination Unreachable

Destination Unreachable error messages[2] are sent by intermediate nodes when they cannot reach the destination of the datagram. Four events are defined for this error message and are described in Table 7.1.

7.1.1.2 Packet Too Big

Packet Too Big error messages[3] are sent by intermediate nodes when they cannot forward the datagram to the next hop because the MTU of the link is smaller than the size of the datagram. The ICMP error message sent back to the source contains the MTU of the next link so the source can adjust its path MTU and resend the datagram with the smaller MTU. Packet Too Big is an essential part of the Path MTU discovery, as discussed in Section 3.6.2. If firewalls in the path prohibit ICMP messages, then the PMTU discovery process does not work. In this case, nodes uses the minimum possible MTU, 1280, even if the whole path supports a larger MTU.

> **Important to remember:**
>
> Firewalls should let ICMP Packet Too Big messages pass through in order to have the Path MTU discovery process work.

---

[2] Identified by the type field value of '1'.
[3] Identified by the type field value of '2'.

Table 7.1 Destination Unreachable message types

Code	Definition	Description
0	No route to destination	Sent by a intermediate node (e.g. router) when it does not find an entry in its routing table for the prefix in the destination address of a datagram.
1	Communication administratively prohibited	Sent by a intermediate node (e.g. firewall) when it has an entry in its routing table but the forwarding is prohibited because of a filtering policy.
3	Address unreachable	Sent by a intermediate node (e.g. router) when it cannot deliver the packet to the destination, but the reason is not related to the previous codes. For example, the router cannot reach the link-layer address of the node.
4	Port unreachable	Sent by the destination node when the packet reached the destination, but there is no transport protocol listener on the destination port.

Note: Some firewall implementations do not send ICMP messages, to hide themselves. The application on the source is then waiting for the timeout, which causes more perceived delays. The correct firewall behavior is to send the ICMP destination unreachable with the 'communication administratively prohibited' code.

7.1.1.3 Time Exceeded

Each intermediate node in the path decrements the Hop Limit field in the IP header by one. Since the source node typically sets the Hop Limit field to a large value, the Hop Limit field in a datagram should not have the value of 1 unless there is a routing loop.

Time Exceeded messages[4] are sent by an intermediate node when it receives a datagram with the Hop Limit field equal to 1 (or zero). The intermediate node discards the original datagram.

Two events are defined for this error message and are described in Table 7.2.

The name 'Time Exceeded' is inherited from the original IPv4 specification [RFC791], where the Time-to-live (TTL) field was defined as the lifetime of the datagram in seconds, rather than the number of hops.

If a destination node receiving fragments does not receive all fragments within 60 seconds, the node replies to the source node with an ICMP Time Exceeded error message with a code value of 1 (fragment reassembly time exceeded).

Table 7.2 Time exceeded message types

Code	Definition	Description
0	Hop limit exceeded	Sent by a intermediate node when it receives a datagram with the Hop Limit field equal to 1 or 0.
1	Fragment reassembly time exceeded	Sent by a destination node when it cannot assemble all fragments; for example, when one fragment is missing after a maximum period of time (default = 60 seconds).

[4] Identified by the type field value of '3'.

7.1.1.4 Parameter Problem

When there is an inconsistency in the IP header of a datagram, the Parameter Problem error message[5] is sent to the source by the node detecting the problem. The ICMP header contains a pointer to the problematic datagram included in the payload of the message. Three events are defined for this error message and are described in Table 7.3.

If a destination node receives a datagram with a Next Header not implemented in the node, then it replies to the source node with an ICMP parameter problem error message.[6]

7.1.2 Informational Messages

ICMP informational messages are identified by a type field value between 128 and 255.

The '`ping`' command (in many IPv6 implementations '`ping6`') sends the ICMP informational Echo request (type = 128) to the target. The target responds with the Echo reply (type = 129). Two fields in the ICMP headers are set by the source to identify uniquely every echo request and reply. Table 7.4 lists the defined ICMP informational messages.

The list of ICMPv6 messages is maintained by the IANA [IANA; 2001].

Table 7.3 Parameter problem message types

Code	Definition	Description
0	Erroneous header field	A header field contains invalid data.
1	Unrecognized Next Header type	The Next Header value is unknown to the node that is processing the header.
2	Unrecognized IPv6 option	An extension header contains an unknown value.

Table 7.4 ICMP informational messages

<table>
<tr><th>ICMP type</th><th>Definition</th><th>Section</th></tr>
<tr><td>128</td><td>Echo request</td><td>This section</td></tr>
<tr><td>129</td><td>Echo reply</td><td>This section</td></tr>
<tr><td>130</td><td>Multicast group membership query</td><td rowspan="3">See Chapter 15</td></tr>
<tr><td>131</td><td>Multicast group membership report</td></tr>
<tr><td>132</td><td>Multicast group membership termination</td></tr>
<tr><td>133</td><td>Router Solicitation</td><td rowspan="2">See Section 5.2.2</td></tr>
<tr><td>134</td><td>Router Advertisement</td></tr>
<tr><td>135</td><td>Neighbor Solicitation</td><td rowspan="2">See Section 6.2</td></tr>
<tr><td>136</td><td>Neighbor Advertisement</td></tr>
<tr><td>137</td><td>Redirect</td><td>See Section 9.3</td></tr>
<tr><td>138</td><td>Router Renumbering</td><td>See Section 9.10</td></tr>
<tr><td>139</td><td>Node Information query</td><td rowspan="2">See Section 8.5</td></tr>
<tr><td>140</td><td>Node Information reply</td></tr>
<tr><td>141</td><td>Inverse Neighbor Solicitation</td><td rowspan="2">See [RFC3122]; used for frame relay networks.</td></tr>
<tr><td>142</td><td>Inverse Neighbor Advertisement</td></tr>
</table>

[5] Identified by the type field value of '4'.

[6] With a code value of 1 (Unrecognized Next Header type).

7.2 Neighbor Discovery

Neighbor Discovery is a control protocol that includes several functions:

- neighbor solicitation and advertisement, previously discussed in Section 6.2;
- router solicitation and advertisement, previously discussed in Section 5.2.2;
- route redirection, discussed in Section 9.3.

All Neighbor Discovery messages are encapsulated using ICMP transport and are identified by specific ICMP types; they use ICMP as their transport protocol, as shown with types 133–136 in Table 7.4.

7.3 Hop Limit Set to 255

All Neighbor Discovery datagrams, used for link scope interactions, have the Hop Limit field in the IP header set to 255. This might be considered as strange since 255 is the maximum value of the Hop limit and permits the packet to be forwarded through 254 routers before being discarded, while it should not leave the link!

However, a router advertisement or a neighbor advertisement is only valid on the same link it is sent, so routers do not forward these packets. One might think that a Hop Limit value of 0 or 1 is the right one to make sure the packet would not be forwarded. But if the value is 0 or 1, the host cannot really verify if the router advertisement was sent by a router on the link or if it comes from a far router by some trick with the Hop Limit being decremented to 0 when it reaches the link. With the source router setting the Hop Limit to 255 in the router advertisement datagram, if the host receives a Hop Limit different than 255 it concludes that the datagram does not come from a router on the link, but has been forwarded from another link: the host discards the router advertisement. This disables a denial-of-service attack coming from outside: the datagram would have a Hop Limit value smaller than 255.

This feature of setting the Hop Limit to 255 is used in many other protocols such as RIP (see Section 9.5).

7.4 Managing ICMP on Hosts and Routers

The ping command sends an ICMP echo request. It is used for troubleshooting.

7.4.1 FreeBSD

To send an ICMP echo request to a node, use '`ping6`', as shown below.

```
% ping6 3ffe:b00:0:1::2
```

When sending to a link-local address, add after the address a '`%`' and the interface name on which the packet should be sent.

```
% ping6 fc80::1%fxp0
```

7.4.2 Linux

To send an ICMP echo request to a node, use 'ping6', as shown below.

```
% ping6 3ffe:b00:0:1::2
```

When sending to a link-local address, add the '-I' option followed by the interface name on which the packet should be sent.

```
% ping6 -I eth0 fe80::1
```

7.4.3 Solaris

To send an ICMP echo request to a node, use 'ping', as shown below.

```
% ping 3ffe:b00:0:1::2
```

When sending to a link-local address, add the '-i' option followed by the interface name on which the packet should be sent.

```
% ping -i le0 fe80::1
```

When a destination host has both IPv4 and IPv6 addresses, use the '-A inet6' to force sending an IPv6 ICMP echo request.

```
% ping -A inet6 host1.example.org
```

7.4.4 Windows

To send an ICMP echo request to a node, use 'ping', as shown below.

```
C> ping 3ffe:b00:0:1::2
```

When sending to a link-local address, add after the address a '%' and the interface index on which the packet should be sent.

```
C> ping fe80::1%3
```

To force the use of IPv6, the '-6' option can be used.

```
C> ping -6 host1.example.org
```

7.4.5 Cisco

To send an ICMP echo request to a node, use 'ping ipv6', as shown below.

```
> ping ipv6 3ffe:b00:1::2
```

Extended ping is also available.

IOS enables rate-limiting for ICMP messages to avoid denial-of-service attacks. The interval between ICMP error messages is set with the 'ipv6 icmp' command. The following sets the interval to 100 milliseconds.

```
configure terminal
ipv6 icmp error-interval 100
```

To troubleshoot ICMP messages, use the debug command.

```
enable
debug ipv6 icmp
```

7.4.6 Hexago

To send an ICMP echo request to a node, use 'ping ipv6', as shown below.

```
> ping ipv6 3ffe:b00:1::2
```

7.4.7 Juniper

To send an ICMP echo request to a node, use 'ping', as shown below.

```
> ping 3ffe:b00:1::2
```

When sending to a link-local address, add the 'interface' option followed by the interface name on which the packet should be sent.

```
> ping fe80::1 interface fe-0/0/1
```

7.5 Summary

ICMPv6 is similar to ICMPv4. Error and informational messages are used by intermediate or destination nodes to inform source nodes about issues concerning sent packets. ICMPv6 is also used as the transport for Neighbor Discovery. ICMPv6 is a transport protocol relative to IPv6.

7.6 References

IANA, ICMPv6 TYPE NUMBERS, June 2001, http://www.iana.org/assignments/icmpv6-parameters

[RFC791] Postel, J., 'Internet Protocol', STD 5, RFC 791, September 1981.

[RFC2463] Conta, A. and Deering, S., 'Internet Control Message Protocol (ICMPv6) for the Internet Protocol Version 6 (IPv6) Specification', RFC 2463, December 1998.

[RFC3122] Conta, A., 'Extensions to IPv6 Neighbor Discovery for Inverse Discovery Specification', RFC 3122, June 2001.

8

Naming with DNS and Selecting an Address

Described simply, the Domain Name System (DNS) is a distributed database across the Internet mapping names and addresses. Modifications to DNS were done to support IPv6 addresses.

This chapter describes the new AAAA record for the hostname to IPv6 address mapping. IPv6 address to hostname is accomplished with a new reverse tree top-level domain: ip6.arpa. The transport of DNS queries is shown for both IPv4 and IPv6. An example is provided to help understand which transport protocol can be used, and where. Some enhanced naming services such as DNS server discovery and node information query are presented. Additionally, since nodes now have an IPv4 address and multiple IPv6 addresses, the selection of the right source and destination address, based on the DNS answers, are described. The chapter ends with configuration examples of major implementations.

8.1 Hostname To IPv6 Address with the AAAA Record

IPv4 DNS defines an address record called 'A', for Address, which maps a hostname to an IPv4 address. The direct equivalent of the A record in IPv6 is the 'AAAA' record. AAAA maps a hostname to an IPv6 address [RFC3596], as shown in Table 8.1. Since IPv6 addresses are four times longer than IPv4 addresses, the name of the IPv6 equivalent record is four times the letter 'A', thus 'AAAA'.

A host may have IPv4 and IPv6 addresses, but may be known by only one name. In this case, the name maps to two records: an A record for the IPv4 address and an AAAA record for the IPv6 address. For example, if the hostname is 'host1.example.org' and the IPv4 and IPv6 addresses are 192.0.2.1 and 3ffe:b00:0:1::1 respectively, then two records will be defined for that host, as illustrated in Table 8.2.

Migrating to IPv6: A Practical Guide to Implementing IPv6 in Mobile and Fixed Networks Marc Blanchet

Table 8.1 A and AAAA record types

Record type	Maps a hostname to
A	An IPv4 address
AAAA	An IPv6 address

Table 8.2 Example of A and AAAA records

Name	Record type	Value
host1.example.org	A	10.1.2.3
host1.example.org	AAAA	3ffe:b00:0:1::1

Table 8.3 Multiple AAAA records

Name	Record type	Value
host1.example.org	AAAA	3ffe:b00:0:1:212:6bff:fe3a:9e9a
host1.example.org	AAAA	2001:1111:30:1:212:6bff:fe3a:9e9a

As shown in the second line of Table 8.2, IPv6 addresses can be written in DNS AAAA records using the compressed form, as described in Section 4.2.1.

If a host has multiple IPv6 addresses, then the hostname maps to multiple AAAA records, as illustrated in Table 8.3.

8.2 IPv6 Address To Hostname

In the DNS, the mapping from an IP address to a hostname, often called reverse mapping, is not bound to the forward mapping (hostname to IP address). A different record, called PTR (PTR for pointer) is used for the reverse mapping. Since the DNS always uses a name as a label, the IP addresses are converted to a pseudo-name under a special top-level domain.

In IPv4, the top-level domain is 'in-addr.arpa' and the address is written in four labels in the reverse order. For example, if the inverse mapping for 192.0.1.2 is host1.example.org, then the DNS record is as described in Table 8.4.

The name in the left part of the record is constructed by inverting the IPv4 address, adding dots between each decimal digit and appending the suffix 'in-addr.arpa'. The right part of the record is the hostname associated with the IP address.

IPv6 uses the same PTR record but constructs the left part of the record by adding dots between each hexadecimal digit of the fully expanded IPv6 address and by using a different

Table 8.4 Example of an IPv4 PTR record

Name	DNS record	Value
2.1.0.192.in-addr.arpa	PTR	host1.example.org

Table 8.5 Example of an IPv6 PTR record

Name	DNS record	Value
1.0.0.0.0.0.0.0.0.0.0.0.0.0.0.0.1.0.0.0.0.0.0.0.0.0.b.0.e.f.f.3.ip6.arpa	PTR	host2.example.org

top-level domain name, 'ip6.arpa'. For example, if the reverse mapping for 3ffe:b00:0:1::1 is host2.example.org, then the DNS record is as described in Table 8.5.

On the left part of the record, the IPv6 address (3ffe:b00:0:1::1) is fully expanded with all zeros (3ffe:0b00:0000:0001:0000:0000:0000:0001), then inverted and dots inserted between every hexadecimal digit. If one had to type this long string on a keyboard, the chances of mistyping it are pretty high. Configuration tools and user interfaces are needed to manage IPv6 in DNS!

The reverse mapping (PTR) is used much less than the forward mapping (A or AAAA), with users employing the forward mapping all the time but most of them never using the reverse mapping. Reverse mapping is mainly used by servers to get a name corresponding to an incoming connection from an IP address. This helps with gathering statistics or troubleshooting, and sometimes basic security. Many organizations and ISPs do not care to enter the reverse mapping for their computers in the DNS which results in a low real value of the PTR record.

8.3 Transport

The DNS is a distributed database, which contains data for both IPv4 and IPv6 addresses. The transport mechanism for sending queries and answers is independent of the requested data type. For example, one could use IPv4 to transport IPv6 address DNS queries; in this case, the data in the database are IPv6 addresses but the transport mechanism is IPv4. Nevertheless, the transport is usually related to the query, since it makes no sense to request IPv6 addresses if the node cannot use IPv6.

Figure 8.1 shows a typical resolution path for the name host1.example.org between a DNS client and the authoritative servers. The process is described in Table 8.6.

In the current IPv4 Internet, all queries, shown as arrows in Figure 8.1, use IPv4 transport. Queries about the IPv6 address of host1.example.org can be sent using IPv4 transport or IPv6 or both along the resolution path.

The transport protocol between the client (A) and the local DNS server (B), shown as step 1 in Figure 8.1, is chosen based on the DNS configuration of the client (A) which identifies the IP address of its local DNS server (B). If the client refers to its local DNS server (B) by its IPv6 address and the server is IPv6 configured and reachable from the client, the client connects to the local DNS server using the IPv6 address and transport.

The transport protocol between the local DNS server (B) and the root servers (C), shown as step 2, is chosen based on the configuration of the local DNS server which identifies the IP address of the root servers. If the root servers have IPv6 addresses, are reachable by IPv6 and the configuration of the local DNS server contains the IPv6 address of the root servers, then the local DNS server will connect to the root servers using the IPv6 address and transport.

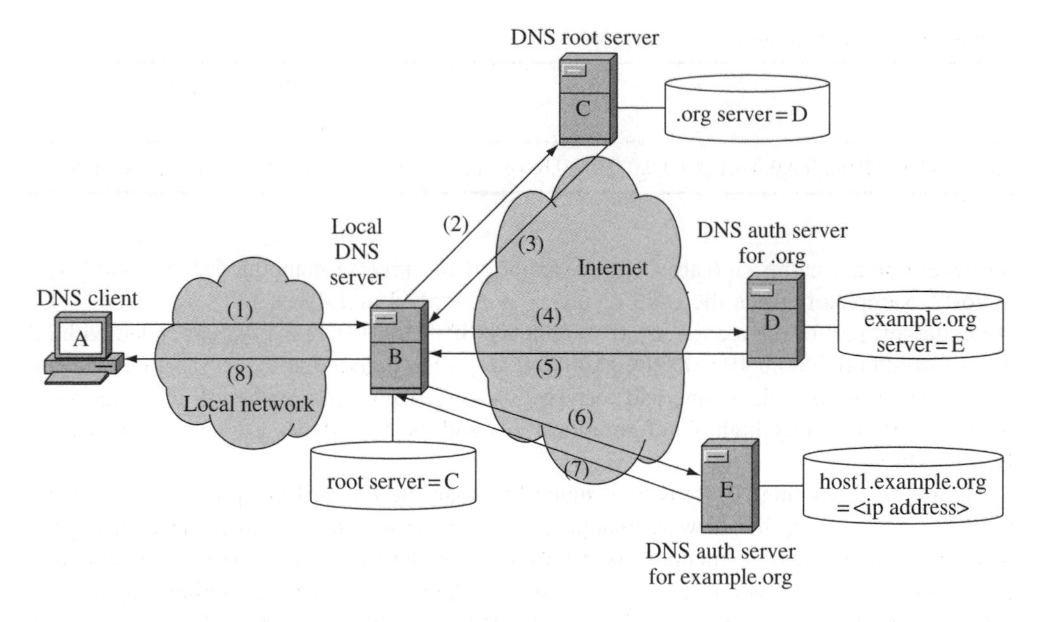

Figure 8.1 DNS resolution process for host1.example.org

Table 8.6 DNS query process

	From	To	Step
1	A	B	The DNS client (A) sends the request for the IP address of host1.example.org to its local DNS server (B).
2	B	C	The local DNS server (B) then sends the query to the root servers (C). The addresses of the root servers are statically configured on the local DNS server (B).
3	C	B	One root server (C) answers back the information about the authoritative server for .org (D).
4	B	D	The local DNS server (B) sends the same query to the authoritative server for .org (D).
5	D	B	The local DNS server (B) receives the answer about the authoritative server for example.org (E) from the authoritative server for .org (D).
6	B	E	The local DNS server (B) sends the query to the authoritative server for example.org (E).
7	E	B	The local DNS server receives the answer for host1.example.org from the authoritative server for example.org (E).
8	B	A	The local DNS server (B) returns the information to the client (A).

The transport protocol between the local DNS server (B) and all other authoritative servers is chosen based on the type of addresses sent by the parent authoritative server. As in step 3 of Figure 8.1, if the root server (C), the parent authoritative server for .org, returns an IPv6 address for the .org authoritative server (D), then the local DNS server (B) will connect to D using the IPv6 address and transport.

To have the full resolution path over IPv6 transport, the client, the local DNS server, the root server and all authoritative servers for all levels of the domain name must have IPv6 addresses and transport. For the client perspective, only the connection to the local DNS server needs to be IPv6, all other connections may be using IPv4 transport.

8.4 DNS Server Discovery

With autoconfiguration, an IPv6 node can boot and configure itself with addresses, prefixes and default routers. However, its DNS server is not configured. At the time of writing, the IETF is still investigating a way [Thaler, 2001] to implement a DNS discovery function where nodes discover the local DNS server without prior static configuration.

8.5 Node Information Query

With link-local addresses and other mechanisms, ad hoc networks are easy to setup with IPv6. In this context, no DNS server is needed and available. An extension of the ICMP protocol [Crawford, 2003] is proposed to enable nodes to query their neighbors for their name in the absence of a DNS service.

The ICMP packet is sent by the node requesting information to its neighbor. It contains the request for:

- the name corresponding to an IPv6 or IPv4 address;
- the IP addresses corresponding to the node.

When querying for a name, the packet is sent to a multicast address formed by FF02:0:0:0:0:2::/96 and a md5 hash of the name for the rightmost 32 bits of the multicast group. A node having the name is listening to this multicast address.

The Kame implementation, included in FreeBSD, NetBSD, OpenBSD and MacOS X, and the USAGI implementation, an IPv6 Linux distribution, have implemented this protocol proposal.

8.6 IP Address Selection

An application must be updated to support IPv6. Inside an IPv4 client application, the function call requesting the DNS A record for a given destination hostname is `gethostbyname()`. This function call is replaced by `getaddrinfo()`, discussed in details in Section 21.4. This new function requests AAAA and A records and returns a list of addresses, in any order and of any address family, to the caller. The application may receive both IPv4 and IPv6 addresses for the same destination. Which IP address the application should use is left to the application.

The application must traverse the list of addresses and try to connect to the targeted addresses in sequence until the connection is successful. If the destination host has IPv4 and IPv6 addresses in the DNS, the typical behavior of applications is to start connecting to IPv6 addresses first. However, giving preference to IPv6 might not always be appropriate.

Moreover, given a destination address, the node has to choose which source address it will use for the connection. A site scoped source address would be a wrong choice for a global destination address.

The mechanism [RFC3484] to choose the source and the destination address matches the source and destination addresses in the longest match possible. For example, a global scoped destination address will be matched with a global scoped source address and a link scoped destination address will be matched with a link scoped source address. Some implementations provide a configuration table to modify the default behavior.

This mechanism is further discussed in the book Web site (http://www.ipv6book.ca).

8.7 Configuring DNS and Address Selection on Hosts and Routers

This section describes the various implementations regarding DNS and address selection. Feel free to skip the implementations in which you are not interested.

8.7.1 Configuring a Unix Client

Unix clients, such as FreeBSD, OpenBSD, NetBSD, MacOSX[1], Linux and Solaris, use the '`/etc/resolv.conf`' file to list the local DNS servers, corresponding to the B server in Figure 8.1. The syntax is:

```
nameserver <ip_address>
```

Multiple lines can be used to specify multiple local DNS servers for redundancy. Note that specifying an IPv6 address means that the client will use IPv6 transport to send the requests to the DNS server, irrelevant of the type of request.

For example, if the local DNS server has 3ffe:b00:0:1::3 as its IPv6 address, the '`/etc/resolv.conf`' file contains:

```
# cat /etc/resolv.conf
nameserver  3ffe:b00:0:1::3
```

8.7.2 Configuring BIND

BIND [BIND] is a popular DNS server implementation. BIND version 4.9.5 and above supports the AAAA record. IPv6 transport is supported in BIND version 8.4.0 and above.

BIND is used in most Unix-based operating systems, such as FreeBSD, OpenBSD, NetBSD, Linux and Solaris. This section describes the syntax of BIND zone files for IPv6 features.

8.7.2.1 AAAA

The hostname to IP address mapping for an IPv4 node is registered in BIND with the A record syntax: '`<name> IN A <ipv4_address>`', as the following example:

```
host1.example.org. IN A 10.1.2.3
```

[1] In fact, MacOSX creates and manages the `resolv.conf` file for compatibility purposes, and the OS resolver process `lookupd` does the name resolution function.

For an IPv6 node, the AAAA record syntax is: '<name> IN A <ipv6_address>', as the following example:

```
host1.example.org. IN AAAA 3ffe:b00:0:1::1
```

The IPv6 address can be written in its full or compressed form.

The examples above use the full name. Usually the record is put in the domain name zone file. For instance, the AAAA record would be put in the 'example.org.zone' file specified in the zone directive of the '/etc/named.conf' file.

```
# cat /etc/named.conf
options {
  directory "/etc/namedb";
};
zone "example.org" IN {
  type master;
  file "example.org.zone";
};
# cat /etc/namedb/example.org.zone
host1  IN AAAA 3ffe:b00:0:1::1
```

8.7.2.2 PTR

The IP address to hostname for an IPv4 node is registered in BIND with the PTR record syntax: 'ip_address_name IN PTR hostname' and uses the in-addr.arpa top-level domain:

```
1.2.0.192.in-addr.arpa IN PTR host1.example.org
```

For an IPv6 node, the same PTR record is used with the IPv6 complete address reversed and the ip6.arpa top-level domain:

```
1.0.0.0.0.0.0.0.0.0.0.0.0.0.0.0.1.0.0.0.0.0.0.0.0.0.b.0.e.f.
f.3.ip6.arpa. IN PTR host1.example.org.
```

The examples above use the full name. Usually the record is put in the ip6.arpa zone file respective to the IPv6 address delegation. For instance, the PTR record would be put in the "3ffe0b000000.zone" file specified in the zone directive of the "/etc/named.conf" file. The assumption is that one owns the 3ffe:0b00:0000::/48 address space and its reverse DNS delegation.

```
# cat /etc/named.conf
options {
  directory "/etc/namedb";
};
zone "0.0.0.0.0.0.b.0.e.f.f.3.ip6.arpa" IN {
 type master;
 file "3ffe0b000000.zone";
};
# cat /etc/namedb/3ffe0b000000.zone
1.0.0.0.0.0.0.0.0.0.0.0.0.0.0.0.1.0.0.0 IN PTR
host1.example.org.
```

8.7.3 *Troubleshooting with Dig*

Dig is a tool used on Unix systems to troubleshoot DNS. To query an A record for host1.example.org, the following command is used:

```
# dig host1.example.org a
```

To query a AAAA record for host1.example.org, the "a" argument is replaced by "aaaa":

```
# dig host1.example.org aaaa
```

To query both records at the same time, the 'any' argument is used:

```
# dig host1.example.org any
```

8.7.4 *FreeBSD*

FreeBSD uses the '/etc/resolv.conf' file for DNS clients, as described in Section 8.7.1, BIND for DNS servers, as described in Section 8.7.2, and dig for troubleshooting as described in Section 8.7.3.

8.7.4.1 Node Information Query

FreeBSD also implements the ICMP node information query by the '-w' option in the ping command. The following example makes an ICMP node information query to all nodes on the link (ff02::1).

```
% ping6 -w ff02::1%fxp0
```

8.7.5 *Linux*

Linux uses the '/etc/resolv.conf' file for DNS clients, as described in Section 8.7.1, BIND for DNS servers, as described in Section 8.7.2, and dig for troubleshooting as described in Section 8.7.3.

8.7.6 *Solaris*

Solaris uses the '/etc/resolv.conf' file for DNS clients, as described in Section 8.7.1, BIND for DNS servers, as described in Section 8.7.2, and dig for troubleshooting as described in Section 8.7.3.

8.7.7 *Windows*

This section describes the specifics of Windows regarding DNS.

8.7.7.1 Static DNS Server Entry

To configure a static DNS server entry, use the 'add dns' subcommand of the 'netsh interface ipv6' command.

```
C> netsh interface ipv6 add dns 3ffe:b00:0:1::3
```

8.7.7.2 Resolution Preference

By default in Windows, if a destination name has both IPv4 and IPv6 addresses in the DNS, then IPv6 addresses are preferred over IPv4 ones.

8.7.7.3 DNS Dynamic Update

The 'Register This Connection's Address In DNS' check box is set by default. The node automatically tries to register the AAAA records for the following addresses of all its interfaces: global or unique-local addresses with EUI-64 or ISATAP. 6to4, temporary addresses and link-local addresses are not registered.

8.7.8 Cisco

While a router does not deal with DNS queries in its forwarding capabilities, the CLI resolves DNS names.

8.7.8.1 Static DNS Server Entry

To configure a static DNS server entry, use the 'ip name-server <address>' command.

```
configure terminal
ip name-server 3ffe:b00:0:1::3
```

8.7.8.2 Resolution

A hostname put in commands such as tftp, ssh or telnet will be resolved first to IPv6 addresses if an AAAA record exists. If not, an A record will be queried.

IOS supports the name to address and address to name resolution when typing a hostname or an address directly after the prompt. By default, the AAAA request is done first on the hostname. If it succeeds, then no A request is done.

8.7.9 Hexago

HexOS is a migration-deployment operating system implementing the tunnel broker mechanism and as such it does not deal with DNS queries in its forwarding function. However, automated registration of tunnel endpoints and automated reverse tree delegation are provided. Full capabilities and configuration for HexOS are presented in Sections 16.2.9 and 16.5.7.

8.7.9.1 Static DNS Server Entry

To configure a static DNS server entry, use the '`ip name-server <address>`' command.

```
configure terminal
ip name-server 3ffe:b00:0:1::3
```

8.7.10 Juniper

While a router does not deal with DNS queries in its forwarding capabilities, the CLI resolves DNS names.

8.7.10.1 Static DNS Server Entry

To configure a static DNS server entry, use the '`name-server <address>`' statement.

```
System {
 name-server {
  3ffe:b00:0:1::3;
 }
}
```

8.7.10.2 Resolution

A hostname put in commands such as ssh or telnet will be resolved first to IPv6 addresses if an AAAA record exists. If not, an A record will be queried.

8.8 Summary

For hostname to IPv6 address mapping, the new AAAA record is used. One hostname can map to a mix of IPv4 and IPv6 addresses using A and AAAA records. IPv6 address to hostname mapping still uses the PTR record but with the new ip6.arpa top-level domain name.

Transport of DNS queries can be done over IPv4 or IPv6 depending on the IP configuration and NS delegation of the parent DNS server for a label. For one client query, the DNS packets may travel over any IP protocol on any segment of the resolution path.

New enhanced services such as DNS server discovery and node information query are available to help the autoconfiguration.

Source and destination address selection is based on an algorithm, where the preferences can be changed if the implementation enables it.

8.9 References

[BIND] ISC BIND, http://www.isc.org/index.pl?/sw/bind/

Crawford, M., 'IPv6 Node Information Queries', draft-ietf-ipngwg-icmp-name-lookups-10, June 2003.

[RFC3484] Draves, R., 'Default Address Selection for Internet Protocol version 6 (IPv6)', RFC 3484, February 2003.

[RFC3596] Thomson, S., Huitema, C., Ksinant, V. and M. Souissi, 'DNS Extensions to Support IP Version 6', RFC 3596, October 2003.
Thaler, D., 'Analysis of DNS Server Discovery Mechanisms for IPv6', Internet Draft draft-ietf-ipngwg-dns-discovery-analysis-00, July 2001.

8.10 Further Reading

Hagino, J., 'Use of ICMPv6 Node Information Query for Reverse DNS Lookup', draft-itojun-ipv6-nodeinfo-revlookup-00 (work in progress), June 2002.
[RFC2672] Crawford, M., 'Non-Terminal DNS Name Redirection', RFC 2672, August 1999.
[RFC2673] Crawford, M., 'Binary Labels in the Domain Name System', RFC 2673, August 1999.
[RFC2874] Crawford, M. and Huitema C., 'DNS Extensions to Support IPv6 Address Aggregation and Renumbering', RFC 2874, July 2000.
[RFC3152] Bush, R., 'Delegation of IP6.ARPA', BCP 49, RFC 3152, August 2001.
[RFC3363] Bush, R., Durand, A., Fink, B., Gudmundsson, O. and Hain T., 'Representing Internet Protocol version 6 (IPv6) Addresses in the Domain Name System (DNS)', RFC 3363, August 2002.
[RFC3364] Austein, R., 'Tradeoffs in Domain Name System (DNS) Support for Internet Protocol version 6 (IPv6)', RFC 3364, August 2002.
[RFC4074] Morishita, Y. and T. Jinmei, 'Common Misbehavior Against DNS Queries for IPv6 Addresses', RFC 4074, May 2005.
[RFC4159] Huston, G. 'Deprecation of "ip6.int"'. RFC 4159, August 2005.

9

Routing

IPv6 did not change the fundamentals of IP routing. It is still based on:

- the longest prefix match;
- the possible use of source routing;
- redirects with ICMP;
- the same routing protocols: Routing Information Protocol (RIP), Open Shortest Path First (OSPF), Intermediate System to Intermediate System (IS-IS) and Border Gateway Protocol (BGP).

No major change in the routing means that a routing expert feels at home when dealing with IPv6 routing. However, there are some changes made to handle routing more efficiently or to make use of IPv6 facilities. This chapter describes these changes, starting with required router addresses, source routing and route redirection. It then presents the routing protocols: RIP, OSPF, IS-IS and BGP. Other topics related to routing such as router renumbering, multihoming and route aggregation are also discussed.

9.1 Required Router Addresses

A router has many configured addresses. Since a router is a node with an additional forwarding function, it inherits all required node addresses as listed in Table 5.9, as well as additional addresses as shown in Table 9.1.

9.2 Source Routing with the Routing Header

Source routing is a way for a source node to influence the path taken by the packet sent to the destination. It does so by including in the header a list of routers that the packet should go through.

Migrating to IPv6: A Practical Guide to Implementing IPv6 in Mobile and Fixed Networks Marc Blanchet

Table 9.1 Required router addresses

Quantity	Address	Requirement	Context
All	required node addresses as defined in Table 5.9		
One	the subnet-router anycast	must be joined	on each forwarding interface
One	the all-routers multicast	must be joined	on each forwarding interface
Any	routing protocols specific multicast	must be joined	on each interface enabled for that routing protocol

IPv4 source routing is handled by the Loose and Strict Source Route options. Each router in the path has to look at the options and see if its IP address is in the list inside the options part of the IP header. This causes all routers in the path to do special processing for those datagrams, which is very slow because the options are not efficiently put on boundaries for fast memory lookup in hardware based processors.

The IPv6 routing header integrates both the IPv4 Loose and Strict options in one function by using the routing extension header, which is processed by only relevant routers involved in the source routing in the path. Not all routers in the path have to look at the routing header, which is a very important improvement for hardware-based fast forwarding.

The routing header contains a list of intermediate nodes, or routers, that need to intercept the datagram and enable special processing for it. Each listed intermediate node is not necessarily the next hop, so not all intermediate nodes in the path need to be listed, but they could be, if one wants the strict source route feature. Intermediate nodes are generally routers, so the word router is used interchangeably.

Figure 9.1 shows an example of the routing header processing where the Network 1, 2, 3 and 4 clouds have many routers, not shown, inside each one. The source node (A) wants to force the path of the packet to the destination node (E) first through routers B and then D.

When the source node (A) sends a datagram with the routing header, it constructs the packet internally with the source address as itself (A), the destination address as the destination node (E), and puts in the routing header the list of routers to go through: first B and then D, as shown in Figure 9.1.

If the source node (A) is using any IPv6 security protection, the protection is calculated on the internal version of the packet.

Before sending the datagram out, as shown in Figure 9.2, the source node (A) puts the IPv6 address of the first router in the list (B = 3ffe:b00:0:3::1) as the destination address of the datagram, and puts the address of the other routers (D = 3ffe:b00:0:6::1) in the routing header, shown as 'in1', and the destination node address (E = 3ffe:b00:0:a::1) at the end of the list of routers in the routing header, shown as 'in2'. As the datagram goes out of the source node (A), its destination address is the first router (B), shown as 'dst'.

Any router inside Network 1 sees the packets as normal and does not inspect the packet further in the extension headers. This is an important advantage of IPv6. Only the router that has the destination address will process the packet specifically.

When the first router (B) receives the datagram, it looks inside the routing header for the next router in the list (D) and puts the next router address (D = 3ffe:b00:0:6::1) as the destination address of the datagram and swaps its own address (B = 3ffe:b00:0:3::1) in the now free slot of D. A counter is kept in the routing header to point to the next hop. Then, all the routers in Network 2 do not look at the packet any more than at a normal IPv6 packet.

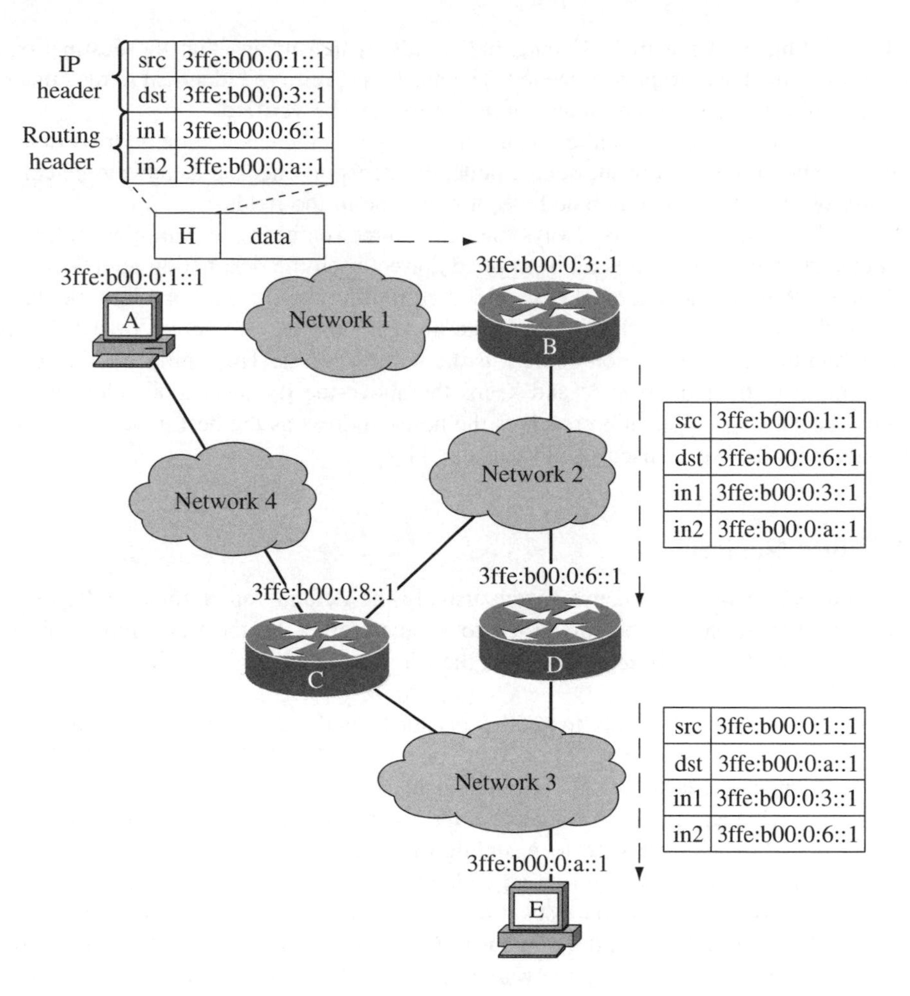

Figure 9.1 Routing header example

IP header	src	3ffe:b00:0:1::1	(A)
	dst	3ffe:b00:0:a::1	(E)
Routing header	in1	3ffe:b00:0:3::1	(B)
	in2	3ffe:b00:0:6::1	(D)

Figure 9.2 Internal Source routing packet

Each router in the list does the same processing as B did until the last router is reached. This last router, D in this example, puts the final destination (E) address (3ffe:b00:0:a::1) as the destination address before sending the datagram. When the destination node (E) receives the datagram, the datagram looks like a normal datagram with the source and destination addresses being the source (A) and destination (E) nodes respectively. However, during the processing of the routing header, the list of routers is preserved in the header, as shown in the last routing header with 'in1' and 'in2' fields, which means that the destination node could look at the list of routers if needed. The header is preserved identical to the constructed one internally inside A, so any integrity protection can be verified.

Since the final destination address is not the one put in the destination address field of the datagram when it goes out of the source node, the transport checksums have to be calculated carefully on the final destination address, not the one in the field.

As the destination address is always the next source router in the path, the routing policy of a network is preserved and can be applied correctly on the destination address.

MobileIPv6 uses a degenerated case of source routing, where only one intermediate node is put in the list. Actually, this intermediate node is the care-of-address of the destination node, while the final destination address is the home address. This optimizes the routing of the datagram to the mobile node and keeps the above-the-IP-stack application unaware of mobility because the application receives the home address as the destination address of the datagram. MobileIPv6 is discussed in Chapter 11.

9.3 Route Redirect

Route redirection is used when a router, usually the default router for a node, receives a packet that should instead be forwarded to another router on the same link. This default router sends an ICMP redirect message to the originating node giving the IP address of the other router.

In Figure 9.3, node A wants to send a packet to node D: node A has B as its default router. Since node A finds that D is outside of the attached link, then A sends the packet, with the destination address of D (3ffe:b00:2::1), to its default router (B). Router B looks in its routing table and find that router C is the right router to reach node D. Then router B sends an ICMP redirect message to A stating that the packet for D should be sent through router C. Node A puts in its own routing table that the route for D is through C. At the same time, router B forwards the initial packet to C. Node A then sends the subsequent packets to D through C. Table 9.2 shows the sequence of packets sent for the ICMP redirect example.

The Redirect message sent by the router should include as much data as possible of the original packet so that the originating node can correlate with the application that sent the packet.

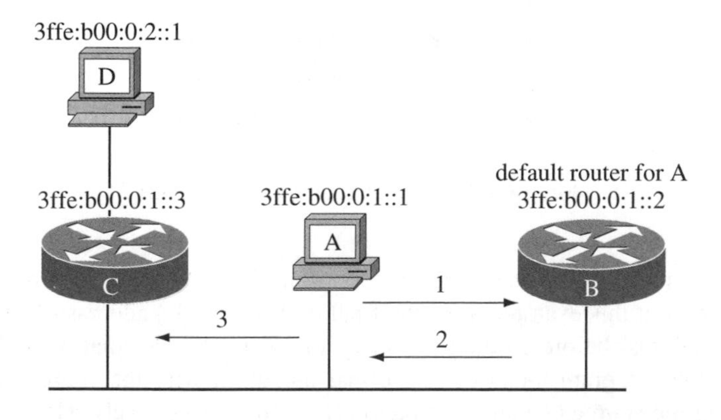

Figure 9.3 ICMP Redirect Example

Table 9.2 ICMP redirect process

From → To	Type of packet	Source and destination address	Description and content
1 A → D	Data packet	Link Src: link-layer address of the source (A) Link Dst: link-layer address of the default router (B) IP Src: address of A (3ffe:b00:0:1::1) IP Dst: address of D (3ffe:b00:0:2::1)	Data packet sent to D from A, through B, the default router for A.
2 B → A	ICMPv6 Type = 137 Hop limit = 255	Link Src: link-layer address of the default router (B) Link Dst: link-layer address of A IP Src: link-local address of B (fe80::2) IP Dst: address of A (3ffe:b00:0:1::1) An option is used to include the originating packet (i.e. 1 above) inside the ICMP data part to help the originating node to correlate the context of the redirect.	ICMP Redirect message sent from B to A, containing the link-local address of C, the right router to reach D.
3 B → D	Data packet	Link Src: link-layer address of the default router (B) Link Dst: link-layer address of the right router (C) IP Src: address of A (3ffe:b00:0:1::1) IP Dst: address of D (3ffe:b00:0:2::1)	Data packet sent to D from B, through C.
4 A → D	2nd data packet	Link Src: link-layer address of the source (A) Link Dst: link-layer address of the right router (C) IP Src: address of A (3ffe:b00:0:1::1) IP Dst: address of D (3ffe:b00:0:2::1)	Data packet sent to D from A, through C.

9.4 Static Routes

Static routes are used to force routing of some prefixes through specific routers. The default route (::/0) is an example of a static route. Static routes have higher preference in a routing table over learned routes from routing protocols. A static route contains the prefix to be routed and the IP address of the router, called the next hop, responsible for routing any packet with a destination inside the given prefix range. Static routing has not changed in IPv6. However, a link-local address should be used as the next hop address.

Differences between IPv4 and IPv6 for static routes:

No difference except that the next hop address should be link-local.

9.5 RIP

RIP (Routing Information Protocol) [RFC1058] is an Interior Gateway Protocol (IGP) for small to medium size networks as the routing protocol. The enhanced version for IPv6, known as RIP next generation (RIPng) [RFC2080, RFC2081], is based on RIP version 2 [RFC1723] and inherits the same generic characteristics:

- Bellman-Ford distance-vector algorithm;
- 30 seconds between updates;
- 180 seconds expiration time for a route not heard;
- fixed metrics;
- network diameter of 15 hops;
- split horizon and poison reverse;
- route tags.

9.5.1 Changes

Many incremental changes for better reliability and for using IPv6 features were introduced in the protocol. Table 9.3 lists the changes to RIP for IPv6.

The link-local source address, the link-scope multicast destination address, the next hop and the hop limit are all link scoped characteristics, which gives additional security and reliability towards any remote rogue router threat.

Table 9.3 RIPng changes and new features

Feature	Description
Advertised Routes	RIPng advertises IPv6 routes composed of IPv6 prefixes with prefix length and metric.
Next Hop	The Next-Hop address is the IPv6 link-local address of the router interface advertising the prefix.
IP Protocol Transport	IPv6 is used to carry RIP datagrams, using UDP as transport protocol.
Source IPv6 Address	The RIP update IPv6 source address is the link-local address of the originating router interface (except when replying to a unicast Request Message from a port other than the RIPng port, when, in this case, the source address is a global valid address; see [RFC2080]).
Destination IPv6 Address	The RIP update destination address is FF02::9, the all-RIP-routers multicast address. Only RIPng routers listen to this multicast address. It is a link-local scope multicast address, which is not forwarded to other links.
Hop Limit = 255	RIP updates have the IPv6 packet Hop Limit set to 255. This enables the listeners to verify if the updates are coming from external rogue routers, as described in Section 7.3.
Port Number = 521	The UDP port is 521, instead of 520 for RIPv1 and 2.
RIPng version = 1	RIPng version number in the RIP packet is 1, which means the first version of RIPng. Since a different transport port is used, the listeners can differentiate RIPv1, RIPv2 and RIPng packets.
Routing Table	The IPv6 routing table is separate from the IPv4 routing table of RIPv1 or 2. The default route is advertised as ::/0.
Authentication	RIPng authentication relies on the underlying IPsec security. Any RIP specific authentication is removed.

Since the routing tables are different for IPv4 routes and IPv6 routes, RIPng deployment is using the 'ships in the night' concept where the routing topologies of IPv4 and IPv6 do not know each other.

Unless otherwise stated, this book uses RIP and RIPng interchangeably to mean RIP for IPv6.

9.6 OSPF

Open Shortest Path First (OSPF) [RFC2328], as version 2, is an IPv4-only IGP. OSPFv3 [RFC2740] is made network protocol independent, similar to IS-IS, so that it can now include IPv6 routes. OSPFv3 shares the same fundamentals as OSPFv2:

- flooding;
- designated router election;
- area support;
- Dijkstra shortest path first calculations;
- on-demand circuit support;
- Stub and NSSA areas;
- multicast extensions (MOSPF).

9.6.1 Changes

Many external changes for better reliability and for using IPv6 features were introduced in the protocol. Table 9.4 lists the major changes made in OSPFv3.

9.6.2 Router ID

In order to make OSPF network protocol independent, the Router and Network Link State Advertisements (LSA) do not have any address. They only carry topology information through the use of the router ID, identifying each router. Since the router ID in OSPFv3 is still a 32 bit value, the implementations either choose one of the IPv4 addresses configured in the router to be the router ID or the router ID will have to be manually configured. Many implementations continue to use the IPv4 syntax for the router ID, even if the semantic is now more generalized to a router ID.

9.6.3 Link-State Database

The link-state database is built with routerIDs and so the topology of the network must be the same for all network protocols. One cannot start a small IPv6 network with tunnels overlaying the IPv4 network and use OSPFv3 to route both protocols, because the two IP networks have different topologies. Use OSPFv2 for IPv4 routing and OSPFv3 for IPv6 routing, which results in two different link-state databases, known as 'ships in the night', which doubles the memory and the SPF calculations. However, having two different link-state databases enables an incremental deployment of IPv6 with a fully-deployed IPv4 network.

Table 9.4 OSPFv3 changes and new features

Feature	Description
Router and network LSAs	These do not have addressing semantics and only carry topology information.
New Intra-Area-Prefix-LSA	This carries IPv6 addresses and prefixes.
Addresses in LSA	These are described as prefix with a prefix length. The default route is ::/0.
Router identification	Router ID is still a 32 bit value, but defines the router, whatever IPv4 or IPv6 address it has. Used in DR, BDR, LSAs, database.
Flooding scopes	Link, area or AS.
Next Hop	The Next-Hop address is the IPv6 link-local address of the router interface advertising the prefix.
New Link-local LSA	This carries the link-local address of the router interface, the prefixes of the link and options.
Runs on per-link basis, instead of a per IP subnet basis	An OSPF interface now connects to a link instead of an IP subnet. Multiple instances on a single link, using an Instance ID in the OSPF packet header, is supported.
Uses IPv6 for transport of OSPF packets	Next Header = 89 to identify an OSPFv3 IPv6 packet.
Source IPv6 Address of OSPF packets	The OSPF packet source address is the link-local address of the originating router interface.
Destination IPv6 address of OSPF packets	All OSPF Routers send Hello packets and listen to FF02::5, the all-ospf-routers link scope multicast address. Designated Router and Backup DR send and listen to FF02::6, the alldrouters link scope multicast address.
Hop Limit = 1 of OSPF packets	1 means link-local scope. However, as with RIPng and Neighbor Discovery, it would have been safer to use a Hop Limit of 255 (see Section 7.3).
OSPF version = 3	Previous versions are equal to 1 or 2.
Authentication is done with IPsec	All OSPF internal authentication data are removed. It now relies on underlying IPsec security to protect the integrity and to provide authentication.

9.7 IS-IS

Intermediate System to Intermediate System (IS-IS) is the OSI IGP protocol. Designed as a network protocol independent routing protocol from the beginning, IS-IS was easily adapted for IPv6 by adding a few type-length-values [Hopps, 2003]. No protocol version has to be incremented to support the new network protocol because there was no change to the IS-IS protocol itself.

9.7.1 Changes

Table 9.5 lists the changes to support IPv6 in IS-IS.

Since IS-IS uses layer 2 encapsulation, no IP address in packets is needed.

IS-IS for IPv6 is considered to be easier to implement compared to OSPFv3.

Table 9.5 Changes to IS-IS to support IPv6

Feature	Description
New IPv6 Reachability TLV (type = 236)	This is an IPv6 route advertisement, containing the IPv6 prefix, the prefix-length, the metric and additional information (advertisement from a higher level and from an external routing protocol).
New IPv6 Interface Address TLV (type = 232)	For Hello packets, it is the link-local address of the router interface. For LSP packets, it is the global address of the router.
IPv6 NLPID (= 142)	It is sent by the router to announce the support of IPv6 routing in IS-IS.

9.7.2 Multi-topology

The use of an integrated link-state database in IS-IS mandates the identical topology of all network protocols it manages. The IPv4 network and the IPv6 network topologies must be identical to an integrated IS-IS.

For environments where IPv6 is deployed incrementally or differently to the IPv4 network topology, integrated IS-IS with IPv4 and IPv6 does not work. Some router vendors have implemented IS-IS with different link-state databases to enable the 'ships in the night' concept, which enables different network topologies for each IP protocol.

9.8 BGP

The Border Gateway Protocol (BGP) [RFC1771] is used to exchange routes between different administrative domains, called autonomous systems (AS). When used between providers to exchange routes, it is called external BGP (eBGP). When used within an administrative domain, it is called internal BGP (iBGP).

The Multiprotocol BGP (MBGP) [RFC2858] is the enhanced version of BGP to support multiple network protocols or address families. For example, MBGP carries VPN and multicast routes as well as IPv4 routes. A new address family in MBGP is defined for IPv6 [RFC2545]; this specific IPv6 version was historically named BGP4+.

9.8.1 Changes

Table 9.6 lists the changes to MBGP for IPv6.

Table 9.6 MBGP changes for IPv6

Feature	Description
Next Hop	An IPv6 address. Global, site and link-scoped are possible.
Network Layer Reachability Information (NLRI)	The announced route. It is expressed as an IPv6 prefix with prefix length.
Address Family	A specific address family is used to identify IPv6 routes.

Most IPv6 router vendors support IPv6 MBGP. It has been used on the IPv6 test backbone, the 6bone, since 1996.

9.8.2 Router ID

BGP uses the Router ID 32 bits field to identify routers between peers. One of the IPv4 addresses of the BGP router is typically used for this Router ID. However, if the BGP router is configured only for IPv6, then a 32 bits value has to be configured manually. This value must be unique among the peers of the router and identifies uniquely the router. Many implementations keep the IPv4 syntax to enter the value, even if the semantics are more generic.

9.8.3 Link-local Addresses for Peering

When routers of multiple AS peer together on a link in an exchange point, a prefix for the link has to be taken out from some organization address space to configure the addresses of the routers on the link. This address space is taken from either one of the AS address spaces or from the exchange point address space. IPv6 does not need to use a specific address space for the link because link-local addresses are available and can be used for BGP peering without harm. Using link-local addresses is simpler than trying to find and choose the right address space from an organization involved in the exchange. Multi-hop BGP, which is build over multiple links, cannot use link-local addresses.

To avoid the problem of changing the peering configuration of the link-local address when the hardware needs to be changed, the link-local address can be manually specified. For example, the fe80::<AS number>:<local ID> addressing structure can be used on a link of an exchange point, where each peer defines a static link-local address which contains its AS number and a peer locally defined number, in case of multiple peering routers of the same AS. Many router vendors support this extension [Manning and Kato, 2001].

9.8.4 Site Scoped Prefixes in Routes

BGP is normally used to exchange global routes between peers. However, the protocol does not restrict the use of site scoped prefixes, which are, by definition, not global but local to a site. Those site scoped prefixes must not be announced in the external peering into the global Internet as discussed in section 23.3.4.

9.9 Tunneling IPv6

In some situations, network managers use tunneling to bypass the normal routing of a network. For example, some distant site has a specific use policy that is different to the policy used in the network in between. This tunneling of IPv6 can be achieved over GRE, IPv4, IPv6, MPLS or others [Conta, Deering, 2002]. Figure 9.4 shows the encapsulation of IPv6 traffic inside an IPv6 packet.

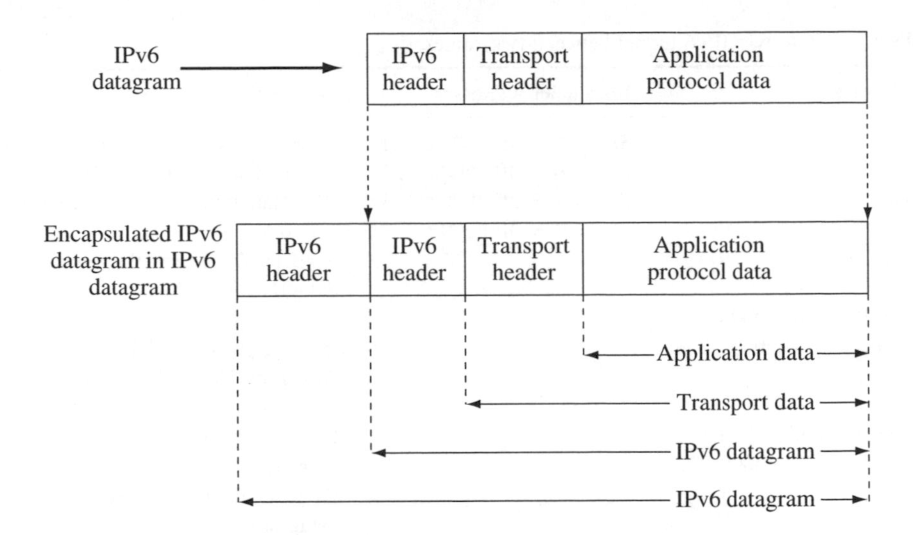

Figure 9.4 IPv6 encapsulation in IPv6

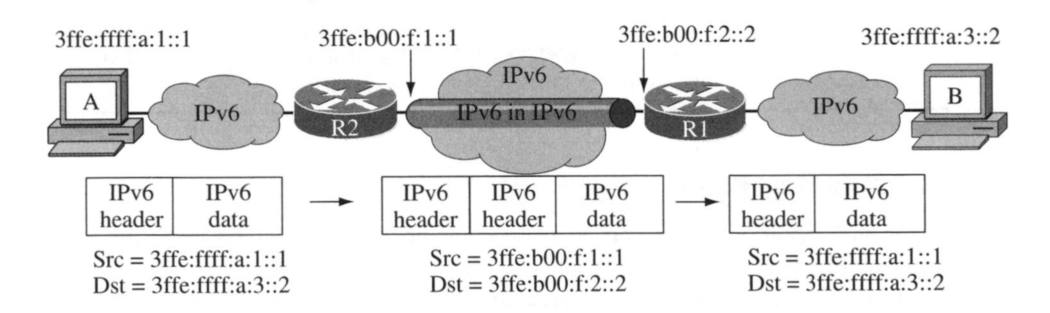

Figure 9.5 IPv6 over IPv6 tunnel between two routers

Figure 9.5 shows an example where node A and B are IPv6 nodes and the tunnel is carried out by the border routers R1 and R2. A sends a packet to B.

The source node A behaves as a normal IPv6 node and has no knowledge of a tunnel in the path to B. However, R2 has a configured tunnel with R1.

As described in Table 9.7, node A sends the IPv6 datagram on its IPv6 network. The source address is A's address (3ffe:ffff:a:1::1) and the destination address is B's address (3ffe:ffff:a:3::2). The IPv6 datagram is forwarded to the R2 router which has a route to the destination network through the tunnel to R1. R2 encapsulates the IPv6 datagram in an IPv6 datagram with the IPv6 source address being R2's address (3ffe:b00:f:1::1) and the destination address being R1's address (3ffe:b00:f:2::2).

A and B have no clue that the datagram was encapsulated over some parts of the path.

Nested tunnels may occur when cascading tunnels are used in the path, causing multiple encapsulations. This is not recommended, especially if fragmentation needs to be applied.

Additional considerations regarding MTU and ICMP with tunnels are discussed in Section 16.2.7. Configuration of IPv6 tunnels can be automated by the use of the Tunnel Setup Protocol, discussed in Section 16.2.9.

Table 9.7 IPv6 over IPv6 tunnel between two routers

IPv6 outer addresses	IPv6 inner addresses	Description
1 None	Src = A = 3ffe:b00:a:1::1 Dst = B = 3ffe:b00:a:3::2	Source A makes an IPv6 datagram for the destination node B.
2 None	Src = A = 3ffe:b00:a:1::1 Dst = B = 3ffe:b00:a:3::2	IPv6 datagram travels on the IPv6 network. Router R2 receives the IPv6 datagram from A.
3 Src = R2 = 3ffe:b00:f:1::1 Dst = R1 = 3ffe:b00:f:2::2		R2 encapsulates the IPv6 datagram into the payload of the IPv6 datagram. Destination of the IPv6 datagram is the end of the tunnel, R1.
4 Src = R2 = 3ffe:b00:f:1::1 Dst = R1 = 3ffe:b00:f:2::2		IPv6 datagram travels on the IPv6 network. R1 receives the encapsulated IPv6 datagram.
5	Src = A = 3ffe:b00:a:1::1 Dst = B = 3ffe:b00:a:3::2	R1 removes the IPv6 header and sends the IPv6 original datagram on its IPv6 network. Destination of the IPv6 datagram is B.
6	Src = A = 3ffe:b00:a:1::1 Dst = B = 3ffe:b00:a:3::2	Destination B receives the IPv6 datagram.

9.10 Renumbering Routers

As discussed in Section 5.3, hosts can be autoconfigured and renumbered using the router advertisements. However, to renumber a full site, routers themselves have to be renumbered, or more appropriately, reconfigured, because the renumbering of routers must be saved permanently in non-volatile memory, so the next reboot will have the new addresses.

A router renumbering protocol [RFC2894] is defined using a new ICMP message (code 138). It uses IPsec for authentication, integrity and has a replay protection scheme. The protocol can be used to modify the configuration of routers for the following items:

- advertised prefixes on links;
- lifetimes (valid and preferred) of the advertised prefixes;
- flags of the router advertisements (onlink, otherconfig);
- router's interface addresses.

Reconfiguration messages include a matching prefix and mask, which enables the change of part of a prefix. For example, when changing provider, a site would change the first 48 bits received from the new provider, but the internal subnet numbers (bits 48 to 64) do not need to be changed or renumbered. A prefix match of 48 bits is used in this case.

The following list shows the steps needed to use the router renumbering protocol in a network.

1. Install security keys in each router.
2. Make a security association between routers.

3. On a network management console or on one of the routers, trigger the reconfiguration message to the first routers.
4. Routers reconfigure themselves automatically and save their new configuration in non-volatile memory.
5. Routers send the reconfiguration message to their neighbors, which propagates the router renumbering. The message is sent to all the router multicast addresses so all routers receive the reconfiguration message.

A test facility is available in the protocol where the actual reconfiguration is not done on the routers receiving the router renumbering requests but, rather, a report is sent back to the sending router with details on what would have been done if the reconfiguration were executed.

In a large network where the number of routers is unknown, it might be a challenge to verify if the renumbering happens effectively on all routers, as stated by the Internet Engineering Steering Group (IESG) in their note written in the protocol RFC [RFC2894].

A good use of this protocol is when a site starts deploying IPv6 locally using the site scoped addresses and later on connects to an IPv6 provider and receives a prefix for the site. The router renumbering protocol can be used to add this new prefix to the configurations of all routers, enabling the use of this new prefix on all hosts on the network that use autoconfiguration. The site scoped addressing might be kept or removed in a subsequent router renumbering action.

Network managers are generally opposed to this kind of protocol, at least with their first impression. This might be your initial view too. In fact, it is scary to have the reconfiguration of routers automated that much. However, it might be useful not only for renumbering but also for changing some router advertisement fields such as the lifetimes on a whole site basis. Few implementations of this protocol are available at the time of writing.

9.11 Internet Routing

An important motivation for the design of IPv6 was to reduce the global routing table size. The aggregation of IGP and EGP IPv6 routes should be better by design, but policies and network engineering decisions over the lifetime of IPv6 might show differently.

In the early 1990s, the global routing table size was seen as a major obstacle to the growth of the Internet, given that core routers might not have the horsepower and memory to handle the global routing table. Recent studies [RFC3221] have shown that the most important obstacle is not the size of the global routing table but its relative instability, caused by the visibility of the site routes in the global routing table. This visibility of site prefixes is used for traffic engineering and for site multihoming, which are two useful features for large networks.

To achieve site multihoming in IPv4, the site prefix is advertised by all the providers of the site. Figure 9.6 shows a site with the 192.0.2.0/24 prefix announced by its two providers to the Internet core.

Table 9.8 lists the entries in the global routing table for this site example.

In the global routing table, the site prefix has two specific entries, one for each provider. This enables the site to be fully redundant if one provider is not reachable, and also permits some load balancing. The example in Figure 9.6 uses portable addresses also called provider independent (PI) addresses, where the site owns the address space.

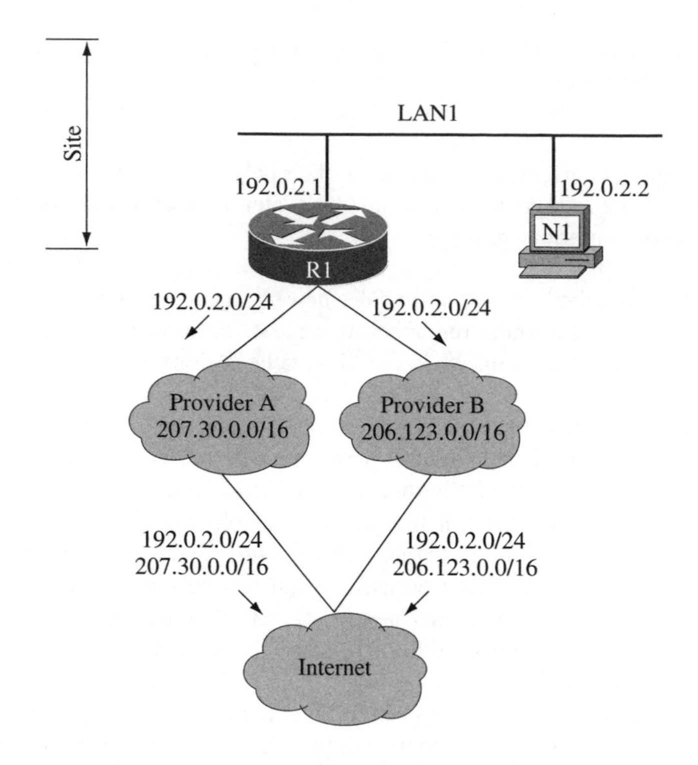

Figure 9.6 Multihoming with provider independent address space

Table 9.8 Route entries in the IPv4 global routing table

Route entry	Origin
192.0.2.0/24	Provider A
192.0.2.0/24	Provider B
206.123.0.0/16	Provider B
207.30.0.0/16	Provider A

If the organization does not own the address space, but instead obtained the address space from one of its providers, then the site specific route is announced by both providers. Figure 9.7 shows an example where the site address space (206.123.31.0/24) is from provider B's address space (206.123.0.0/16). Provider A has to announce the 206.123.31.0/24 site address space. If provider B does not announce the specific route (206.123.31.0/24), then the longest match rule of routing results in all incoming traffic to the site going through provider A, which announces a longer prefix (/24) than provider B (/16) for the site address space. Provider B has to announce the site address space (206.123.31.0/24) as well as the larger address space (206.123.0.0/16).

Table 9.9 lists the entries in the global routing table for this site example.

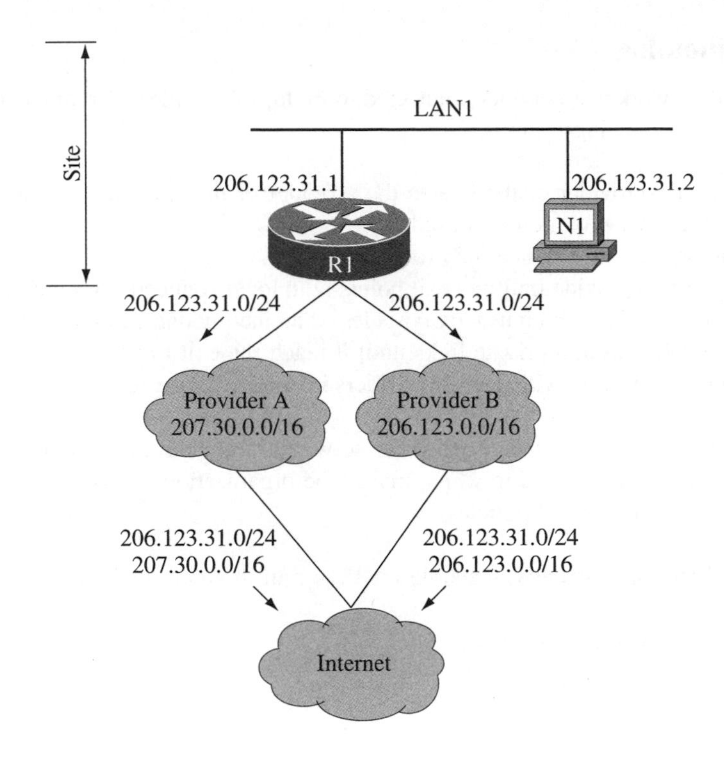

Figure 9.7 Multihoming with provider assigned addresses

Table 9.9 Route entries in the IPv4 global routing table

Route entry	Origin
206.123.31.0/24	Provider A
206.123.31.0/24	Provider B
206.123.0.0/16	Provider B
207.30.0.0/16	Provider A

From the Internet core aggregation point of view, these two examples show that multihoming with or without provider independent address space causes specific routes to be announced in the global routing table.

Any slight instability of the site reachability generates route withdrawals and insertions, causing many exchanges between the BGP routers of the Internet. Any multihoming solution that exposes the site prefixes in the global routing table creates this instability.

Aggregation is useful to achieve the smallest global routing table as well as for stability of the whole Internet. However, current traffic engineering and multihoming techniques in IPv4 break aggregation: using the same techniques in IPv6 also breaks aggregation.

9.12 Multihoming

A multihomed network is a network connected to multiple providers. Organizations use this technique for the following reasons:

- Redundancy: if a provider or the link to the provider or the exit router to the provider is down, then the other provider(s) is used.
- Load balancing: to load balance the providers links.
- Local exit point: to provide multiple exit points with local connectivity for a large network over multiple locations, which usually costs less than having one exit point that aggregates the Internet traffic through private links until it reach the exit router.
- Less transit costs: peering with Tier-2 providers is usually cost-effective because no transit costs are involved.
- Better response time: by peering with other networks nearer to the target customer base.
- Provider-independent addressing which makes the organization independent of providers.
- Policies: internal or external policies.

Many techniques are currently available for IPv6 multihoming, including:

- provider independent address space;
- multiple prefixes;
- cross-tunnels at site exit routers.

Among the other techniques being discussed in the IPv6 community is propagation using router renumbering and advertisements.

This section will describe the above techniques.

9.12.1 Provider Independent Address Space

As in IPv4, there is no technical barrier to using provider independent address space for IPv6 multihoming. In this case, the address space registries would give address space to organizations that are multihomed. Figure 9.8 shows an example where the site has the 3ffe:b00:1::/48 provider independent address space.

The site is connected to provider A which has the 2001:410::/32 address space and to provider B which has the 2001:5c0::/32 address space. Both receive the 3ffe:b00:1::/48 announcement from the site and redistribute and announce that route to the Internet core.

This solution has the same drawbacks as in IPv4 multihoming with provider independent address space, such as multiple entries in the routing table for each site. This means instability, route flapping and high requirements for core router vendors to implement large route databases in routers. If all sites are multihomed, then at least four entries exist in the global routing table for each site. Billions of sites will be a challenge, if ever possible.

A recent proposal [Roberts, 2005] is to provide IPv6 provider independent address space only to enterprises which already have an autonomous system (AS) number. Since the only reason for an enterprise to get an AS number is to be multihomed, which usually requires significant costly infrastructure, then those enterprises could apply to get a provider independent address space directly from the registries. Since the number of enterprises in this situation is low compared to the total number of enterprises, this proposal is worth considering.

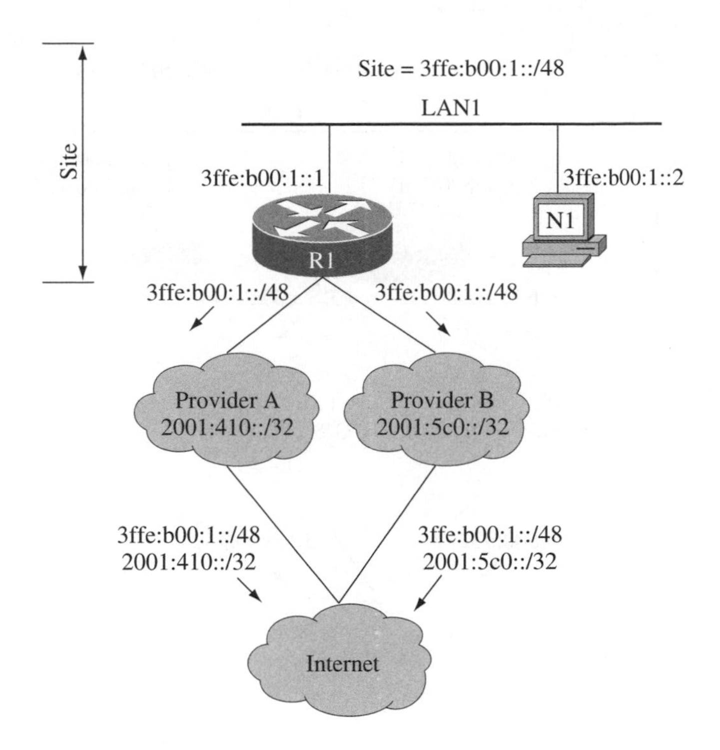

Figure 9.8 Multihoming example with PI IPv6 addresses

Currently, the policy for address allocation [RIPE267] implemented by address registries does not permit the allocation of IPv6 provider independent address space, which makes this solution theoretical only. However, no technical barrier prevents the deployment of this solution if required in the future. The address allocation policy is discussed in Section 23.2.

9.12.2 Multiple Prefixes

The IPv6 basic solution to the multihoming problem is to use multiple prefixes inside a site. Each provider gives a /48 to the site. The site propagates each address space inside the network so that each host has multiple addresses, one per provider. Figure 9.9 shows an example with two providers (A and B) and one site consisting of a border router R1, a LAN and the node N1. Provider A has the 3ffe:b00::/32 address space from which it assigns 3ffe:b00:1::/48 to the site. Provider B has the 3ffe:c00::/32 address space from which it assigns 3ffe:c00:2::/48 to the site. N1 has two addresses on the LAN: 3ffe:b00:1::2 and 3ffe:c00:2::2.

From the Internet core perspective, aggregation is intact so the site is not seen in the core. On the host side, the node N1 has to choose the right source address. The address selection mechanism, discussed in Section 8.6, is used, but has limitations in the way it always chooses the best source address. Moreover, when providers and organization network managers are filtering based on source or destination addresses in their network, the multiple prefix gives much more filter management to the extent that operators might not deploy the multiple prefixes solution only for that reason.

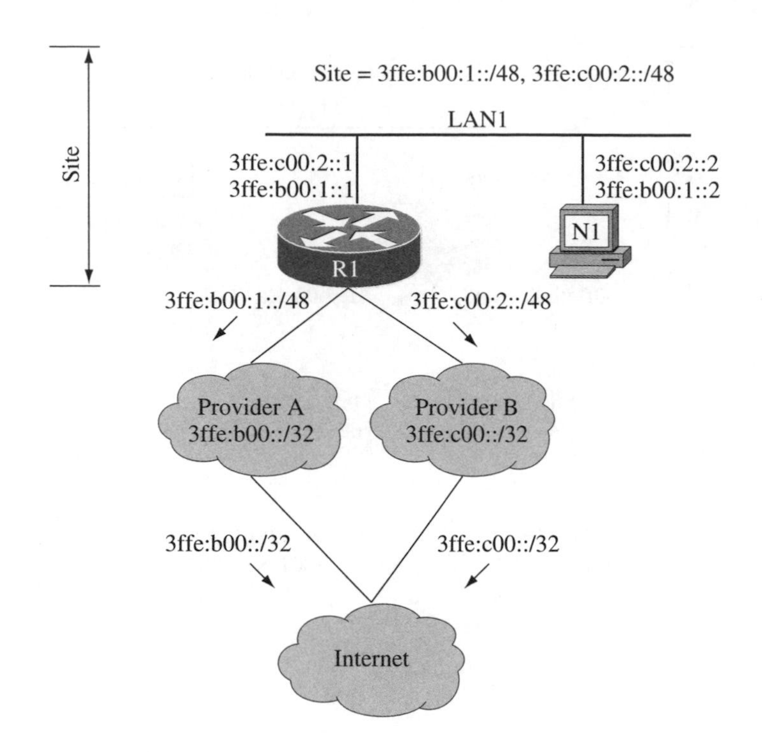

Figure 9.9 Multiple prefix multihoming example

9.12.3 Cross-tunnels at Site Exit Routers

When multiple routers and links are used to connect the providers to the site, cross-tunnels [RFC3178] can be used to keep the aggregation of address space at the core while having a link failure mechanism at the site level. Figure 9.10 shows an example of a site that has two providers (A and B), connected to R1 and R2 respectively. Each provider assigns an address space to the site: 3ffe:b00:1::/48 and 2001:5c0:1::/48. The assigned address space (for example, 3ffe:b00:1::/48) is announced to provider (A) by the site on the primary link (R1 to RA). A backup link, usually done by a tunnel, is established between the provider router (RA) and the other site router (R2), where the assigned address space (3ffe:b00:1::/48) is announced with a lower preference. With this preference change, the primary link is always used unless it is not working. In this latter case, the backup link is used.

This solution works only for short distance links, since the two backup link costs would be prohibitive on long-haul. Moreover, since multiple prefixes are used in the site, it exhibits the same issues as the multiple prefixes solution.

9.12.4 Propagation using Router Renumbering and Advertisements

Router advertisements, as discussed in Section 5.2.2.1, are used to send prefixes to nodes. Router Renumbering, as discussed in Section 9.10, is used to reconfigure the routers. If a site is multihomed and received multiple prefixes from its providers, then the two mechanisms

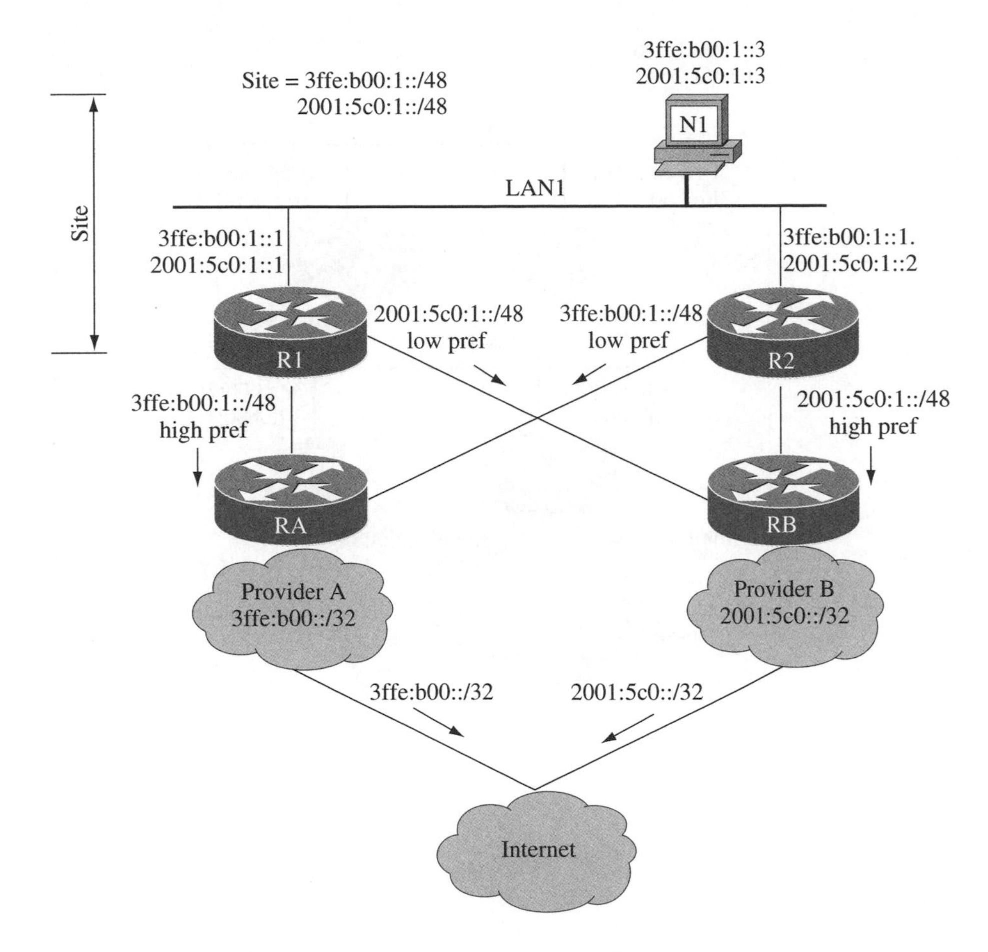

Figure 9.10 Cross-tunnels IPv6 multihoming example

can be used for a better multihoming solution. When a provider link goes down, the border router starts a router renumbering inside the network to withdraw the prefix of that provider. The router advertisements are updated based on the router renumbering changes, which makes the hosts withdraw the faulty prefix.

Figure 9.11 shows an example of a site with the 3ffe:b00:1::/48 prefix allocated from provider A 3ffe:b00::/32 address space and the 3ffe:c00:2::/48 prefix allocated from provider B 3ffe:c00::/32 address space. Routers R1 and R2 are configured for the router renumbering protocol and advertise the two prefixes on LAN1 and LAN2. Nodes N1 and N2 each have two addresses from the two prefixes.

When the link to provider B becomes down as shown in Figure 9.12, R1 withdraws the 3ffe:c00:2::/48 prefix in the router renumbering and router advertisements. R2 receives the router renumbering message and consequently withdraws the 3ffe:c00:2::/48 in its router advertisements. N1 and N2 withdraw the prefix and now have only one prefix configured.

This solution does not break the aggregation and provides link failure recovery. Effectively, it renumbers the whole network when a provider link is down. However, if the provider

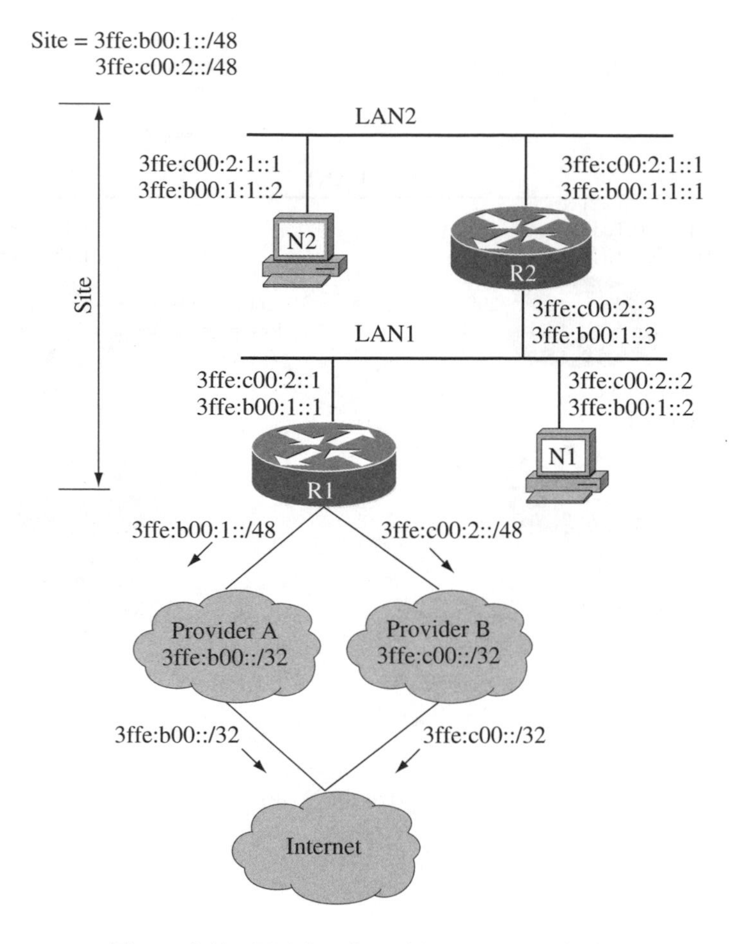

Figure 9.11 Multihoming with router renumbering

link is unstable, the convergence of the propagation can make the network unreliable. As discussed in Section 9.10, the router renumbering protocol in large networks is still subject to debate.

9.12.5 Multihoming Work Progress

IPv6 multihoming can use the IPv4 multihoming techniques. However, the current issues of large routing table and instability are not resolved, and will be more important as the Internet continues to grow. IPv6 is an opportunity to make multihoming more efficient [RFC3582]. Some IPv6-only techniques such as multiple prefixes are available today, others are being discussed. Please consult the book web site (http://www.ipv6book.ca) for the latest developments.

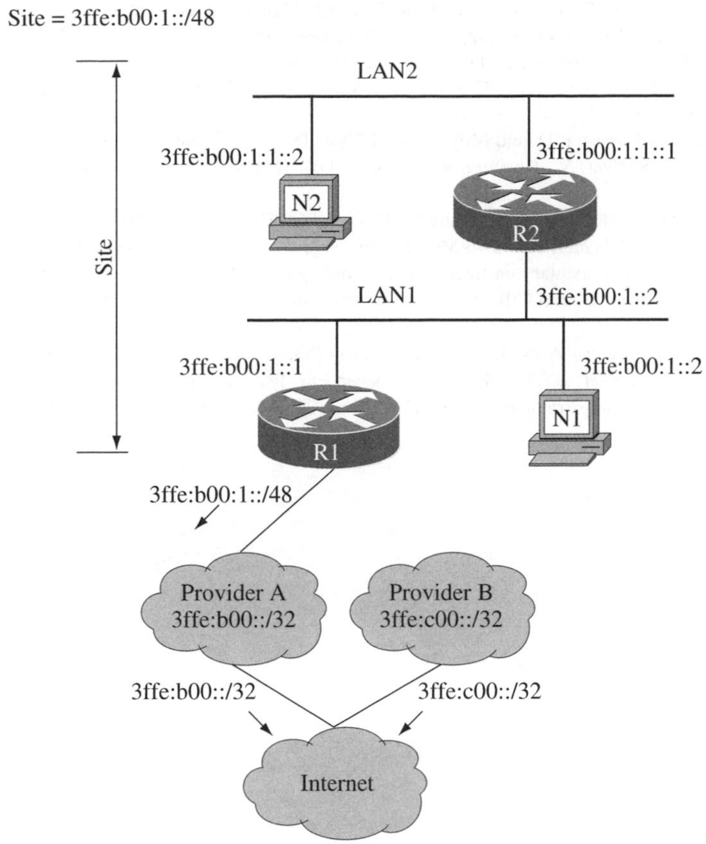

Figure 9.12 Multihomed Link failure with RR and RA

9.13 Summary

Routing has not changed that much in IPv6: route redirect works similar to IPv4 although source routing is done differently by using the routing extension header. The same routing protocol mechanisms are used but upgraded versions of the protocols are necessary. Router renumbering is defined to reconfigure routers in the event of prefix changes. Multihoming is an important topic in the next phases of the Internet and IPv6 has many solutions in the works.

9.14 References

Conta, A. and Deering S., 'Generic Packet Tunneling in IPv6 Specification', draft-ietf-ipv6-tunnel-v02-00 (work in progress), July 2002.

Hopps, C., 'Routing IPv6 with IS-IS', Internet-Draft draft-ietf-isis-ipv6-05, January 2003.

Manning, B. and Kato, A., 'BGP4+ Peering Using IPv6 Link-local Address', draft-kato-bgp-ipv6-link-local-00 (work in progress), September 2001.

[RFC1058] Hedrick, C., 'Routing Information Protocol', RFC 1058, June 1988.

[RFC1723] Malkin, G., 'RIP Version 2 – Carrying Additional Information', STD 56, RFC 1723, November 1994.

[RFC1771] Rekhter, Y. and Li, T., 'A Border Gateway Protocol 4 (BGP-4)', RFC 1771, March 1995.

[RFC2080] Malkin, G. and Minnear, R., 'RIPng for IPv6', RFC 2080, January 1997.
[RFC2081] Malkin, G., 'RIPng Protocol Applicability Statement', RFC 2081, January 1997.
[RFC2328] Moy, J., 'OSPF Version 2', STD 54, RFC 2328, April 1998.
[RFC2545] Marques, P. and Dupont, F., 'Use of BGP-4 Multiprotocol Extensions for IPv6 Inter-Domain Routing', RFC 2545, March 1999.
[RFC2740] Coltun, R., Ferguson, D. and Moy, J., 'OSPF for IPv6', RFC 2740, December 1999.
[RFC2858] Bates, T., Rekhter, Y., Chandra, R. and Katz, D., 'Multiprotocol Extensions for BGP-4', RFC 2858, June 2000.
[RFC2894] Crawford, M., 'Router Renumbering for IPv6', RFC 2894, August 2000.
[RFC3178] Hagino, J. and Snyder, H., 'IPv6 Multihoming Support at Site Exit Routers', RFC 3178, October 2001.
[RFC3221] Huston, G., 'Commentary on Inter-Domain Routing in the Internet', RFC 3221, December 2001.
[RFC3582] Abley, J., Black, B. and Gill, V., 'Goals for IPv6 Site-Multihoming Architectures', RFC 3582, August 2003.
[RIPE267] RIPE, 'IPv6 Address Allocation and Assignment Policy', RIPE-267, January 2003.
Roberts L., 'Policy Proposal 2005-1: Provider Independent IPv6 Assignments for End-sites', April 2005, http://www.arin.net/meetings/ minutes/ARIN_XV/PDF/tue/2005-1.pdf and http://www.arin.net/policy/proposals/2005_1.html

10

Configuring Routing

This chapter presents the routing configuration of FreeBSD, Linux, Solaris and Windows, acting as hosts and routers, and Cisco, Hexago, Juniper and Zebra, acting as routers. It includes the most important commands and configuration statements while not trying to replicate the vendor documentation, nor starting from zero to configure routers. The goal is to provide templates and basic configurations to bootstrap the learning of these vendors OS and to act as a reference for occasional uses. The reader is expected to have some knowledge of the respective vendor's OS.

For each vendor, sections on forwarding, static routes, route redirect, Routing Information Protocol(RIP), Open Shortest Path First (OSPF), Intermediate System to Intermediate System (IS-IS), Border Gateway Protocol (BGP) and troubleshooting are presented.

If configuring routers is not your primary interest, you can skip this chapter.

10.1 Considerations on Using Autoconfiguration for Router Interfaces

Router interface addresses are often referenced within routing protocols and router configurations. Since interface autoconfiguration is based on the MAC address, the EUI-64 made addresses are dependent on the hardware interface. A replacement of a faulty hardware interface generates a new autoconfigured address. To avoid reconfiguration, one should avoid interface autoconfiguration on routers and instead configure a static address on router interfaces. This advice also applies to server interface addresses.

Advice on how to configure IPv6 addresses on router interfaces:

Do not use autoconfiguration for router interfaces: configure static addresses instead.

Migrating to IPv6: A Practical Guide to Implementing IPv6 in Mobile and Fixed Networks Marc Blanchet

10.2 FreeBSD

This section discusses routing on FreeBSD, as a host and as a router.

10.2.1 Forwarding

To configure FreeBSD to forward packets between interfaces, set the 'ipv6_gateway_enable' variable to 'yes' in the '/etc/rc.conf' file.

```
# cat /etc/rc.conf
ipv6_gateway_enable="yes"
```

At boot time, this variable sets the kernel variable 'net.inet6.ip6.forwarding' to 1. To verify this variable, use the 'sysctl' command, as shown below.

```
# sysctl net.inet6.ip6.forwarding
net.inet6.ip6.forwarding: 1
```

To set the variable, use 'sysctl net.inet6.ip6.forwarding=1'.

Setting FreeBSD as a router disables listening of router advertisements, because a router should not listen to router advertisements of other routers.

Forwarding should be set when multiple interfaces are used and when traffic is forwarded between these interfaces. This applies to logical and physical interfaces. For example, a host with one physical interface and one tunnel interface acting as a router should set the forwarding mode.

10.2.2 Static Routes

To install a route, use the 'route add' command. The following installs an IPv6 default route targeted to fe80::2.

```
# route add -inet6 default fe80::2
```

To permanently install a static default route, set the 'ipv6_defaultrouter' variable in the '/etc/rc.conf' file to the address of the router, as shown below.

```
# cat /etc/rc.conf
ipv6_defaultrouter="fe80::2"
```

At boot time, the rc script executes the following command, based on the content of the previous variable.

```
# route add -inet6 default fe80::2
```

A FreeBSD host using the autoconfiguration process should receive a default route in the router advertisements, which means this static default route setting is not necessary in autoconfiguration mode.

A static route is set using the 'ipv6_static_routes' and 'ipv6_route' variables in '/etc/rc.conf'. The example below shows how to set the 3ffe:b00:1::/48 route to fe80::2.

```
# cat /etc/rc.conf
ipv6_static_routes="1"
ipv6_route_1="3ffe:b00:1:: -prefixlen 48 fe80::2"
```

The '1' string could be any string defined in the 'ipv6_static_routes' variable and referenced in the name of the 'ipv6_route' variable that defines the specific route. The syntax of the 'ipv6_route' variable is the 'route' command syntax without the 'add' argument. At boot time, these variables execute the following route command.

```
# route add -inet6 3ffe:b00:1:: -prefixlen 48 fe80::2
```

Some of the IPv6 prefixes are not suitable for routing. For example, the internal IPv4-mapped address, such as ::ffff:0.0.0.0/96, should not appear on the wire. To ensure this behavior, the '-reject' argument is used in the route command to send an ICMP unreachable message for this route instead of installing it in the kernel. At boot time, the '/etc/rc.network6' script executes the following command to reject these routes.

```
# route add -inet6 ::ffff:0.0.0.0 -prefixlen 96 ::1 -reject
```

10.2.3 Route Redirect

The 'net.inet6.ip6.redirect' kernel variable is used to control if FreeBSD accepts route redirects ICMP messages. It should be set to 1 in most cases and is the default.

```
# sysctl net.inet6.ip6.redirect
net.inet6.ip6.redirect: 1
```

10.2.4 RIP

The RIPv6 routing process is implemented in the 'route6d' daemon. It is started at boot time by setting the 'ipv6_router_enable' to yes in the '/etc/rc.conf' file.

```
# cat /etc/rc.conf
ipv6_router_enable="yes"
```

At boot time, this variable executes the 'route6d' command with the 'ipv6_router_flags' variable used to set options to the 'route6d' process.

Route advertisements of site scoped prefixes are disabled by default. This is a precaution of the stack developers because site scoped prefixes should not be advertised outside the site. For all internal routers, the variable 'ipv6_router_flags' should be set to '-l' to advertise site scoped prefixes, as shown below.

```
# cat /etc/rc.conf
ipv6_router_flags="-l"
```

10.2.5 *Troubleshooting*

To troubleshoot routes, use the '`netstat -r`' command. To restrict the output to IPv6 routes, use the '`-f inet6`' argument.

```
# netstat -f inet6 -rn
```

The '`traceroute6`' command is used to identify the path taken from the source to the destination. It is the IPv6 version of traceroute.

```
# traceroute6 3ffe:b00:1:1::1
```

10.3 Linux

This section discusses routing on Linux, as a host and as a router. Some features discussed below are configured as variables in the kernel using either the '`sysctl`' command to set the variable, or putting the value in the corresponding file in the proc filesystem. For example, to enable the IPv6 forwarding, use either

```
# sysctl -w net.ipv6.conf.all.forwarding=1
```

or

```
# echo "1" > /proc/sys/net/ipv6/conf/all/forwarding
```

10.3.1 *Forwarding*

To configure Linux to forward packets between interfaces, set the '`net.ipv6.conf.all.forwarding`' variable to '`1`' as shown below.

```
# sysctl -w net.ipv6.conf.all.forwarding=1
```

10.3.2 *Static Routes*

To add a static route, use the '`route add`' command. The following example installs the 3ffe:b00:1::/48 static route pointing to fe80::2.

```
#route -A inet6 add 3ffe:b00:1::/48 gw fe80::2
```

To delete a route, use the '`route del`' command. The keyword '`gw`' used in the add command is replaced by '`via`'. The following example deletes the 3ffe:b00:1::/48 route.

```
#route -A inet6 del 3ffe:b00:1::/48 via fe80::2
```

The current Linux kernel does not support the IPv6 default route (::/0). To make a default route, use the current allocated address space. The equivalent of the default route is the 2000:/3 route, since it is the defined unicast address range in the addressing architecture (see Section 4.3.1).

```
#route -A inet6 add 2000::/3 gw fe80::2
```

If using unique-local addressing, add also a route for fc00::/16 or fd00::/16.

10.3.3 Route Redirect

By default, route redirects are accepted when Linux is not forwarding and refused when Linux is forwarding. To set the acceptance of route redirects, set the sysctl '`net.ipv6.conf.all.accept_redirects`' variable to '`1`', as shown below.

```
# sysctl -w net.ipv6.conf.all.accept_redirects=1
```

10.3.4 Troubleshooting

To troubleshoot IPv6 routes, use the '`route -A inet6`' command without any additional argument.

```
# route -A inet6
```

The IPv6 routes can also be seen using the '`netstat`' command and by looking in the '`/proc`' filesystem, as shown below.

```
# more /proc/net/ipv6_route
```

The '`traceroute6`' command is used to identify the path taken from the source to the destination. It is the IPv6 version of traceroute.

```
# traceroute6 3ffe:b00:1:1::1
```

10.4 Solaris

This section discusses routing on Solaris, as a host.

10.4.1 Static Routes

To install a route, use the '`route add`' command. The following installs an IPv6 default route targeted to fe80::2.

```
# route add -inet6 default fe80::2
```

10.4.2 Troubleshooting

To troubleshoot, use the '`netstat`' command. To restrict the output to IPv6 routes, use the '`-f inet6`' and '`-r`' arguments.

```
# netstat -f inet6 -rn
```

The 'traceroute' command is used to identify the path taken from the source to the destination.

```
# traceroute 3ffe:b00:1:1::1
```

When the argument is a domain name, use the '-A inet6' option to force IPv6.

10.5 Windows

This section discusses routing on Windows, as a host and as a router. All statements are under the 'netsh interface ipv6' command.

10.5.1 Forwarding

By default, IPv6 forwarding in Windows is off. To enable IPv6 forwarding, use the 'set forwarding' statement, as shown below.

```
C> netsh interface ipv6 set interface "Local Area Connection"
forwarding=enabled
```

This command has to be done on each interface by replacing the 'Local Area Connection' with the appropriate interface name.

10.5.2 Static Routes

To add a static route, use the 'add route' statement and specify the route prefix, the interface and the next hop. The following example installs a static default route (::/0) on the first LAN interface ('Local Area Connection'), targeted to fe80::1.

```
C> netsh interface ipv6 add route ::/0 "Local Area Connection"
nexthop=fe80::1 publish=yes
```

To delete a route, use the 'delete route' statement.

```
C> netsh interface ipv6 delete route ::/0 "Local Area Connection"
```

10.5.3 Troubleshooting

To list the IPv6 routes, use the 'show routes' command, as shown below.

```
C> netsh interface ipv6 show routes
```

To traceroute an IPv6 path, use the 'tracert' command with a destination IPv6 address, as shown below.

```
C> tracert 3ffe:b00:1:1::1
```

When the argument is a domain name, use the '−6' option to force IPv6.

To display latency and packet loss of a path, use the 'pathping' command, as shown below.

```
C> pathping 3ffe:b00:1:1::1
```

When the argument is a domain name, use the '−6' option to force IPv6.

10.6 Cisco

This section describes the routing configuration of Cisco routers. In each example, 'configure terminal' or 'enable' is shown to distinguish in which context the commands are used. All levels are shown using indentation. To save redundant explanations, only changes are shown. For example, the command 'ipv6 unicast-routing' is only shown once, even if it should be included in all IPv6 Cisco configuration.

Most IPv6 commands are similar to the IPv4 commands, by replacing the 'ip' keyword with 'ipv6'.

10.6.1 IPv6 Forwarding

If IPv6 forwarding is not explicitly configured, a Cisco IOS router behaves as an IPv6 host and does not forward IPv6 packets. To configure IPv6 forwarding, use the 'ipv6 unicast-routing' command.

```
configure terminal
ipv6 unicast-routing
```

10.6.2 Cisco Express Forwarding

To enable IPv6 Cisco Express Forwarding (CEF) use the 'ipv6 cef' statement, but IPv4 CEF must be enabled first using the 'ip cef' statement. IPv6 packets with global or site scoped source and destination addresses are CEF switched. IPv6 packets with link-local addresses are not CEF switched.

The following example shows the commands to enable IPv4 and IPv6 CEF.

```
configure terminal
ip cef
ipv6 cef
```

The following example shows the commands to enable IPv4 and IPv6 distributed CEF.

```
configure terminal
ip cef distributed
ipv6 cef distributed
```

10.6.3 Prefix Lists

Prefix lists are used to filter routes in routing protocols. The 'ipv6 prefix-list' command is used with a name to define a prefix list. Actions such as deny or permit follow the statement and matching rules are applied.

The following example defines a prefix list named NO6TO4 with the matching rule to deny the advertisement of 6to4 (see Section 16.2.5) routes in RIP. 6to4 routes are defined as 2002::/16 or greater than 16.

```
configure terminal
ipv6 prefix-list NO6TO4 deny 2002::/16 ge 16
ipv6 router rip R1
  distribute-list prefix-list no6to4 in Ethernet 0
```

10.6.4 Static Routes

IPv6 static routes are configured using the 'ipv6 route' command. The following example installs the 3ffe:b00:0:1::/64 static route targeted to the fe80::2 router.

```
configure terminal
ipv6 route 3ffe:b00:0:1::/64 fe80::2
```

This other example installs a route pointing to the first Ethernet interface.

```
configure terminal
ipv6 route 3ffe:b00:0:1::/64 ethernet 0
```

Static routes are listed using the 'show ipv6 route static' command.

```
enable
show ipv6 route static
```

10.6.5 Route Redirect

By default, route redirect is enabled on Cisco IOS. The 'no ipv6 redirects' command disables it.

```
configure terminal
no ipv6 redirects
```

10.6.6 RIP

RIP for IPv6 is enabled with the 'ipv6 router rip' command. RIP processes are defined with a unique name. The RIP advertisements occur when the process is attached to an interface, using the 'ipv6 rip enable' command under the interface mode.

The following example configures RIP for IPv6 with the process name R1. It is enabled on the first Ethernet interface.

```
configure terminal
ipv6 router rip R1
interface ethernet 0
  ipv6 rip R1 enable
```

To inject a default route into RIP advertisements, causing this router to be seen as a default router, use the '`ipv6 rip default-information originate`' statement under the interface mode.

```
configure terminal
interface ethernet 0
  ipv6 rip R1 default-information originate
```

To redistribute static routes in RIP advertisements, use the '`redistribute static`' statement under the '`ipv6 router rip`' mode.

```
configure terminal
ipv6 router rip R1
  redistribute static
```

RIP information is shown by the '`show ipv6 rip`' command in enable mode. Details about the routing database are listed with the '`database`' argument.

```
enable
show ipv6 rip
```

To debug RIP, use the '`debug ipv6 rip`' command in enable mode.

```
enable
debug ipv6 rip
```

10.6.7 OSPF

OSPFv3 commands are similar to their OSPF IPv4 counterparts. OSPFv3 is enabled on a per interface basis, instead of a network command as in OSFPv2.

To enable the OSPF routing daemon, use the '`ipv6 router ospf`' statement. This enters a sub-mode for additional OSPF configuration statements. To start advertising OSPF on an interface, use the '`ipv6 ospf`' in the interface mode.

The following example configures OSPF for IPv6 applied to the first Ethernet interface.

```
configure terminal
ipv6 router ospf
interface ethernet 0
  ipv6 ospf
```

To redistribute BGP routes in OSPF, use the '`redistribute bgp`' command with the AS number of the BGP process, as shown below.

```
configure terminal
router bgp 65001
ipv6 router ospf
  redistribute bgp 65001
```

OSPF information is shown by the '`show ipv6 ospf`' command in enable mode.

```
enable
show ipv6 ospf
```

To debug OSPF, use the '`debug ipv6 ospf`' command in enable mode.

```
enable
debug ipv6 ospf
```

To reinitialize the SPF calculations, use the '`clear ipv6 ospf force-spf`' command in enable mode.

```
enable
clear ipv6 ospf force-spf
```

10.6.8 IS-IS

Different to OSPF and RIP, IS-IS carries IPv4 and IPv6 routes in the same protocol and process, so very few commands are added for IPv6.

The '`address-family ipv6`' statement is used to enable IPv6 in the IS-IS routing engine. The '`ipv6 router isis`' statement is used under the interface mode to enable IS-IS on that interface. The following example shows the IS-IS process named AREA1 with the IPv6 address family and assigned to the first Ethernet interface.

```
configure terminal
router isis AREA1
  net 49.001.0000.0000.000c.00
  address-family ipv6
interface Ethernet0
  ipv6 router isis AREA1
```

To redistribute BGP routes in IS-IS, use the '`redistribute bgp`' statement under the '`address-family ipv6`' mode, as shown below.

```
configure terminal
router isis AREA1
  address-family ipv6
    redistribute bgp 65000
```

To inject a default route in the IS-IS advertisements, causing this router to advertise as a default router, use the '`default-information originate`' under the '`address-family ipv6`' mode.

```
configure terminal
router isis AREA1
  address-family ipv6
    default-information originate
```

The default behavior of IS-IS is to check adjacencies for the same set of protocols. When the IPv6 IS-IS topology is not identical to the IPv4 IS-IS topology (see Section 9.7.2), the adjacency check fails. This happens when some neighbors have IPv4, others have IPv4 and IPv6 and others might have IPv6 only. All routers within an area must support the same set of IP protocols.

To remove this adjacency check to enable multi-topology, use the '`no adjacency-check`' under the '`address-family ipv6`' mode, as shown below.

```
configure terminal
router isis AREA1
  address-family ipv6
    no adjacency-check
```

IS-IS information is shown by the '`show isis`' command in enable mode.

```
enable
show isis AREA1
```

10.6.9 BGP

To enable the BGP routing daemon, use the '`router bgp`' statement followed by the AS number. This enters a sub-mode for additional BGP configuration statements.

BGP configuration is done in two parts: establishing the peering and exchanging routes. The peering is established with the '`neighbor remote-as`' statement. The IPv6 route exchange is activated under the '`address-family ipv6`' statement using the '`neighbor activate`' statement.

The following example configures a BGP for IPv6 routing daemon with AS 65001, identified by the 192.0.2.1 router-id (see Section 9.8.2), peering with 3ffe:c00:1:1::1 in AS 65002. Under the IPv6 address family, it activates the exchange of routes with the neighbor and advertises the 3ffe:b00::/24 route to the neighbor.

```
configure terminal
router bgp 65001
 bgp router-id 192.0.2.1
 neighbor 3ffe:c00:1:1::1 remote-as 65002
 address-family ipv6
  neighbor 3ffe:c00:1:1::1 activate
  network 3ffe:b00::/24
```

When exchanging IPv6 routes only, use the '`no bgp default ipv4-unicast`' statement under '`router bgp`', since by default, IPv4 routes are exchanged.

```
configure terminal
router bgp 65001
  no bgp default ipv4-unicast
```

In enable mode, use the 'show bgp ipv6' command to list the details of the BGP process.

```
enable
show bgp ipv6
```

To troubleshoot BGP IPv6 transactions, use the 'debug bgp ipv6' command in enable mode.

```
enable
debug bgp ipv6
```

10.6.10 Troubleshooting

To troubleshoot IPv6 routes, use the 'show ipv6 route' command in enable mode.

```
enable
show ipv6 route
```

10.7 Hexago

This section describes the routing configuration of the Hexago Migration Brokers.

HexOS configuration syntax is similar to other vendors' syntax. In each example, 'configure terminal' or 'enable' is shown to distinguish in which context the commands are used. All levels are shown using indentation. Exit commands are not shown for simplicity.

HexOS have IPv6 forwarding enabled by default.

IPv6 static routes are configured using the 'ipv6 route' command. The following example installs the 3ffe:b00:0:1::/64 static route targeted to the fe80::2 router.

```
configure terminal
ipv6 route 3ffe:b00:0:1::/64 fe80::2
```

Static routes are listed using the 'show ipv6 route static' command.

```
enable
show ipv6 route static
```

10.8 Juniper

JunOS identifies the IPv6 routing information base (RIB) with 'inet6.X', instead of 'inet.X' for IPv4. The first IPv6 table is named 'inet6.0'.

10.8.1 Martian Routes

JunOS identifies invalid routes as martians. When received by the router, these routes are discarded. For IPv6, the predefined martians are:

- the loopback address (::1);
- the reserved and unassigned prefixes;
- the link-local unicast prefix (fe80::/16).

To add a martian route, the '`martians`' statement must be used. In the following example, the unique-local prefix range (any route which starts with fd00::/8 with a prefix length from 8 to 128) is added as a martian route.

```
routing-options {
  rib inet6.0 {
    martians {
      fd00::/8 orlonger;
    }
  }
}
```

10.8.2 Router ID

When the router is an IPv6-only router (ie. where no IPv4 address configuration is provided), a 32 bit router ID is configured with the '`router-id`' statement. The router ID is used in routing protocols to identify the router. The router-id statement takes an IPv4-formatted 32 bit value as argument. The following example shows the router-id defined for this router as 192.0.2.1.

```
routing-options {
 router-id 192.0.2.1;
}
```

10.8.3 Static Routes

To install a static route, use the '`route`' statement under '`routing-options rib inet6.0 static`' level. The following example installs the 3ffe:b00:0:3::/64 static route pointing to the 3ffe:b00:0:1::3 next hop.

```
routing-options {
 rib inet6.0 {
  static {
   route 3ffe:b00:0:3::/64 next-hop 3ffe:b00:0:1::3;
  }
 }
}
```

10.8.4 RIP

RIP for IPv6 is configured under the '`ripng`' statement under '`protocols`'. Most statements in `ripng` are identical to their RIP counterpart, so these are not described. The following RIP statements are not used in RIPng:

- '`authentication-(key|type)`' statements because authentication is now done with IPSEC;
- '`rib-group`' statement because routing table groups are not supported.

RIPng must be specified for each interface using the '`neighbor`' statement under '`group`'. The following example shows a router configured with RIPng on its fe-0/0/1 interface under the group named 'LAN'.

```
protocols {
 ripng {
  group LAN {
   neighbor fe-0/0/1;
  }
 }
}
```

Most '`show rip`' and '`clear rip`' commands are similar for RIPng using the '`ripng`' keyword instead of '`rip`'. The following commands are specific to RIPng:

- `clear ripng statistics`
- `show ripng statistics`
- `clear ripng general-statistics`
- `show ripng general-statistics`
- `show ripng neighbor`
- `show route protocol ripng table inet6`

The following example shows the command to list the RIP IPv6 routes.

```
show route protocol ripng table inet6
```

10.8.5 OSPF

OSPFv3 is configured under the '`ospf3`' statement at the same level as the '`ospf`' statement, under '`protocols`'. Most statements in '`ospf3`' are identical to their OSPFv2 counterpart, so these are not described. The following OSFPv2 statements are not used in OSPFv3:

- '`authentication-(key|type)`' statements because authentication is now done with IPSEC;
- traffic engineering features.

OSPFv3 must be activated on a per interface basis. The following example shows a router configured with OSPFv3 on the fe-0/0/1 interface for area 0.

```
protocols {
 ospf3 {
  area 0 {
   interface fe-0/0/1;
  }
 }
}
```

Most 'show ospf' and 'clear ospf' commands are similar in OSPFv3 using the 'ospf3' keyword instead of 'ospf'. The following commands are specific to OSPFv3:

- `clear ospf3 database`
- `show ospf3 database`
- `show ospf3 interface`
- `clear ospf neighbor`
- `show ospf3 neighbor`
- `show ospf3 route`

The following example lists the IPv6 OSPF routes.

```
show ospf3 route
```

10.8.6 IS-IS

Since it is a network protocol independent routing protocol, IS-IS configuration on JunOS does not need any changes to statements to support IPv6. By default, all enabled network protocols on interfaces are advertised in IS-IS unless specifically disabled.

The following example shows IS-IS configured on the fe-0/0/1 interface with a network entity title (NET) address of 49.0001.0000.0000.000c.00.

```
interfaces {
 fe-0/0/1 {
  unit 0 {
   family iso {
    address 49.0001.0000.0000.000c.00;
   }
  }
 }
}
protocols {
 isis {
  interface fe-0/0/1;
 }
}
```

This configuration carries both IPv4 and IPv6 routes.

To disable the routing of IPv4 or IPv6 inside IS-IS, use the 'no-ipv4-routing' and 'no-ipv6-routing' statements respectively. The following example shows a router with no IPv4 IS-IS routing.

```
protocols {
 isis {
  interface fe-0/0/1;
  no-ipv4-routing;
 }
}
```

As of JunOS 5.5, the IPv4 and IPv6 network topologies must be identical.

Most '`show isis`' and '`clear isis`' commands are used as they are. To see the IP routes in the IS-IS routing table, use the '`show isis routes`' command as shown below.

```
show isis routes
```

10.8.7 BGP

JunOS BGP is configured under the '`protocols bgp`' level, where a peer group is defined using the '`group`' statement.

Each BGP address family routes are installed into different routing tables.

The following example shows the basic BGP configuration for a router in AS 65001 having an eBGP peering, using the '`type external`' statement, with the 3ffe:b00:1:1::2 neighbor in AS 65002. Unicast IPv6 routes are exchanged using the '`family inet6 unicast`' statement.

```
routing-options {
 autonomous-system 65001;
}
protocols {
 bgp {
  group R2 {
   peer-as 65002;
   type external;
   neighbor 3ffe:b00:1:1::2 {
    family inet6 {
     unicast;
    }
   }
  }
 }
}
```

To carry unicast IPv4 routes in the same BGP session as well, the '`family inet`' statement is added at the same level as the '`family inet6`' statement, as shown below.

```
neighbor 3ffe:b00:1:1::2 {
 family inet6 {
  unicast;
 }
 family inet {
  unicast;
 }
```

To enable IPv6 and IPv4 route exchanges over an IPv4 BGP session, the IPv4-compatible addresses are used in the interface level. This precludes the need to configure IPv6 static routes.

The following example shows the 192.0.2.1 router within AS 65001. BGP peering is done over IPv4, forced with the '`local-address 192.0.2.1`' statement, with the 192.0.2.2 router within AS 65002. IPv4 and IPv6 routes are exchanged because the '`family inet`' and '`family inet6`' statements are used and the fe-0/0/1 interface is configured with an IPv4-compatible IPv6 address (::192.0.2.1).

```
protocols {
 bgp {
  group R2 {
   peer-as 65002;
   type external;
   neighbor 192.0.2.2;
   local-address 192.0.2.1;
   family inet {
    unicast;
   }
   family inet6 {
    unicast;
   }
  }
 }
}
interfaces {
 fe-0/0/1 {
  unit 0 {
   family inet {
    address 192.0.2.1/24;
   }
   family inet6 {
    address ::192.0.2.1/126;
   }
  }
 }
}
```

Note the use of /126 in the IPv4-compatible IPv6 address, since this is a tunnel interface. Tunnels are discussed in Section 16.2. In this case, the BGP next-hop is the IPv4-compatible IPv6 prefix (::192.0.2.1).

Peering can be established with link-local neighbor addresses. Since link-local addresses are ambiguous depending on the interface, the interface must be specified with the '`local-interface`' statement under '`neighbor`', as shown below.

```
protocols {
 bgp {
  group R2 {
   peer-as 65002;
   type external;
```

```
    neighbor fe80::1 {
     local-interface fe-0/0/1;
    }
    family inet6 {
     unicast;
    }
   }
  }
}
```

When IPv6-only peering is done, the router-id is specified in the IPv4 address format, using the 'routing-options router-id' statement.

```
routing-options {
 router-id 192.0.2.1;
}
```

Most 'show bgp' and 'clear bgp' commands can be used as with IPv4. To see the status of BGP neighbors, use the 'show bgp neighbor'.

```
show bgp neighbor
```

10.8.8 Troubleshooting

Most show commands accepting an IPv4 prefix or address also support an IPv6 prefix or address, such as 'show route', as shown below.

```
show route 3ffe:b00:0:1::/64
```

When an address family is specified, use 'inet' for IPv4 and 'inet6' for IPv6.

```
show route forwarding-table family inet6
```

When a RIB table is specified, use 'inet.X' for IPv4 and 'inet6.X' for IPv6, where X stands for the identifier of the table.

```
show route table inet6.0
```

10.9 Zebra

This section describes the routing configuration of Zebra v0.93b, an open-source routing daemon. It runs on most Unix variants such as FreeBSD, Linux and Solaris. It implements RIP, OSPF and BGP for IPv4 and IPv6. Zebra implements one daemon per routing protocol, a daemon to manage the interfaces and kernel routes named 'zebra', and a CLI daemon named 'vtysh' to configure these daemons in an integrated way.

Zebra configuration syntax is based on Cisco CLI syntax. In each example, 'configure terminal' or 'enable' is shown to distinguish in which context the commands are used. All levels are shown using indentation. Exit commands are not shown for simplicity.

10.9.1 Static Routes

To install a static route, use the '`ipv6 route`' command. The following example shows the 3ffe:b00:0:1::/64 static route targeted to the fe80::2 router.

```
configure terminal
ipv6 route 3ffe:b00:1::/64 fe80::2
```

10.9.2 RIP

The daemon implementing RIP for IPv6 is '`ripngd`'. Its configuration file is '`/usr/local/etc/ripngd.conf`'. To enable the RIPng routing daemon, use the '`router ripng`' statement. This enters a sub-mode for additional RIPng configuration statements, such as the '`network`' command to announce a route. The following example shows a RIPng routing daemon activated announcing the 3ffe:b00:1::/64 route.

```
configure terminal
router ripng
 network 3ffe:b00:1::/64
```

In enable mode, use the '`show ipv6 ripng`' command to list the details of the RIPng process.

```
enable
show ipv6 ripng
```

To see debugging details, use the '`debug ripng`' command in enable mode. The command has options such as '`events`', '`packet`' and '`zebra`'.

```
enable
debug ripng packet
```

10.9.3 OSPF

The daemon implementing OSPF for IPv6 is '`ospf6d`'. Its configuration file is '`/usr/local/etc/ospf6d.conf`'. To enable the OSPF routing daemon, use the '`router ospf6`' statement. This enters a sub-mode for additional OSPF configuration statements.

The '`interface`' sub-command is used to enable OSPF advertisements on that interface.

The following example configures an OSPF for IPv6 routing daemon, identified by the 192.0.2.1 router-id (see Section 9.6.2), advertising on the '`fxp0`' interface with area 0 (written in IPv4 address notation), and redistributing static routes in OSPF.

```
configure terminal
router ospf6
  router-id 192.0.2.1
  interface fxp0 area 0.0.0.0
  redistribute static
```

In enable mode, use the 'show ipv6 ospf6' command to list the details of the OSPF process. Options such as 'database', 'interface' or 'neighbor' are available for additional details.

```
enable
show ipv6 ospf6 [database | interface | neighbor]
```

To list the OSPF routes, use the 'show ipv6 route ospf6' command.

```
enable
show ipv6 route ospf6
```

10.9.4 BGP

The daemon implementing BGP for IPv4 and IPv6 is 'bgpd'. Its configuration file is '/usr/local/etc/bgpd.conf'. To enable the BGP routing daemon, use the 'router bgp' statement followed by the AS number. This enters a sub-mode for additional BGP configuration statements.

BGP configuration is done in two parts: establishing the peering and exchanging routes. The peering is established with the 'neighbor remote-as' statement, and optionally the 'router-id' statement when IPv6-only peering is done. The IPv6 route exchange is activated under the 'address-family ipv6' statement using the 'neighbor activate' statement.

The following example configures a BGP for IPv6 routing daemon with AS 65001, identified by the 192.0.2.1 router-id (see Section 9.8.2), peering with 3ffe:c00:1:1::1 in AS 65002. Under the IPv6 address family, it activates the exchange of routes with the neighbor and advertises the 3ffe:b00::/24 route to the neighbor.

```
configure terminal
router bgp 65001
 bgp router-id 192.0.2.1
 neighbor 3ffe:c00:1:1::1 remote-as 65002
 address-family ipv6
  neighbor 3ffe:c00:1:1::1 activate
  network 3ffe:b00::/24
```

When exchanging IPv6 routes only, use the 'no bgp default ipv4-unicast' statement under 'router bgp', since by default, IPv4 routes are exchanged.

```
configure terminal
router bgp 65001
  no bgp default ipv4-unicast
```

In enable mode, use the 'show ipv6 bgp' command to list the details of the BGP process. This command has options such as 'community', 'neighbors' and 'summary'.

```
enable
show ipv6 bgp
```

To see debugging details, use the 'debug bgp' command in enable mode. This command has options such as 'events', 'filters' and 'updates'.

```
enable
debug bgp updates
```

10.9.5 Troubleshooting

To troubleshoot IPv6 routes, use the 'show ipv6 route' command.

```
enable
show ipv6 route
```

10.10 Summary

This chapter describes the routing configuration statements and commands of the following operating systems: FreeBSD, Linux, Solaris, Windows, Cisco, Hexago, Juniper and Zebra. Template commands and examples are given for each OS to help the reader find the right commands.

10.11 Further Reading

FreeBSD Handbook, The FreeBSD Handbook Project, 2003, http://www.freebsd.org/handbook/index.html
Linux IPv6 HowTO, Peter Bieringer, 2003, http://www.bieringer.de/linux/IPv6
IPv6 Administration Guide – Solaris 9, Sun Microsystems, August 2003.
Windows IPv6 Documentation, Microsoft, 2003, http://www.microsoft.com/ipv6
IPv6 for Cisco IOS Command Reference, Cisco, 2003.
Migration Broker Configuration Guide – HexOS 3.0, Hexago, 2004.
Migration Broker Command Reference – HexOS 3.0, Hexago, 2004.
JunOS Internet Software Configuration Guide – Routing Protocols – JunOS 5.5, Juniper, 2002.

11

Mobility

In the late 1970s, there were no micro-computers or laptops, only computers taking up a significant space in a room and having no easy way to move. This is the era in which IPv4 was designed, so no provision was made for mobile computers. Nowadays, mobile computers are not only laptops, but cell phones, personal digital assistant (PDA), pagers, airplanes, cars, tanks, etc.

The handling of mobility is accomplished at many layers as shown in Figure 11.1.

With the wireless LAN standard IEEE 802.11 [IEEE80211], a mobile node can move from one access point to another without any disruption of the IP traffic, because the mobility aspects are handled by the 802.11 protocol and by the access point devices. However, while moving, if a mobile node changes or needs to change its IP address because its new point of attachment is on another IP subnet or network, then the IP traffic is disrupted.

In the context of IP, mobility is defined by the node changing its IP address while moving. If it moves without changing IP address, there is no IP mobility involved. If it moves from an Ethernet LAN to another Ethernet LAN and changes IP address, then IP mobility is involved. If it moves from a wireless LAN to a PPP connection and changes IP address, then IP mobility is involved.

MobileIP is an IP protocol designed to sustain the IP connections while the IP address changes. MobileIP is designed to work for IPv4 and IPv6, although it is optimized for IPv6. The mobile node is known by its peers with a permanent address and an agent is mandated to forward the traffic to the current location of the mobile node. The agent knows the location of the mobile node because each time the mobile node moves, it registers its new location, i.e. the new IP address, to its agent.

This chapter describes MobileIP with a focus on MobileIPv6. It starts with a general overview using the phone call forwarding metaphor and then describes MobileIP in detail.

Migrating to IPv6: A Practical Guide to Implementing IPv6 in Mobile and Fixed Networks Marc Blanchet

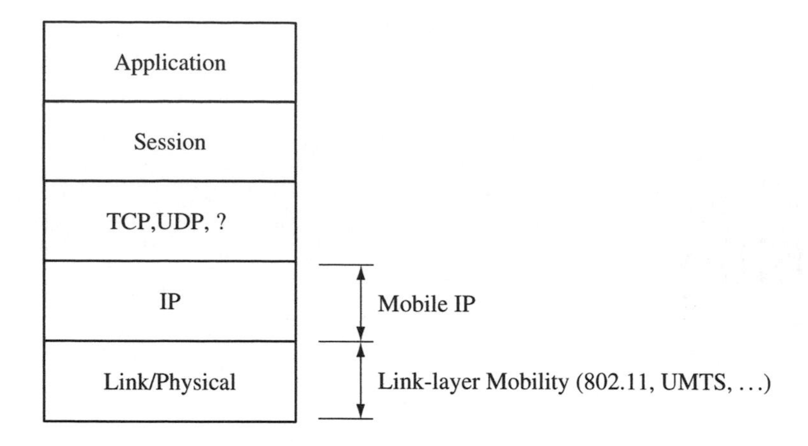

Figure 11.1 Mobility at different layers

11.1 Overview

MobileIP process is very similar to the call forwarding function in telephony, as illustrated in Figure 11.2 and Table 11.1.

Alice knows Bob's primary telephone number (1). If Bob moves to another location (3) where he has another phone number, then he configures the switch to forward the calls coming for its primary number to its remote phone number (4). Alice still knows Bob's

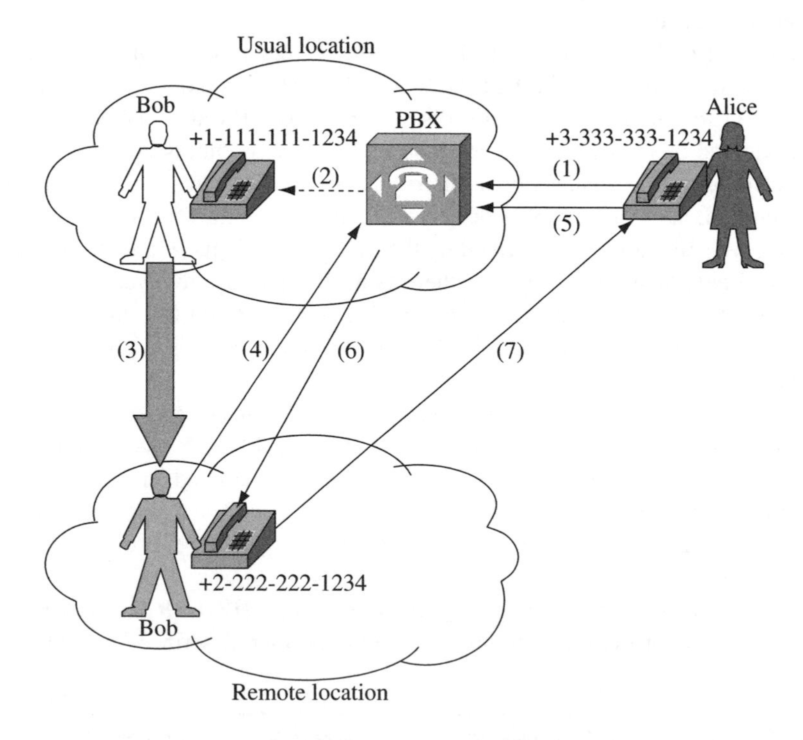

Figure 11.2 Call forwarding

Table 11.1 Equivalence between MobileIP and call forwarding

Call Forwarding Function	MobileIP Function
Primary phone number	Home IP address
Primary switch	Home agent (HA)
Process to register the new phone number	Binding update (BU)
New phone number	Care-of-address (CoA)
Correspondent	Correspondent node (CN)
Mobile user	Mobile node (MN)

primary telephone number, but an agent, the switch, takes care of forwarding Bob's calls to him (6). If Bob moves again, he only needs to change the configuration of the switch to forward calls to the new phone number, without having to tell Alice and all his correspondents about where he is now.

11.1.1 MobileIP Terminology

In the MobileIP context and terminology,

- the primary telephone number is the node home IP address;
- the switch is the home agent (HA), which is usually a router on the home network;
- the process to register your new telephone number is the binding update (BU);
- the correspondent (Alice) is the correspondent node;
- and the new phone number is the Care-of-address (CoA).

The word 'home' in MobileIP terminology is not related to the physical home where one lives, but is defined as the permanent place where the permanent IP address resides and where a permanent agent forwards the traffic directed to the permanent IP address. For an employee, the home network is typically the organization network where the permanent IP address is assigned and a router on that network takes care of the home agent function.

11.1.2 Basic MobileIP Process

In a nutshell, the MobileIP process is described by the following steps, illustrated in Figure 11.3:

1. A mobile node uses its permanent address, the home address, when it is located on its home network. If a node sends a datagram to the mobile node (step 1 in Figure 11.3), it sends it to the home address of the mobile node.
2. The mobile node receives the datagram on the home network.
3. When a mobile node is visiting a network, away from its home network, it acquires by some means from the visiting network a temporary IP address, called the care-of-address (CoA).
4. The mobile node registers this CoA to its home agent.

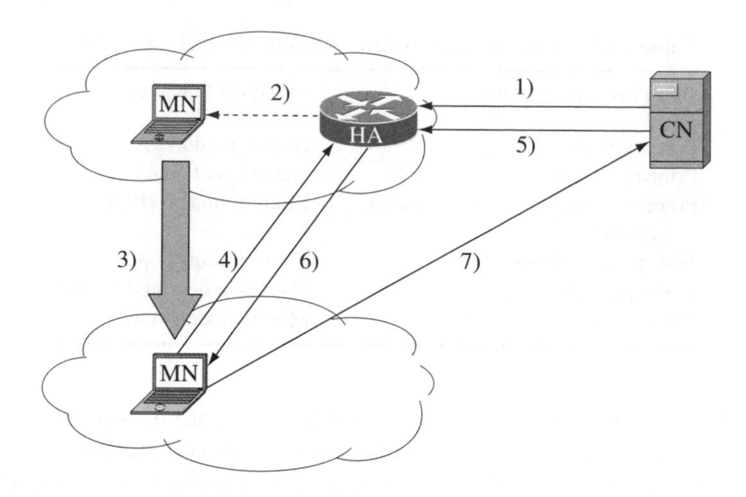

Figure 11.3 MobileIP basic process

5. The correspondent node knows the mobile node only by its home address, thus it sends a datagram with the destination address as the home address of the mobile node.
6. The datagram is routed to the home network where the home agent intercepts and forwards the datagram to the mobile node based on the home agent knowledge of the mobile node care-of-address, this address being located in the visiting network. The mobile node receives the datagram.
7. When the mobile node sends a datagram to a correspondent node, the source address of the datagram is the home address of the mobile node and it sends it directly to the correspondent node.

The correspondent node has no knowledge of the mobility of the (mobile) node. The effort of handling mobility is entirely done by the mobile node's own infrastructure: the home agent and the mobile node itself.

11.1.3 Triangle Routing

To continue with the call forwarding metaphor, when Bob calls Alice while he is away from his primary phone, he uses another phone with another phone number, as illustrated in Figure 11.4.

If Bob does not tell Alice the new phone number (+2-222-222-1234 in Figure 11.4), she still calls Bob's primary phone number (+1-111-111-1234) and the switch forwards the call to Bob's current phone number. This may cost Bob a lot since he will be paying for the long distance call between the primary phone number (+1-111-111-1234) and the current phone number (+2-222-222-1234).

In the MobileIP context, this behavior is called triangle routing and is the basic behavior in absence of any optimization, as illustrated in Figure 11.5.

The mobile node sends datagrams directly to its correspondent nodes (step 3 in Figure 11.5), by setting the home address as the source address of the packets. The correspondent nodes still send their datagrams to the home address (step 1) and the home agent forwards them

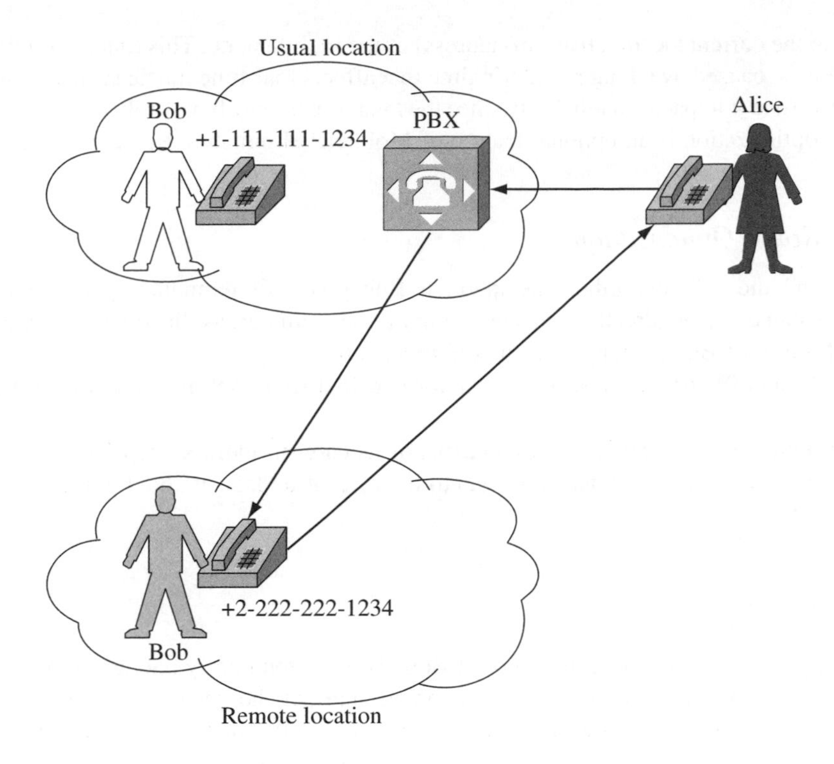

Figure 11.4 Call forwarding triangle

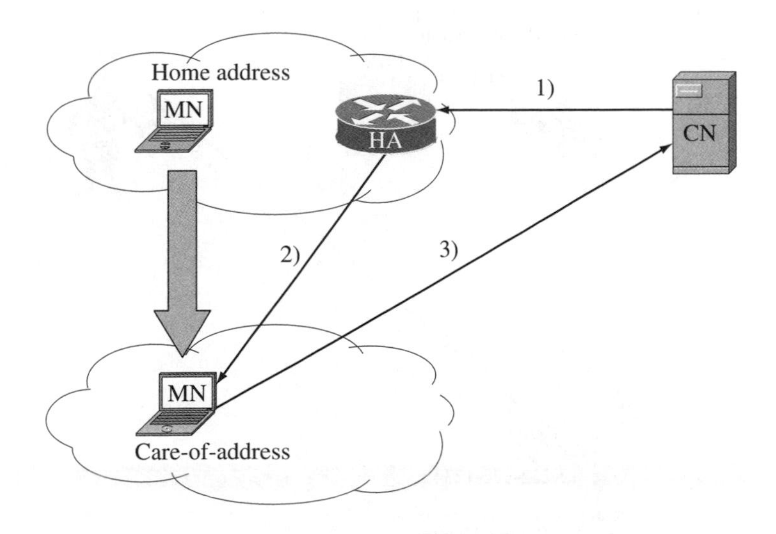

Figure 11.5 MobileIP triangle routing

(step 2) to the current location (care-of-address) of the mobile node. This triangle routing may create delays, caused by a long round trip time that affects real-time traffic such as voice over IP (VoIP). Also it impacts on reliability since the longer path may have broken links.

Route optimization is an optional feature of MobileIP eliminating triangle routing.

11.1.4 Route Optimization

Again using the call forwarding metaphor, if Bob gives his temporary phone number to Alice, she can call Bob directly on his temporary phone number, as illustrated in Figure 11.6. This will not cost Bob a penny since it is a direct call.

In the MobileIP context, this same behavior is called route optimization, as illustrated in Figure 11.7.

The mobile node informs its correspondents of its care-of-address (step 1 in Figure 11.7), so correspondent nodes send datagrams (2) to the care-of-address without going through the home agent.

11.1.5 Handoff

Continuing the call forwarding metaphor, if Bob gives his temporary phone number to Alice, then whenever Bob moves again to a new phone number, he needs to tell Alice the new phone number as well as register the new phone number to the switch controlling the primary phone number. While Bob is telling them the new phone number, some of the correspondents still call him on his previous temporary phone number and Bob is not there, causing a black hole as illustrated in Figure 11.8.

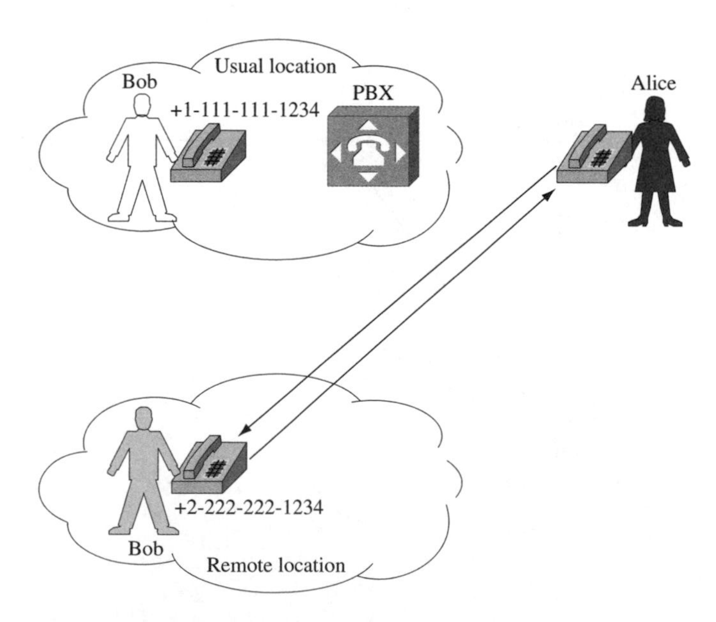

Figure 11.6 Call forwarding direct call

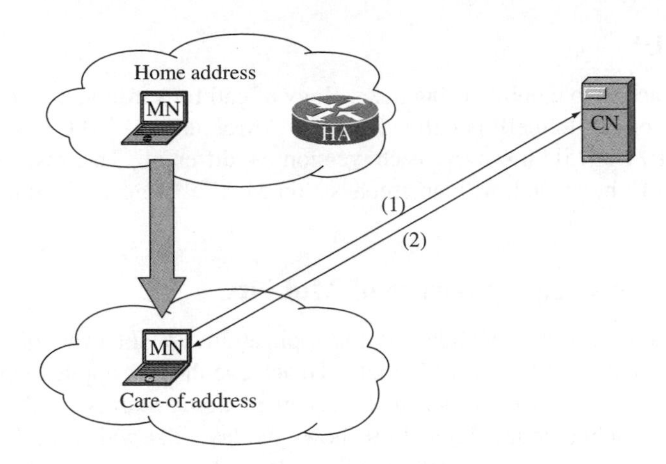

Figure 11.7 MobileIP route optimization

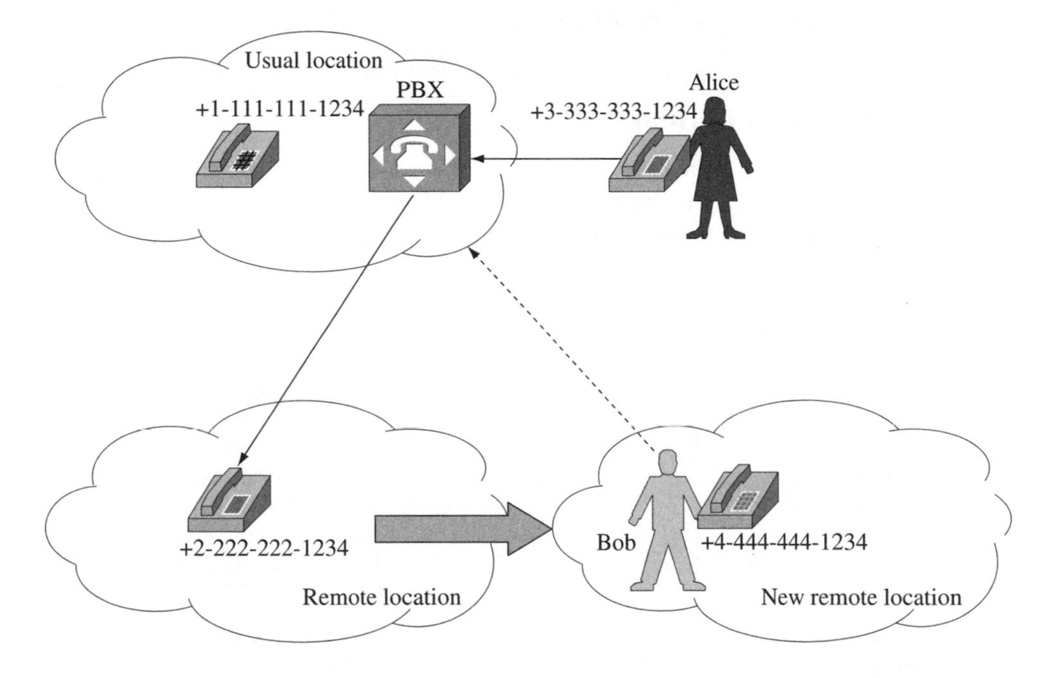

Figure 11.8 Call forwarding miss

In the MobileIP context, this is called the handoff problem, where correspondents send datagrams to a previous care-of-address after the mobile node has already gone. Many solutions are possible depending on the context of the handoff.

If Bob does not give his temporary phone numbers to his correspondents, then they will still call the primary phone number. Therefore, the only handoff problem remaining in this case concerns the time during which Bob is using a new temporary phone and the primary switch has not yet been set to forward calls to the new phone number.

11.2 MobileIP

After this introduction to mobility using the analogy of call forwarding, this section describes how mobileIP works. MobileIP is defined for IPv4 (MobileIPv4) [RFC3344] and for IPv6 (MobileIPv6) [RFC3775], however, each version is different. The rest of this chapter describes MobileIP in general, with an emphasis on MobileIPv6 specific functions.

11.3 Applications are not Aware of Mobility

An important design criteria of MobileIP is that applications are not aware of mobility, therefore they are not modified to support MobileIP. To achieve this, the application on the mobile node always uses and sees the permanent address, named 'home address', as the source address of packets on the mobile node. Figure 11.9 shows the layers at which the IP addresses are managed for mobility purposes, when the mobile node sends a packet to the destination node.

MobileIP manages the changes of the source address in the IP stack of both peers, restoring the permanent address as the source address of the connection when the correspondent node application receives the packet from the stack.

11.4 Mobile Node is at Home

MobileIP home agents (HA) are specially configured routers implementing MobileIP. Periodically, they send router advertisements with specific MobileIP data to tell mobile nodes that a home agent is available on the network and advertising a prefix, as shown in Figure 11.10.

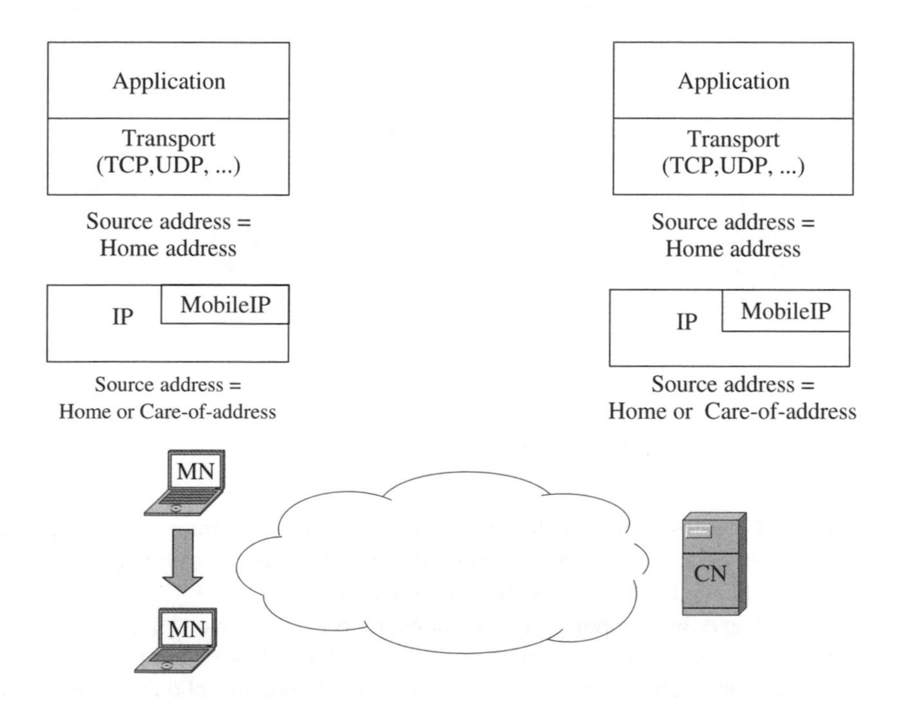

Figure 11.9 Application and IP stack view of mobileIP

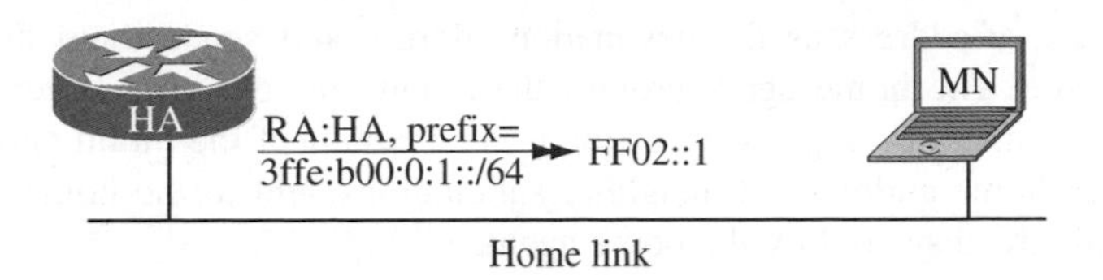

Figure 11.10 Home agent advertising

The mobile node receives these advertisements and compares the received network prefix with its previously saved home prefix. If they are identical, then the mobile node knows it is on its home link. The home address is the permanent address used for all connections, either while at home or visiting another network.

While at home, a mobile node behaves as a normal IP node. No mobility specific processing is done. The mobile node sends its packets using the home address as the source address.

11.5 Mobile Node is away from Home

A mobile node needs to know when it has moved to start using the MobileIP functions. At the MobileIP level, a node moves when its IP address changes.[1] To find when it has moved, a node compares its saved home address with a new acquired address. When the addresses are equal, it is at home, otherwise it is visiting another network. A mobile node visiting a network acquires an IPv6 address by autoconfiguration or DHCPv6.

Correspondent nodes (CN) know the mobile node by its home address. By default, the correspondent node sends datagrams to the home address, as shown in Figure 11.11.

The home agent, which proxies the mobile node while it is away, intercepts the packets addressed to the mobile node home address, encapsulates them in an IPv6 packet with the

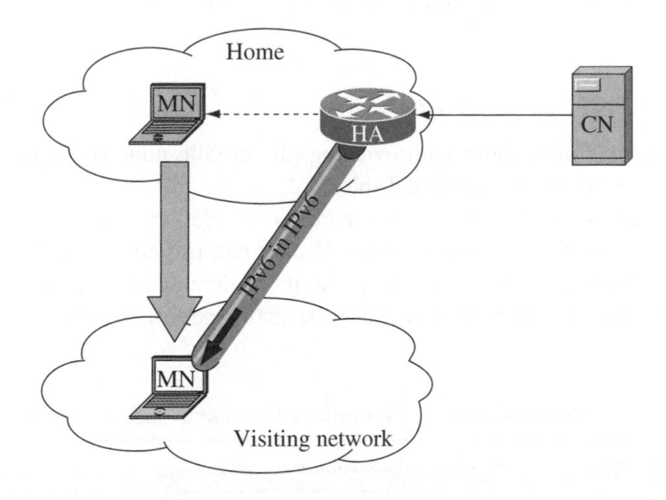

Figure 11.11 Correspondent node sends a packet to the home address

[1] Recall that when a mobile node may move from one access point to another one without changing IP address, the mobility functions are taken care of at the layers below IP and do not involve MobileIP.

mobile node care-of-address as the destination address and sends them the mobile node at the visiting site. The home agent proxies the mobile node home address by sending a neighbor advertisement message (see Section 6.2) on behalf of the mobile node, advertising the mobile node home address. All neighbor solicitations sent to the home address by any node on the link are answered by the home agent.

11.5.1 Mobile Node Registering to the Home Agent

After acquiring its new IP address and discovering it is visiting, the mobile node registers its new IP address to its home agent, through a secured message named the Binding Update (BU). The binding update is protected by IPsec (see Section 11.8), which means that a security association was pre-established between the mobile node and the home agent prior to the binding update. After the binding update, the home agent can then forward any packets received for the mobile node to its new IP address, as shown in Figure 11.11.

11.5.2 Mobile Node Registering to the Correspondent Nodes

The Binding Update (BU) is also sent to some of its correspondent nodes to inform them of the new IP address of the mobile node. After receiving the binding update, the correspondent nodes send packets directly to the new location of the mobile node without going through the home agent. The binding update also contains a lifetime of the new IP address, which is kept in the cache of the correspondent node. The correspondent node sends a binding acknowledgement to the mobile node, acknowledging the reception of the binding update. The binding update message format is described in Section 11.11.

The binding updates to the correspondent nodes require the correspondent node to implement the return routability procedure, described in Section 11.8.2.

11.5.3 Mobile Node Sending Packets

After sending the binding updates while visiting, the mobile node sends packets directly to its correspondent nodes as described in Table 11.2.

The care-of-address used as the source address enables the packet to go through the visiting network without any filtering issues. If the visiting network is filtering the source address of the outgoing packets by verifying that the source address belongs to the network, then the packet is not filtered if the source address is the care-of-address of the mobile node.

Table 11.2 IPv6 header of a packet sent by the mobile node while visiting

Field	Description
Source address	Mobile node Care-of-address
Destination address	Correspondent node address
Home address destination option extension header	Contains the home address of the mobile node.

When the packet arrives at the correspondent node, the packet source address is replaced by the home address included in the destination option header before pushing the packet in the application stack. The application sees the home address of the mobile node as the source address, which is the permanent address of the mobile node.

11.5.4 Correspondent Node Sending Packets to the Mobile Node

By default, the correspondent node sends the packets to the mobile node using its home address as destination address, as shown in Table 11.3.

The packet is intercepted by the home agent and encapsulated to the current location of the mobile node, as shown Figure 11.11.

If a binding update tells the correspondent node of the new IP address of the mobile node, the care-of-address, then the correspondent node sends the packet to the care-of-address, as shown in Figure 11.12.

The packet also contains a routing header which encapsulates the home address of the mobile node as the next hop. The IPv6 header of a packet sent by the correspondent node is shown in Table 11.4.

When the mobile node receives this packet, the IP stack sees the routing header and processes it by replacing the destination address with the first (and only one) address in the routing header (routing header processing is described in Section 9.2). Thus the IPv6 header presented to the application in the mobile node is shown in Table 11.5.

Table 11.3 IPv6 header of a packet sent by the CN to MN HA

Field	Description
Source address	Correspondent node address
Destination address	Mobile node Home address

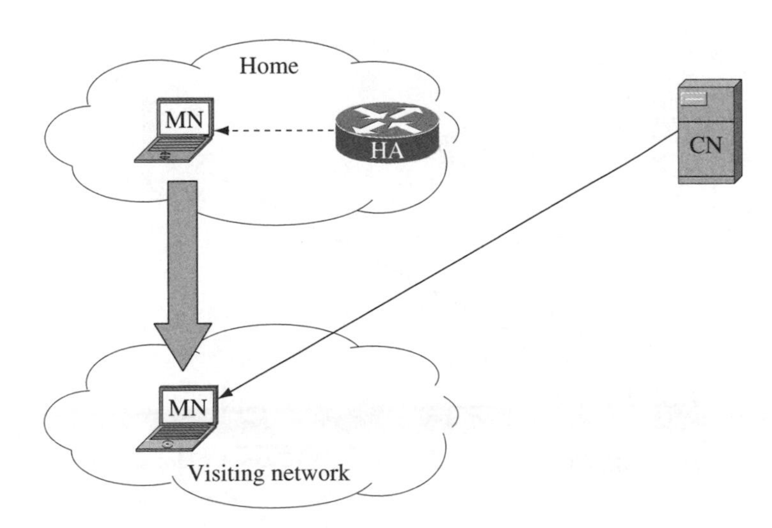

Figure 11.12 Correspondent node sending to MN CoA

Table 11.4 IPv6 header of a packet sent by the CN with MN CoA

Field	Description
Source address	Correspondent node address
Destination address	Mobile node care-of-address
Routing header	Contains the home address of the mobile node as the next hop.

Table 11.5 IPv6 header of a packet presented to the application

Field	Description
Source address	Correspondent node address
Destination address	Mobile node home address

11.6 Mobile Node is Moving Again

When the mobile node moves again, it acquires a new IP address on the new visiting network and sends the binding updates to its home agent and correspondent nodes, as described above. However, there is a time gap between when the mobile node has acquired its new IP address, named care-of-address 2, but the correspondent node has not yet received the binding update. During that gap, the correspondent nodes send the traffic to the previous IP address, care-of-address 1, which is no longer valid, thus packets are discarded, as shown in Figure 11.13.

This is the handoff problem: while moving, traffic is sent to the wrong place. Solutions to this issue are discussed in Section 11.10.1.

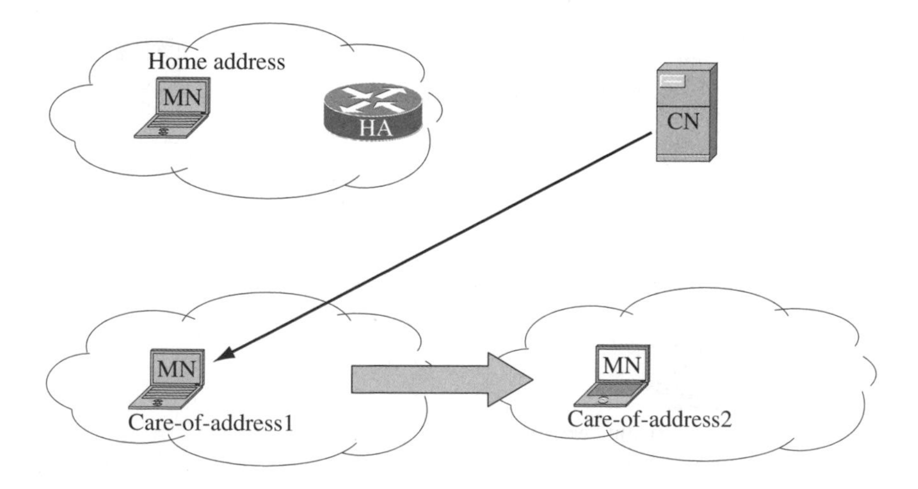

Figure 11.13 Mobile node moving again

11.7 Mobile Node Comes Back Home

When the mobile node returns to its home network, it does the same process as if moving again: it gets an address in the network and, compares it with its home address. When back home, the comparison is positive and the mobile node no longer uses the mobileIP functions to send traffic to its correspondent node. However, it will send the binding updates to its home agent and correspondent nodes to inform them of the new location.

11.8 Securing the Binding Update

MobileIP is essentially a host routing protocol, designed to modify the normal routing for a specific host. Since it changes the usual way of sending a datagram to a host, MobileIP could be an opportunity to create new passive or active attacks, denial of service or more generally security threats [Mankin, 2001].

The binding update, which informs a correspondent node of the new care-of-address, is implicitly authenticated by the correspondent node, verifying that it does not come from a rogue node. To authenticate the update, the correspondent node and the mobile node need to have a previously established security association. Since there is no global public-key infrastructure yet on the global Internet, the other way is to share a secret.

11.8.1 Security Association with Home Agent

The interaction of the mobile node with its home agent includes the re-routing of the traffic addressed to the mobile node to another location, its visiting address. A weak or non-existent security association between the mobile node and its home agent would enable a rogue node to send binding updates to the home agent redirecting traffic destined for the mobile node to a rogue location. For this reason, IPsec ESP [RFC3776] in transport mode is used between the mobile node and its home agent to secure the MobileIPv6 control messages such as the binding update.

11.8.2 Return Routability Procedure with Correspondent Nodes

After a mobile node informs its correspondents of its new IP address, the correspondents send the traffic directly to the new location. If a rogue node can send rogue IP addresses to the correspondent nodes, then all the traffic to the mobile node is redirected to a rogue location. To avoid these issues, a security association between the mobile node and each of its correspondent nodes, similar to the one used with its home agent, could be used. This was actually proposed in the early days of the mobileIPv6 definition. However, it is impractical because the correspondent nodes of mobile nodes are public Web servers, portals, e-mail servers, etc., all of which have no intrinsic security association possible with nodes. How can a well-known portal or search engine can maintain millions of security associations and manage and encrypt this level of traffic? A lightweight process done by the mobile node and the correspondent nodes through the home agent enables the correspondent node to establish that the mobile node really owns the home address it claims to, while not creating a costly non-scalable security association. This process, called the return routability procedure, is shown in Figure 11.14.

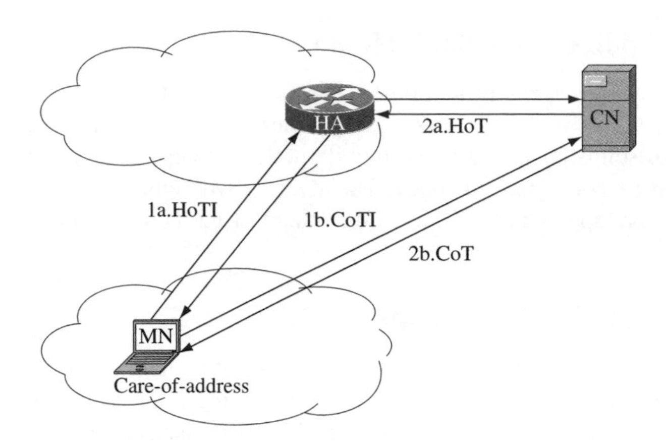

Figure 11.14 Return routability procedure

The goal of the return routability procedure is to create an association between the correspondent node and the mobile node where the correspondent node has a way to verify that the binding updates of the mobile node really come from the mobile node that owns the claimed home IP address.

The process involves the mobile node sending random generated keys sent through the home agent and replied to by the correspondent node. The cryptographic process creates a cache in the correspondent node to authenticate future binding updates.

Table 11.6 lists the steps in the procedure shown in Figure 11.14. Details of the cryptographic processing, cookies and keys are discussed in the MobileIPv6 specification [RFC3775].

Table 11.6 Return routability procedure steps

From → To	Message	Description and content
0		Correspondent node generates a random key once and nonces regularly.
1a MN → HA → CN	Home Test Init (HoTI)	The mobile node generates a home init cookie and sends it to the correspondent node through the home agent.
1b MN → CN	Care-of Test Init (CoTI)	The mobile node generates a care-of init cookie and sends it directly to the correspondent node without going through the home agent.
2a CN → HA → MN	Home Test (HoT)	The correspondent node replies to the home test init (HoTI) message by sending this message to the mobile node through the home agent. The message contains the home init cookie and a hash, called home keygen token, based on the correspondent node random key, a nonce and the home address.

2b	CN → MN	Care-of Test (CoT)	The correspondent node replies to the care-of test init (CoTI) message by sending this message directly to the mobile node without going through the home agent. The message contains the care-of init cookie and a hash, called care-of keygen token, based on the correspondent node random key, a nonce and the care-of-address.
3			Mobile node and correspondent node now share some secret used by the correspondent node to verify the origin of future binding updates.

Hashes based on the exchanged information are then used in the binding updates and acknowledgments between the mobile node and the correspondent node to verify the validity of these messages.

This return routability procedure requires the correspondent nodes to implement this algorithm and to maintain a cache of mobile nodes. IPv6 nodes are not required to implement this algorithm: in this case, the init messages sent by the mobile node in step 1 will be refused by the correspondent node not implementing the return routability procedure and an ICMP error message will be sent back to the mobile node. The mobile node may cache that information not to send further init messages to this correspondent node. When a correspondent node does not support the return routability procedure, the traffic from the correspondent node to the mobile node is sent to the home address and the home agent tunnels the traffic to the current care-of-address of the mobile node.

11.9 Correspondent Node is Not MobileIP Aware

MobileIPv6 has many optimizations relying on some minimal MobileIPv6 processing done by the correspondent node. During the course of the IETF standards work for IPv6 and MobileIP, the requirement that any IPv6 correspondent must implement some MobileIP functionality was dropped from required to optional. The previous sections explain MobileIPv6 with the assumption that the correspondent node is MobileIPv6 aware. This section describes the differences when a correspondent node is not MobileIPv6 aware.

A mobileIP node implementation manages its interaction using a cache which includes a list of correspondent nodes, home agents, states of bindings, etc. This cache is also used to handle the case when some of its correspondent nodes are not implementing MobileIP.

A correspondent node that is not implementing mobileIPv6 does not understand or use the mobileIP messaging, such as binding updates, home address destination option and return routability messages.

11.9.1 Mobile Node Registering to the Correspondent Node

Section 11.5.2 describes the process where the mobile node registers with the correspondent node by starting the return routability procedure, described in Section 11.8.2. Without prior knowledge of the correspondent node, the mobile node starts the return routability procedure by sending the Home test Init (HoTI) and Care-of Test Init (CoTI) messages to the correspondent node. When the correspondent node does not understand these mobileIP messages which are included in the mobility extension header, the correspondent node sends an ICMP error message[2] to the mobile node, stating its ignorance of such an extension header. Figure 11.15 shows the return routability procedure started by the mobile node and the ICMP error message reply by the correspondent node.

The mobile node then updates its cache identifying that this correspondent node is not implementing mobileIP and therefore should not receive binding updates and return routability messages in the future.

11.9.2 Mobile Node Sending Packets

Section 11.5.3 describes the process of the visiting mobile node sending packets to its correspondent nodes. If a correspondent node is not implementing MobileIPv6, its stack cannot process the Home-Address destination option sent by the mobile node. This home address destination option means the source address of the packet is the care-of-address of the mobile node. Recall that an application is not aware of mobileIP because it always receives the mobile node home address as the source address of the packet. The application receiving the care-of-address as the source address while the connection was established with the home address, simply discards the packet.

After the failure of the return routability procedure used before the mobile node sends packets directly to its correspondent nodes, the mobile node puts in its cache the status that this correspondent node does not implement mobileIPv6.

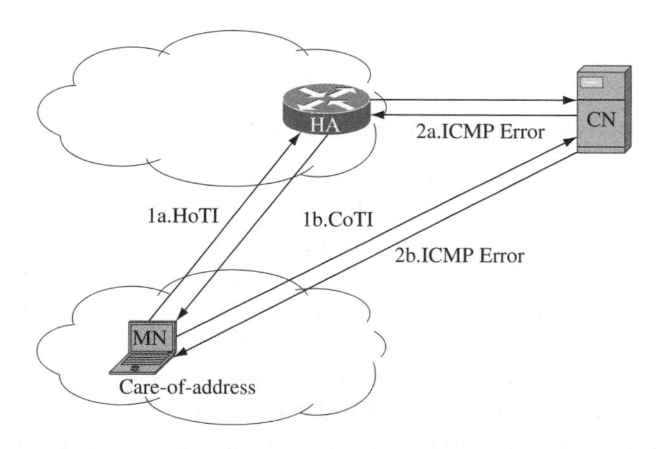

Figure 11.15 ICMP error message sent by the CN

[2] Parameter problem, code 1: see Table 7.3.

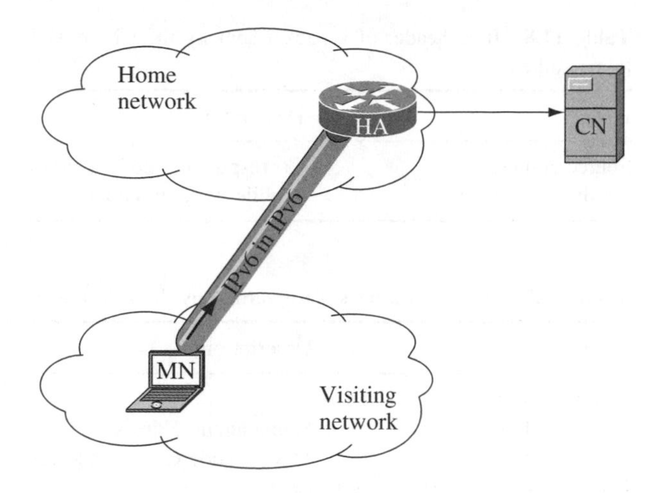

Figure 11.16 Mobile node sending packets through home agent

Instead of sending a packet directly to the correspondent node, which would reject it, the mobile node encapsulates its IPv6 packet to the correspondent node in another IPv6 packet to the home agent, building a tunnel back to the home agent, as shown in Figure 11.16.

The inner packet has the mobileIP home address as its source address. The home agent decapsulates the packet and forwards the inner packet to the correspondent node. Table 11.7 shows the headers of the packet sent by the mobile node to the correspondent node through the home agent. The outer packet to the home agent is protected by IPsec.

The correspondent node receives the decapsulated packet as if it was coming from the home address of the mobile node.

11.9.3 Correspondent Node Sending Packets to the Mobile Node

Section 11.5.4 describes how a correspondent node sends packets directly to the mobile node, by using the mobile node care-of-address as the destination address and by using the routing header. This is applicable when the correspondent node supports the basic mobileIPv6 functionality needed for correspondent nodes. If the correspondent node does not support mobileIPv6, then it has no knowledge of the mobile node care-of-address and only knows the mobile node by its home address. In this case, the correspondent node sends the packet to the home address, as shown in Table 11.8.

Table 11.7 Header of a packet sent by the MN to CN through HA

Field	Description
Outer Packet	
Source address	Mobile node care-of-address
Destination address	Home agent address
Inner Packet	
Source address	Mobile node home address
Destination address	Correspondent node address

Table 11.8 IPv6 header of a packet sent by the CN to MN home address

Field	Description
Source address	Correspondent node address
Destination address	Mobile node home address

Table 11.9 Header of a packet forwarded by the HA to MN

Field	Description
Outer Packet	
Source address	Home agent address
Destination address	Mobile node Care-of-address
IPsec extension header	ESP
Inner Packet	
Source address	Correspondent node address
Destination address	Mobile node home address

The home agent intercepts the packet and encapsulates it in an IPv6 packet to the mobile node care-of-address, as shown in Figure 11.11. The header of the encapsulated packet is shown in Table 11.9.

The home agent should protect the encapsulated packet with IPsec ESP because the home agent and the mobile node have a security association.

11.10 Advanced Features

MobileIP has many more features. As the intent of this chapter is to get an overview of MobileIPv6, this section discusses some issues and features without trying to be comprehensive. The interested reader should refer to a MobileIPv6 book e.g., [Soliman, 2004].

11.10.1 Fast Handoff

As discussed previously, after the mobile node has changed location but before its correspondent nodes have received the binding update, there is a gap when the traffic is sent to the previous location, as shown in Figure 11.13. This handoff situation is solved in different ways, depending on the requirements of the applications and the layer 2 infrastructure. Two ways to solve the problem are described here. At the time of writing this book, no solution has been yet standardized at the IETF for fast handoff.

One solution is where the mobile node requests to a router in the previous network to intercept all traffic going to the previous location of the mobile node, encapsulate the traffic and forward it to the new location, as shown in Figure 11.17.

A router in the previous network, named 'foreign agent', is offering the service on behalf of the mobile node. There are also some security issues of a router forwarding traffic of another node, without previous authentication.

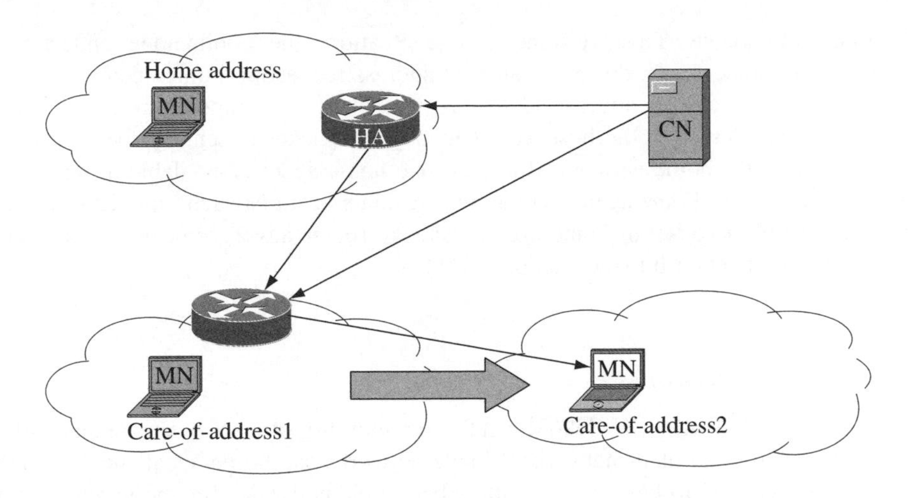

Figure 11.17 Relaying from previous location

Another way to solve this handoff problem is by having the home agent copy the packets to both locations, as shown in Figure 11.18.

This is especially useful when the mobile node is attached to multiple links at the same time. The key advantage is less delay for the mobile nodes to get the datagrams from the correspondent node.

11.10.2 Home Agent is Not Reachable

The home agent of a mobile node might become unavailable, for different reasons, such as hardware or software failure of the home agent device, failure of the link, loss of connectivity

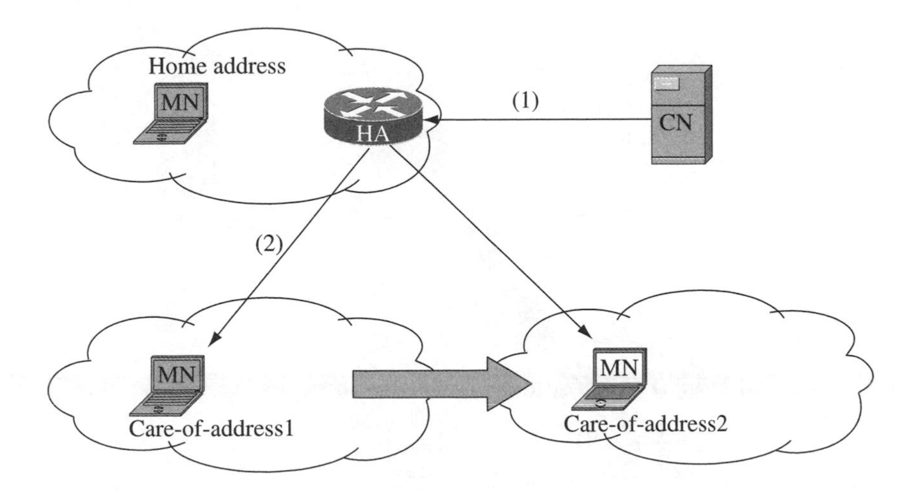

Figure 11.18 Home agent copying packets to both locations

to the home network, etc. To solve some of these situations, the mobile node sends a request message to the Home Agent discovery anycast address requesting home agents to respond. The Home agent discovery anycast address is formed by the home address prefix and a reserved anycast address in the host part. This anycast packet is sent by the mobile node and forwarded to the home network. If at least one home agent is available, it responds by sending the list of live home agents on the link. A home agent can send the list of all home agents on the link because all home agents listen to router advertisements to look for the home agent identification bit (see Section 11.11).

11.10.3 Mobile Networks

Network mobility (NEMO) [RFC3963] handles the mobility of a set of computers with one router connecting to the other networks. Mobile networks can be personal such as a laptop with Wifi and PDA or airplane, car or military battlefield networks, linking wireless devices together.

As shown in Figure 11.19, NEMO is implemented by building an IPv6-in-IPv6 tunnel between the mobile router (MR) and the home agent (HA). This tunnel carries IPv6 traffic from and to the mobile network (MN1, MN2), by encapsulating the original IPv6 packets into IPv6 packets carried between the home agent (HA) and the mobile router (MR).

The binding updates are used to exchange the information of the prefix(es) of the mobile network, the care-of-address and a mobile router flag to build the tunnel. The home agent then creates the tunnel interface and assigns all the mobile network prefixes to this tunnel interface. The outer IPv6 addresses of the tunnel are the mobile router care-of-address and the home agent address. Apart from the tunnel establishment and the routing of the mobile network prefixes, the mobile router is acting like a normal mobile node.

In the NEMO basic architecture, only the mobile router (MR) must support MobileIPv6 and NEMO. The mobile nodes (MN1, MN2) within the mobile network do not have to implement either MobileIPv6 or NEMO.

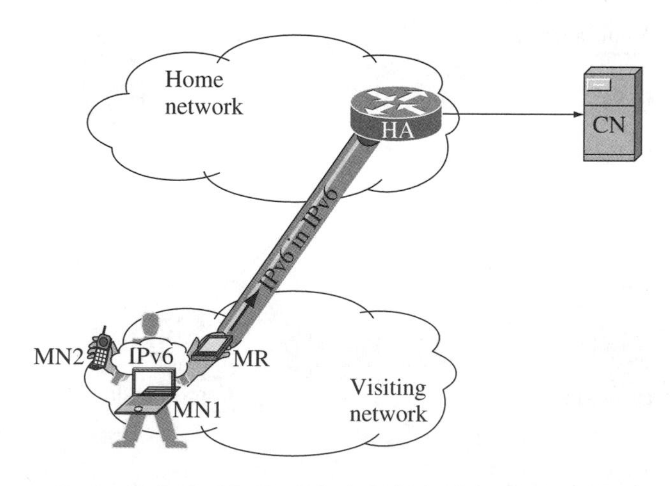

Figure 11.19 NEMO basic architecture

11.11 MobileIP Messaging

This section describes the MobileIPv6 messages and their fields. This is for reference to help the reader who wants to understand more of the details, while not being comprehensive. This section can be skipped.

11.11.1 Mobility Extension Header

MobileIPv6 uses a specific extension header, named the Mobility header, for the bindings messaging. It is inserted in the chain of extension headers after the IPv6 header. The mobility header fields are listed in Table 11.10.

Messages defined within the mobility header are listed in Table 11.11.

11.11.2 Home Address Destination Option Extension Header

The home address destination option message is an IPv6 extension header used by the mobile node, while visiting, to inform the recipient, its home agent and the correspondent nodes, of its home address.

11.11.3 Type 2 Routing Header

The type 2 routing header is an IPv6 routing header (see Section 9.2) used by the correspondent node to send packets directly to the mobile node, without going through the home agent. The type 2 mandates the use of only one hop source routing, while type 1 may include many hops for complex source routing. The address put by the correspondent node in the type 2 routing header is the home address of the mobile node, while the IPv6 destination address of the packet is the mobile node care-of-address. Such a packet is routed directly to the mobile node, but the mobile node internal IPv6 stack processes the routing header and replaces the destination address with the home address taken in the routing header. The applications inside the mobile node will then see the home address as the destination address, which makes them unaware of mobility in general, and of the direct path routing used. The type 2 routing header contains the fields listed in Table 11.12.

11.11.4 ICMP Messages

MobileIPv6 uses the ICMP messages listed in Table 11.13.

Table 11.10 Mobility header fields

Name	Size	Description
Payload protocol	8 bits	Identifies the next header after this one. Should be set to none (59). Reserved for future use.
Header length	8 bits	Length of the mobility header in units of 8 octets
Mobility Header Type	8 bits	Identifies the type of message.
Checksum	16 bits	Checksum of the mobility header.
Data	variable	Content of the message.

Table 11.11 Mobility header message types

Name	Mobility header type	Description
Binding Refresh Request	0	Sent by correspondent nodes to mobile node to request a binding refresh.
Home Test Init (HoTI)	1	Sent by mobile node to initiate a return routability procedure. The message contains the Home Init Cookie, a 64 bit random token generated by the mobile node.
Care-of Test Init (CoTI)	2	Sent by the mobile node to initiate a return routability procedure. The message contains the Care-of Init cookie, a 64 bit random token generated by the mobile node.
Home Test (HoT)	3	Sent by the correspondent node to the mobile node. The message contains the Home Init Cookie and the Home Keygen Token.
Care-of Test (CoT)	4	Sent by the correspondent node to the mobile node. The message contains the Care-of Init cookie and the Care-of Keygen Token.
Binding Update (BU)	5	Sent by the mobile node to the home agent and correspondent nodes. The message contains some flags, a sequence number and the lifetime of the binding. A BU may contain the following options: binding authorization data, nonce indices and alternate care-of-address.
Binding Acknowledgment (BA)	6	Sent by the home agent and the correspondent node to acknowledge the reception of the binding update. The message contains some flags, a status number stating the acceptance/refusal of the BU, a sequence number (copied from the BU sequence number) and the lifetime of the binding granted by the sending node. A BA may contain the following options: binding authorization data and binding refresh advice.
Binding error	7	Sent by the correspondent node to signal an error. The message contains a status number and the home address destination received from the mobile node.

Table 11.12 Type 2 routing header fields

Name	Size	Value	Description
Next header	8 bits		Identifies the next header.
Header extension length	8 bits	2	Length of the mobility header in units of 8 octets.
Routing type	8 bits	2	Identifies the type of message.
Segments left	8 bits		Number of hops. This is a degenerated case of the more general routing header case.
Home address	128 bits		Home address of the mobile node.

Table 11.13 ICMP Mobility message types

Name	Description
Home agent address discovery request	Sent by the mobile node to the home agent's anycast address on its own home subnet prefix. This messages enables the mobile node to find the home agents.
Home agent address discovery reply	Sent by the home agents responding to the mobile node home agent address discovery request. It contains the home agent's address.
Prefix solicitation	Sent by the mobile node to its home agent. It requests the prefix used at home.
Prefix advertisement	Sent by the home agent to advertise the home prefix to the mobile node, in response to its prefix solicitation request.

11.11.5 Neighbor Discovery

MobileIPv6 makes some changes to the neighbor discovery protocol, as listed in Table 11.14.

Additionally, to help the mobile node detect their movement, home agents should send router advertisements as fast as every 50 milliseconds. Compared to minutes in the non-mobile case, this rate of sending router advertisement means a burden for all nodes on

Table 11.14 MobileIPv6 changes to router advertisements

Name	Description
Home agent flag	A flag in the router advertisement is used by the home agents to identify themselves as home agents.
Router address flag in the prefix advertisement	A flag is added to prefix information to identify the home agent global address. This is used by the dynamic home agent discovery.
Advertisement interval	This value is sent by the home agents to tell the nodes the frequency of sending the router advertisements. This helps the mobile node to calculate the maximum time to wait until it could safely decide it is out of reach of the home agent, thus away from its home network.
Home agent information	Optional information sent by the home agent to the mobile node, to influence its use of the home address and the home agent lifetime.

that network to process these packets. Care should be taken to use these values only on subnets which have mobile nodes requiring this rate for fast movement detection.

11.12 Deployment Considerations

Considerations on managing security in networks using MobileIPv6 and on using IPv4 and IPv6 mobility are discussed in this section.

11.12.1 Enterprise Network with Mobile Nodes on Most Links

An enterprise has a large network where mobile nodes have their primary point of attachment on most links. A typical case is where laptops are used by a large number of the users, they connect on their own office LAN, and there are many LANs. One then needs to provide the home agent capability on all these LANs. Providing that capability on most or all LANs might be seen by some network managers as a complex management task. Moreover, it also means that all these home agents must be reachable from the outside, e.g. the Internet, when the mobile nodes are visiting networks. This implies opening the firewall to let mobileIP and IPv6 packets going through the internal network up to the home agents, without too much control. This will be difficult to filter correctly and might be an important security risk that some organizations might not want to take.

One other way to solve this issue is to dedicate a /64 prefix for a virtual LAN, assigned to a home agent which is located at the border of the network and reachable from the outside and the inside, as shown in Figure 11.20.

The mobile node gets its prefix (3ffe:b00:0:1::/64) and home address on the home virtual LAN (A) by static configuration. When the mobile node is inside on its office LAN (B), it is not located on the LAN which advertises its home prefix. The mobile node is considered away from its home LAN and will behave as if it is visiting a network. When the mobile node goes outside (C), it is visiting too. So the mobile node is always 'away' from home,

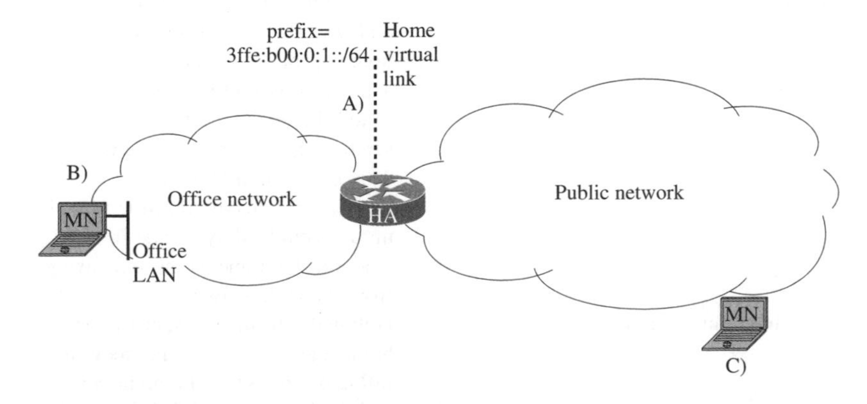

Figure 11.20 Home virtual link

even when it is connected to its office LAN. This approach has its own drawbacks, but solves the security issues discussed above.

11.12.2 Security Considerations

A network that attaches foreign mobile nodes (e.g. visiting) needs to open its firewall to let many special types of IPv6 packets pass through. These types of packets are:

- IPv6 in IPv6 packets;
- IPv6 packets with mobility header;
- IPv6 packets with home address destination option;
- ICMPv6 mobility packets;
- IPv6 packets with routing header.

These types of packets might present a security risk that an organization might not want to take. In this case, the visiting mobile nodes will not be able to use the mobileIPv6 functionality on that visiting network.

11.12.3 IP Version Centric

MobileIPv4 is for IPv4 mobility; MobileIPv6 is for IPv6 mobility. A mobile node implementing MobileIPv6 means the visiting network must have IPv6 availability. The TSP tunnel broker, a migration tool discussed in Section 16.2.9, provides mobility for both versions of the protocol, as discussed below.

11.12.4 Ubiquitous IP

Since MobileIP is IP version centric, a mobile node that is visiting IPv4-only networks receiving a global IPv4 address, IPv4-only networks receiving a private address, and IPv6-only networks, do not have a single mobility solution for all these cases. However, the TSP tunnel broker, as discussed in Section 16.2.9, does provide ubiquitous IP, by enabling both IP protocols to be seamlessly available all the time.

Figure 11.21 shows such a case where the mobile network is moving from IPv4 to IPv4 with NAT to IPv6 and the TSP tunnel broker provides connectivity for both protocols in all the cases. The TSP tunnel broker is similar to the home agent functionality.

11.13 Configuring Mobility

This section describes how to configure MobileIPv6 and TSP tunnel broker on the platforms.

11.13.1 FreeBSD

FreeBSD has an implementation of MobileIPv6 as part of the Kame IPv6 stack and supports TSP tunnel broker.

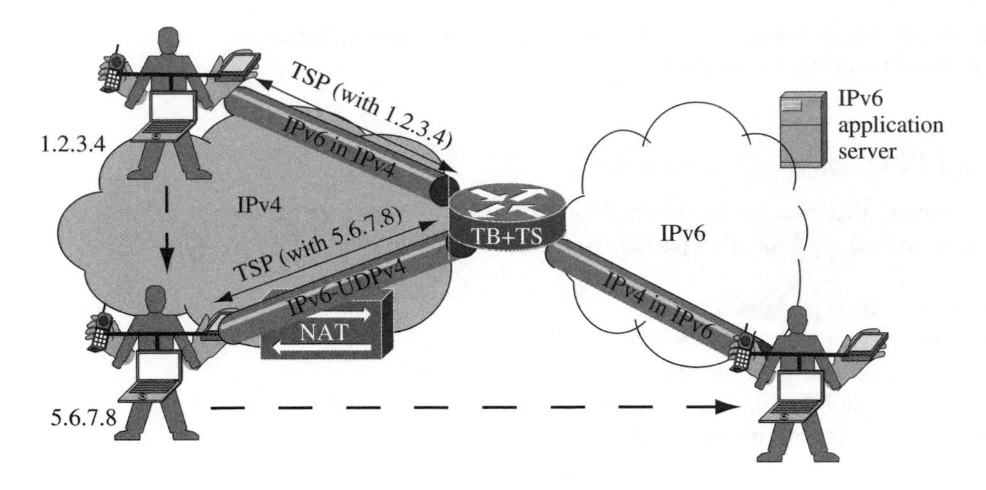

Figure 11.21 TSP tunnel broker as mobility server

11.13.1.1 MobileIPv6

By default, the FreeBSD kernel is not compiled for MobileIPv6. Table 11.15 lists the lines to be added to the kernel configuration file.

Modify the /etc/rc.conf to enable MobileIPv6 and configure interfaces appropriately.

To configure a home agent, use the following lines.

```
# cat /etc/rc.conf
ipv6_mobile_enable="YES"
ipv6_mobile_nodetype="home_agent"
ipv6_mobile_home_prefixes="3ffe:b00:0:1::/64"
ipv6_mobile_home_link="fxp0"
```

Since a home agent is a router, forwarding (ipv6_gateway_enable) must also be configured, as discussed in Section 10.2.

As a home agent, the router advertisement interval, configured in the /etc/rtadvd.conf file, is much shorter than the normal case. The following lines define the interval between 4 and 6 seconds.

```
# cat /etc/rtadvd.conf
fxp0:
  :maxinterval#6:mininterval#4:
```

Table 11.15 FreeBSD kernel config for MobileIPv6

Kernel config line	Description
`options MIP6`	To enable MobileIPv6 processing. Must be present in any MobileIPv6 functionality: mobile node, correspondent node and home agent.
`options MIP6_MOBILE_NODE`	To enable the MobileIPv6 mobile node functionality.
`options MIP6_HOME_AGENT`	To enable the MobileIPv6 home agent functionality.

To configure a mobile node, use the following lines.

```
# cat /etc/rc.conf
ipv6_mobile_enable="YES"
ipv6_mobile_nodetype="mobile_node"
ipv6_mobile_home_prefixes="3ffe:b00:0:1::/64"
```

The Kame MobileIPv6 toolkit contains commands as listed in Table 11.16.

IPsec security associations must be set between the home agent and the mobile node. These configuration statements are not discussed here.

11.13.1.2 TSP tunnel broker client

The TSP tunnel broker client configuration on FreeBSD is discussed in Section 16.5.2.3. Specific mobility configuration parameters are discussed in Section 11.13.5.

11.13.2 Linux

Linux supports MobileIPv6 and TSP tunnel broker.

11.13.2.1 MobileIPv6

MobileIPv6 on Linux is developed by HUT [Helsinki University of Technology]. It is configured by the `network-mip6.conf` file. Table 11.17 lists the key statements of this configuration file.

A mobile node configuration file is shown below.

```
# cat /etc/network-mip6.conf
FUNCTIONALITY=mn
HOMEDEV=mip6mnha1
HOMEADDRESS=3ffe:b00:0:1::/128
HOMEAGENT=3ffe:b00:0:1::1/64
```

Table 11.16 FreeBSD MobileIPv6 commands

Command	Description
`had`	Dynamic home agent discovery protocol daemon
`mip6control`	A control program for MobileIPv6 functions
`mip6stat`	Statistics of packets related to MobileIPv6
`mip6makeconfig.sh`	Tool to generate IPsec template configurations
`mip6seccontrol.sh`	To set IPsec parameters

Table 11.17 Linux network-mip6.conf MobileIPv6 configuration file statements

Statement name	Possible values	Description
FUNCTIONALITY	cn	Define the node as correspondent node
	ha	Define the node as home agent
	mn	Define the node as mobile node
DEBUGLEVEL	*Number*	Define the level of debug messages
HOMEDEV	*Device name*	Define the interface name on which the home address is set
HOMEADDRESS	*IPv6 address/128*	Sets the home address of the mobile node
HOMEAGENT	*IPv6 address/64*	Sets the home agent address for the mobile node

The `mipdiag` command is used for troubleshooting.

11.13.2.2 TSP tunnel broker client

The TSP tunnel broker client configuration on Linux is discussed in Section 16.5.3.4. Specific mobility configuration parameters are discussed in Section 11.13.5.

11.13.3 Solaris

The TSP tunnel broker client configuration on Solaris is discussed in Section 16.5.4.3. Specific mobility configuration parameters are discussed in Section 11.13.5.

11.13.4 Windows

The TSP tunnel broker client configuration on Windows is discussed in Section 16.5.5.5. Specific mobility configuration parameters are discussed in Section 11.13.5.

11.13.5 Hexago

The Hexago TSP tunnel broker client can be used to provide pretty good mobility, as discussed in Section 11.12.4. Some parameter default values should be changed for fast handoff. The `tspc.conf` file parameters listed in Table 11.18 should be modified by decreasing the values to provide a faster handoff.

Table 11.18 TSP client tspc.conf statements for mobility

Variable	Default value	Possible values	Description
`retry_delay`	0	*number*	When a TSP connection fails, the number of seconds to wait before retrying to connect to the tunnel broker.
`keepalive`	`yes`	`yes no`	When set to 'yes', the client sends keepalives to keep the tunnel active. It is especially useful to have the NAT keep its mapping to maintain a sustainable tunnel over UDP.
`keepalive_interval`	`30`	*number*	This interval in seconds should be smaller than the NAT mapping timeout for UDP. The client sends a keepalive to the broker every keepalive_interval seconds. The broker may force a higher value than the client wants, given an expected load on the broker by a provider.

11.14 Summary

MobileIP is used to sustain the IP communications while changing the IP address. MobileIPv6 is more deployable and optimized than MobileIPv4, such as direct communication between the mobile node and the correspondent node.

11.15 References

Helsinki University of Technology, MIPL Mobile IPv6 for Linux, http://www.mipl.mediapoli.com.

[IEEE80211] IEEE 802.11, 1999 Edition (ISO/IEC 8802-11: 1999) IEEE Standards for Information Technology – Telecommunications and Information Exchange between Systems – Local and Metropolitan Area Network – Specific Requirements – Part 11: Wireless LAN Medium Access Control (MAC) and Physical Layer (PHY) Specifications.

Mankin, A., 'Threat Models introduced by Mobile IPv6 and Requirements for Security in Mobile IPv6', Internet-Draft draft-team-mobileip-mipv6-sec-reqts-00, July 2001.

[RFC3344] Perkins, C., 'IP Mobility Support for IPv4', RFC 3344, August 2002.

[RFC3775] Johnson, D., Perkins, C. and Arkko, J., 'Mobility Support in IPv6', RFC 3775, June 2004.

[RFC3776] Arkko, J., Devarapalli, V. and Dupont, F., 'Using IPsec to Protect Mobile IPv6 Signaling Between Mobile Nodes and Home Agents', RFC 3776, June 2004.

[RFC3963] Devarapalli, V., Wakikawa, R., Petrescu, A. and Thubert, P., 'Network Mobility (NEMO) Basic Support Protocol', RFC 3963, January 2005.

Soliman, H., *MobileIPv6*, Addison-Wesley, April 2004.

12

Wireless IP

IP is now being used over wireless links, making it even more ubiquitous than before. Since the IP protocol suite, including TCP, is link-layer independent, it was not designed for the specificities of wireless links. The initial deployment of IP over satellite links [RFC2488] showed many inefficiencies. This chapter discusses the challenges and modifications to the IP protocol suite to accommodate the new wireless networks, such as Wifi [IEEE80211] and 3G [3GPP, 3GPP2].

12.1 Characteristics of Wireless Links

A recent trend in wireless and cellular networks, like Bluetooth and 3G, is to use the IP protocol as the layer 3 protocol, which gives ubiquitous, seamless and universal connectivity. However, IP is not optimized for wireless links.

Wireless links exhibit some specific characteristics:

- Limited bandwidth: the radio spectrum is a rare and costly resource where the bandwidth available on radio links is usually much smaller than wired links. 2.5G networks have 10–20 kilobit per second (kbps) uplink and 10–40 kbps downlink. 3G networks have 64 kbps uplink and 384 kbps downlink for their minimal grade service.
- Higher error rate: because of the expensive radio spectrum, cellular networks design trade offs tend to prefer higher error rates to minimize the use of the spectrum. The typical bit-error rate (BER) is $1e^{-3}$ and can be up to $1e^{-2}$ in severe conditions [RFC3095].
- Latency and long link round trip time: typical 100–200 ms [RFC3095] round trip time in radio networks is pretty long compared to wired networks. The RTT could be as high as one second.

Migrating to IPv6: A Practical Guide to Implementing IPv6 in Mobile and Fixed Networks Marc Blanchet

- Changes in link characteristics: a mobile device moves, thus changes its distance to the base station. The wireless protocol then adapts and changes the link characteristics, such as decreased bandwidth. The resulting bandwidth-delay product in 3G networks is similar to satellite systems [RFC2488], known to have sub-optimal TCP performance.

These characteristics, and the dynamic nature of wireless links, result in sub-optimal TCP performance on wireless links because of the following issues:

- IP header length costs too much bandwidth.
- TCP flow control makes non-optimal assumptions about network congestion.

These two issues are discussed in the following sections.

12.2 Header Compression over Limited Bandwidth Link Layers

Most real-time protocols for voice and video over IP use the Real Time protocol (RTP) [RFC3550] over UDP. A RTP datagram over UDP over IPv4 has a combined header size of 40 octets: 20 octets for the IPv4 header, 8 octets for the UDP header and 12 octets for the RTP header. Over IPv6, the combined header size is 60 octets, since the IPv6 header is 20 octets longer than the IPv4 one and UDP and RTP header sizes are the same. If the voice traffic rate is 50 datagrams per second, then the IPv6, UDP and RTP combined headers (60 octets) consume $50 \times 60 \times 8 = 24\,\text{kbit/second}$, while the voice traffic with GSM encoding uses 13 kbit/second. Considering that a typical voice over IP datagram might have 15–20 octets of data, the header tax is very high, especially considering the limited bandwidth.

Header compression [RFC2507] is needed for the following reasons (note that the compression of the data portion of the voice or video datagrams is already handled by the application protocols):

- improved response time;
- bandwidth usage efficiency for small packets;
- lower latency for smaller packets;
- decreased overhead of header;
- reduced packet loss.

A header compression protocol [RFC2508] compresses the RTP-UDP-IPv4 headers to 4 octets with an UDP checksum, 2 octets without, using a modified version of the known Van Jacobson compression scheme [RFC1144]. However, it only works well for low link round trip time and suffers a bad domino effect of losing many datagrams if an error happens in one compressed datagram [RFC3096]. A more optimized protocol, the Robust Header Compression (RoHC)[RFC3095] is designed for wireless links and used in WCDMA, EDGE and CDMA2000 wireless networks.

This section describes the overall technique of header compression without going into the details[RFC2508, RFC3095].

Header compression [RFC3095] is accomplished by removing the redundancy of header information for a specific flow of datagrams on a point-to-point link. Most wireless network technologies behave as point-to-point links. A specific flow can be characterized by the same source and destination addresses, the same source and destination ports, and possibly some session context fields, such as the Synchronization Source (SSRC) field in RTP which identifies the RTP session. Multiple datagrams of the same flow contain the same header fields, are redundant, and therefore can be eliminated, effectively compressing the header for the second and subsequent datagrams. Additionally, when some header fields are changed but the new values are predictable, the compression algorithm may predict them.

A context id is used to identify the context of a specific compression state. A ROHC packet with a specific link-layer number is used for the control part of the compression. A simplified view of the compression process is as follows:

1. Send the full header with a new context identifier for the first datagram.
2. Send a compressed header with the context identifier for the subsequent datagrams.
3. When a compressed header field changes, either resend the full header or change to a new context identifier.

Tables 12.1, 12.2 and 12.3 list the characteristics of the fields for IPv4, IPv6 and UDP headers, respectively.

Table 12.1 IPv4 header compression

Name	Use for compression
Version	Static
Header Length	Eliminated
TOS	Dynamic
Total Length	Eliminated
Identification	Dynamic
Flags	Dynamic
Fragment Offset	Eliminated
Time to Live	Dynamic
Protocol	Static
Checksum	Eliminated
Source Address	Static
Destination Address	Static

Table 12.2 IPv6 header compression

Name	Use for compression
Version	Static
Traffic Class	Dynamic
Flow Label	Static
Payload Length	Eliminated
Next Header	Static
Hop Limit	Dynamic
Source Address	Static
Destination Address	Static

Table 12.3 UDP header compression

Name	Use for compression
Source Port	Static
Destination Port	Static
Checksum	Dynamic
Length	Eliminated

Some fields are static on the point-to-point link, some are dynamic and some are not relevant and can be eliminated. For example, a typical RTP/UDP/IPv4 packet has static fields of 25 octets and dynamic fields of 15 octets. A typical RTP/UDP/IPv6 packet has static fields of 49 octets and dynamic fields of 11 octets. Therefore, the IPv6 compressed headers are smaller than their IPv4 counterpart because fewer octets are dynamic.

12.3 TCP Behavior over Wireless

TCP design assumes a small bit error rate, typical to wired networks. The flow control is based on network congestion, not on the bit error rate of the link layer. TCP slows down the data rate in case of failures, so each independent TCP connection, decreasing its data rate, contributes to decreasing the congestion.

On a wireless link, the bit error rate is much higher and usually causes the majority of packet loss. TCP behavior on these links reacts as if there was network congestion, by decreasing the data rate and backing off, which creates poor performance, even if there is no network congestion.

When the TCP connection is traversing a wireless and a wired network as shown in Figure 12.1, the bit error rate on the wireless link and the congestion control on the wired network compete and double the problem in the TCP back off algorithm.

Many solutions have been proposed to solve this issue. One method is to intercept the TCP connection with a TCP proxy at the border of the wireless link, implementing a modified algorithm to take care of the high bit error rate on the wireless link, as shown in Figure 12.2.

The TCP proxy is seen as the source of the connection by the destination. However, this method breaks the end-to-end connection model of TCP/IP and therefore is not recommended. Moreover, there is no IETF standard for this mechanism.

The combination of a larger congestion window size [RFC3390], Selective Acknowledgments (SACK) [RFC2018] and ECN, as described in Section 19.2.1, and specific TCP configuration values are recommended [RFC3481] for better TCP performance over wireless links.

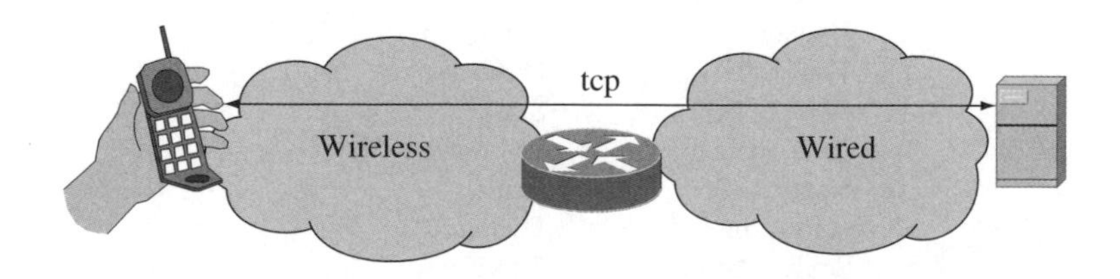

Figure 12.1 TCP over wireless and wired networks

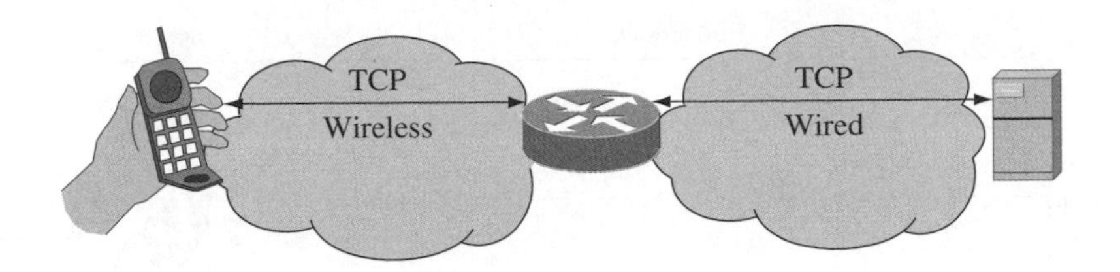

Figure 12.2 Split TCP

12.4 3GPP

The third generation partnership program, 3GPP, is the standard [3GPP] for GSM/GPRS-based 3G networks. Only a limited IP view is described here.

From an IP view [RFC3314], the User Equipment (UE) such as a cell phone has a point-to-point link to the Gateway GPRS Support Node (GGSN), its next hop router, over the General Packet Radio Services (GPRS) air network, as shown in Figure 12.3.

The point-to-point link uses the Packet Data Protocol (PDP) within a negotiated context in the underlying infrastructure. The point-to-point link is assigned a /64 prefix.

Figure 12.4 shows the protocol stack of the GPRS infrastructure which includes the Universal Terrestrial Radio Access Network (UTRAN) and the Serving GPRS Support Node (SGSN), delivering the layers below IP. The whole 3GPP protocol stack and interactions are not discussed here.[1]

The User Equipment, such as the cell phone, establishes a PDP context with the GPRS infrastructure, creating a point-to-point link to the GGSN. The GGSN is an IP router forwarding IP packets to the Internet.

A Terminal Equipment (TE), such as a laptop, is connected over PPP to the Mobile Termination (MT), the cell phone, as shown in Figure 12.5.

This PPP connection creates another PDP context on the GPRS air network for carrying the IP traffic of the TE through the network. Figure 12.6 shows a modified version of the Figure 12.4 protocol stack, where the laptop (TE) IP traffic is shown. Note that the cell phone can still send IP traffic, but will use its own PDP context (shown in Figure 12.6).

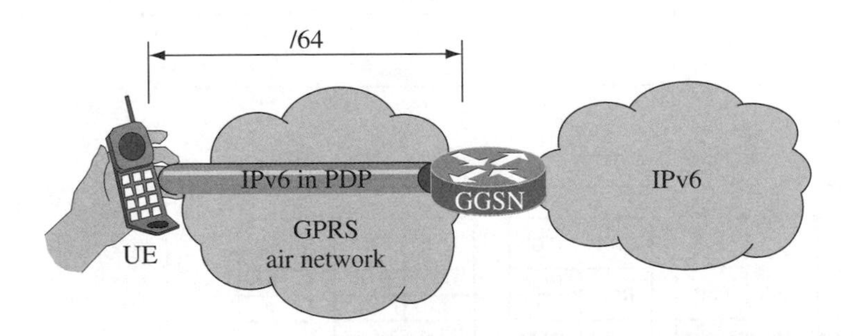

Figure 12.3 IPv6 link on 3GPP

[1] From Figure 12.4, note that the UTRAN, SGSN and GGSN modules use UDP/IP between themselves to carry the above IP packets.

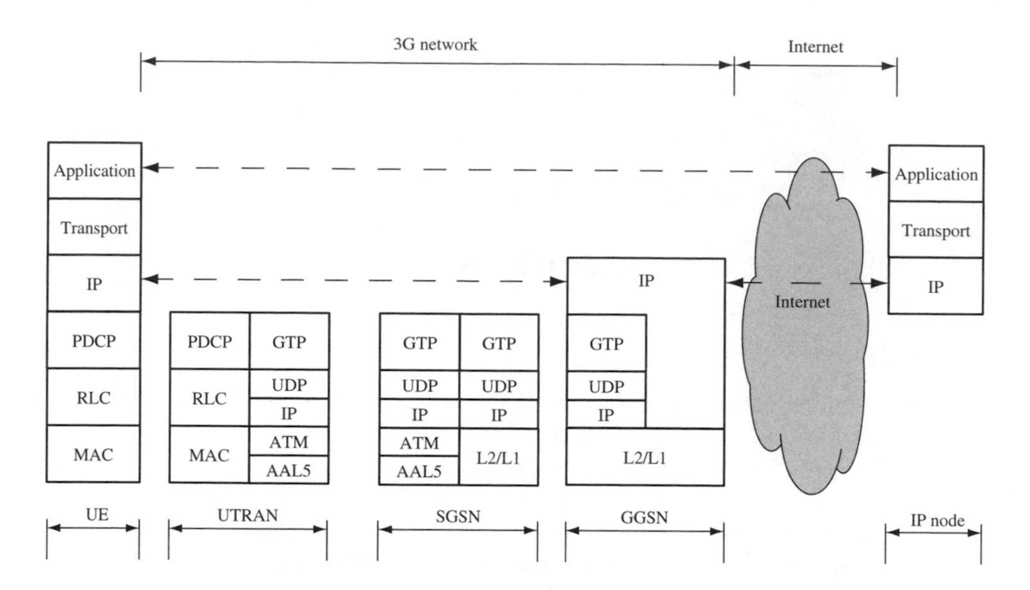

Figure 12.4 3GPP protocol stack

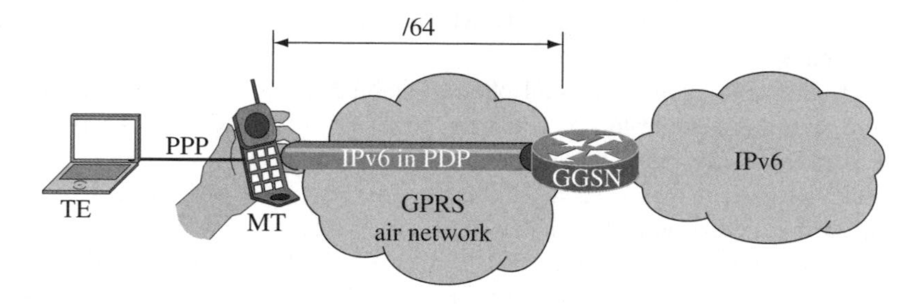

Figure 12.5 IPv6 link on 3GPP with TE

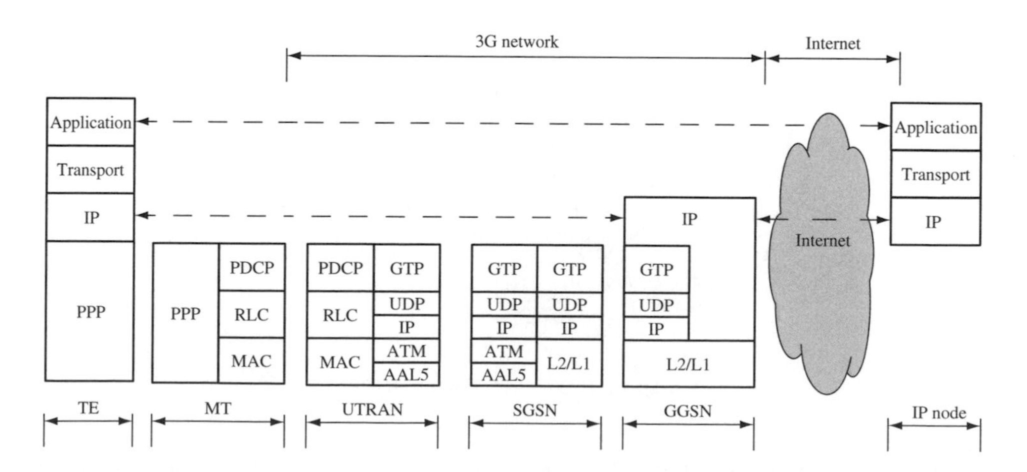

Figure 12.6 3GPP protocol stack with TE

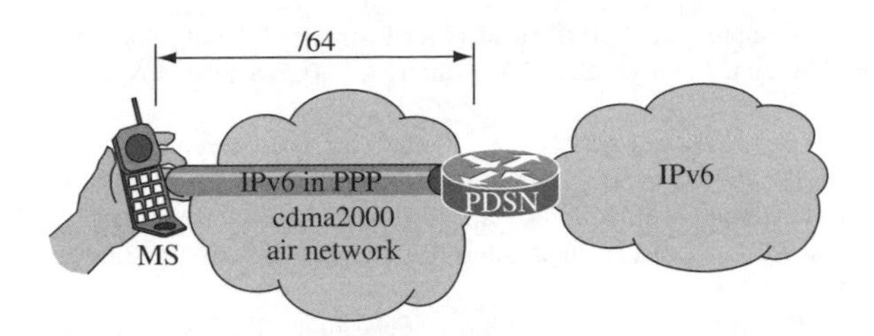

Figure 12.7 3GPP2 IPv6 in PPP

12.5 3GPP2

The third generation partnership program 2, 3GPP2, is the standard [3GPP2] for CDMA2000-based 3G networks. It has a different architecture than 3GPP. Only a limited IP view is described here.

As shown in Figure 12.7, the Mobile Station (MS) uses PPP with its access server, the Packet Data Serving node (PDSN), to access the IPv6 network, for the IPv6 connectivity.[2] This PPP channel uses the radio network as its link-layer.

As discussed in Section 6.7, PPP provides a channel for IPv4 and a channel for IPv6 on the same PPP link if both ends support both IP protocols. Negotiation of the IPv6 link-local addresses is first done inside PPP. When link-local addresses are assigned on each side, the PDSN sends router advertisements on the link, advertising a /64 prefix assigned to this PPP link from a PDSN IPv6 prefix pool. The mobile station then autoconfigures using the received /64 prefix.

The 3GPP2 mobile station can use IP header compression with the PDSN as discussed in Section 12.2.

12.6 Summary

Wireless links introduce new link-layer characteristics such as limited bandwidth, higher error rate and latency. The IP protocol stack is modified to accommodate these changes, by compressing the IP and transport headers and by optimizing TCP. The 3G wireless networks, 3GPP and 3GPP2, introduce new architectures where the handset gets a /64 prefix over the point-to-point link. These large subjects have had whole books devoted to them: this chapter provides only a limited perspective on these issues.

12.7 References

[3GPP] The Third Generation Partnership Project (3GPP), http://www.3gpp.org

[3GPP2] The Third Generation Partnership Project 2 (3GPP2), http://www.3gpp2.org

[IEEE80211] IEEE 802.11, 1999 Edition (ISO/IEC 8802-11: 1999) IEEE Standards for Information Technology – Telecommunications and Information Exchange between Systems – Local and Metropolitan Area Network – Specific Requirements – Part 11: Wireless LAN Medium Access Control (MAC) and Physical Layer (PHY) Specifications.

---

[2] However, the 3GPP2 roadmap is based on IPv4 and MobileIPv4. IPv6 appeared more recently and there is no description of the use of MobileIPv6.

[RFC1144] Jacobson, V., 'Compressing TCP/IP Headers for Low-speed Serial Links', RFC 1144, February 1990.
[RFC2018] Mathis, M., Mahdavi, J., Floyd, S. and Romanow, A., 'TCP Selective Acknowledgment Options', RFC 2018, October 1996.
[RFC2488] Allman, M., Glover, D. and Sanchez, L., 'Enhancing TCP Over Satellite Channels using Standard Mechanisms', BCP 28, RFC 2488, January 1999.
[RFC2507] Degermark, M., Nordgren, B. and Pink, S., 'IP Header Compression', RFC 2507, February 1999.
[RFC2508] Casner, S. and Jacobson, V., 'Compressing IP/UDP/RTP Headers for Low-Speed Serial Links', RFC 2508, February 1999.
[RFC3095] Bormann, C., Burmeister, C., Degermark, M., Fukushima, H., Hannu, H., Jonsson, L-E., Hakenberg, R., Koren, T., Le, K., Liu, Z., Martensson, A., Miyazaki, A., Svanbro, K., Wiebke, T., Yoshimura, T. and Zheng, H., 'RObust Header Compression (ROHC): Framework and Four Profiles: RTP, UDP, ESP, and Uncompressed', RFC 3095, July 2001.
[RFC3096] Degermark, M., 'Requirements for Robust IP/UDP/RTP Header Compression', RFC 3096, July 2001.
[RFC3314] Wasserman, M., 'Recommendations for IPv6 in Third Generation Partnership Project (3GPP) Standards', RFC 3314, September 2002.
[RFC3390] Allman, M., Floyd, S. and C. Partridge, 'Increasing TCP's Initial Window', RFC 3390, October 2002.
[RFC3481] Inamura, H., Montenegro, G., Ludwig, R., Gurtov, A. and Khafizov, F., 'TCP over Second (2.5G) and Third (3G) Generation Wireless Networks', BCP 71, RFC 3481, February 2003.
[RFC3550] Schulzrinne, H., Casner, S., Frederick, R. and V. Jacobson, 'RTP: A Transport Protocol for Real-Time Applications', RFC 3550, July 2003.

12.8 Further Reading

[RFC3168] Ramakrishnan, K., Floyd, S. and Black, D., 'The Addition of Explicit Congestion Notification (ECN) to IP', RFC 3168, September 2001.

13

Security

Computers, networks, operating systems, applications, users, policies and protocols are components forming a complex system, where they interact with each other. Security should be applied to each component separately, as well as to all components as a system. Perfect security does not exist: security is about managing risks.

For the purposes of this book, the discussion will be focused on the specifics of IPv6 security, and should not be considered by any means to be comprehensive on network security.

Figure 13.1 identifies very simply where some of the security solutions discussed in this chapter fit in the networking stack.

IP security (IPsec) works at the IP layer, SSL/TLS at the session layer, SSH at the session/application layer and Secure Neighbor Discovery (ND) at the frontier between IP and the link-layer. These topics as well as others, such as temporary addresses and network address translation, are discussed in this chapter. Is IPv6 more secure than IPv4? Keep reading until the end of the chapter. . . .

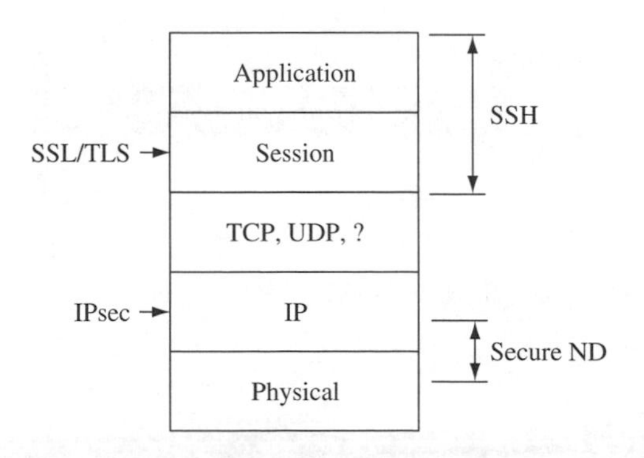

Figure 13.1 Layers and security

13.1 IP Security (IPsec)

When the IETF was working on requirements for IPv6, security at the IP level was identified as a key requirement [RFC1752]: IP security was to be mandatory for IPv6. This work started a separate working group on IP security (IPsec). IPsec [RFC2401] is defined and designed to be used for both IPv4 and IPv6. This is good for the security of currently deployed IPv4 networks, but bad for IPv6 since it decreases the traction towards IPv6 based on security needs.

IPsec protects the IP layer communications. It is usually handled in the kernel of operating systems, and applications may or may not be aware of the IP security established between two nodes.

IPsec has two encapsulation modes: transport and tunnel. When two IPsec peers need to secure their communication, they first start by establishing a security association. Then they use either or both the Authenticated Header or the Encapsulating Security Payload services.

13.1.1 IPsec Transport and Tunnel Modes

As noted above, IPsec has two modes of encapsulation: transport and tunnel. Figure 13.2 shows nodes N1 and N2 establishing a secure IP connection using IPsec, which illustrates the transport mode: end-to-end security.

Figure 13.3 shows the tunnel mode, where N1 establishes a secure IP connection with the Virtual Private Network (VPN) server, not with N2. N1 encapsulates its traffic to N2 in a secure IP connection towards the VPN server. The VPN server decapsulates the secured traffic and forwards it to N2, which receives the traffic unsecured from N1. A secure tunnel is built from N1 to the VPN server.

Modes can be combined in many different ways. Figure 13.4 shows the tunnel mode used between N1 and the VPN server, and the transport mode used between N1 and N2. Every packet going from N1 will have two IPsec encapsulations: one to the VPN server and one for N2.

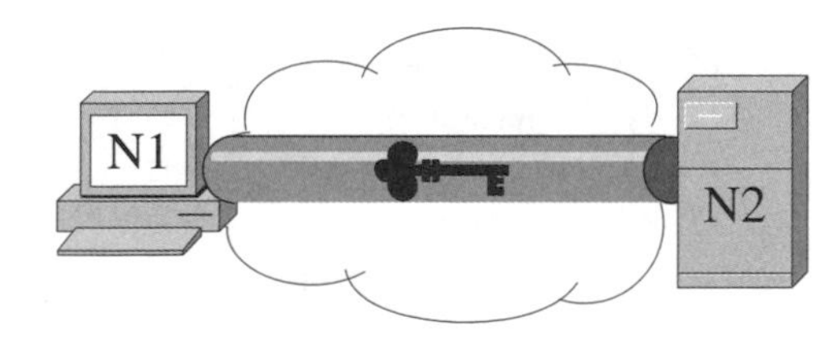

Figure 13.2 IPsec transport mode between two end nodes

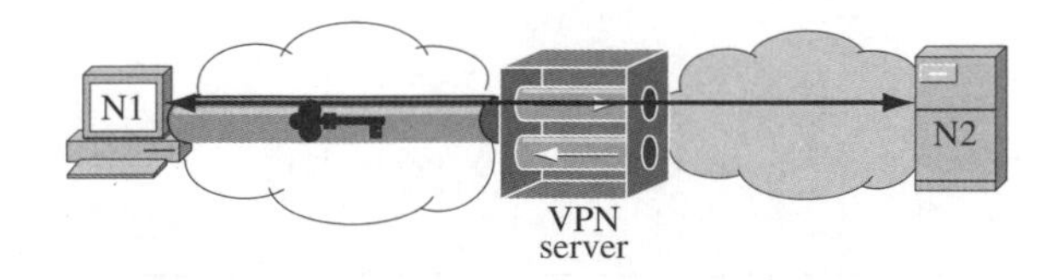

Figure 13.3 IPsec tunnel mode from a node to a VPN server

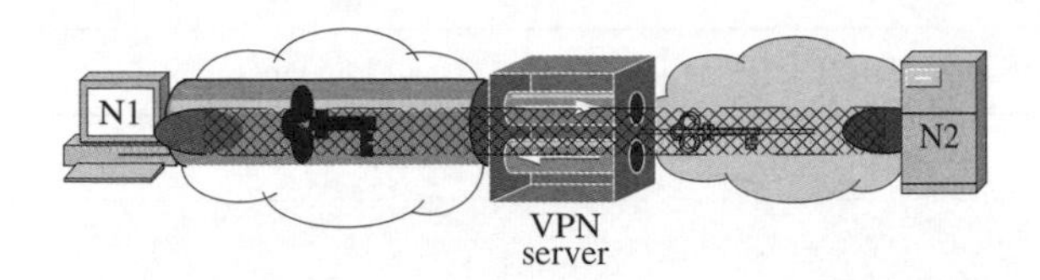

Figure 13.4 IPsec tunnel and transport modes used simultaneously

> **Important to remember:**
>
> IPsec has two modes of encapsulation: transport (between two end nodes) and tunnel (between gateways).

13.1.2 Establishing a Security Association

For each pair of IPsec peers, a security association is established prior to sending secured packets. This involves keys exchanges and cryptographic algorithms negotiations[1] using the Internet Key Exchange (IKE) protocol [RFC2409]. IKE uses the IP addresses to identify the keys used in the exchange, which makes IKE aware of the IP protocol and addresses. Since IKE has nothing special regarding IPv6, it is not detailed here.

13.1.3 AH Header

IPsec has two security services: Authenticated Header and Encapsulating Security Payload. The IPsec Authentication Header (AH) [RFC2402] provides:

- integrity of the whole packet;
- authentication of the source;
- replay protection.

Figure 13.5 shows the IPv6 packet, where the gray fields are protected by AH and the white are not. Fields in the header that change values in a unpredictable way in the path cannot be protected. The traffic class field could be changed by diffserv routers. The flow label could be changed between different QoS domains. The hop limit is decremented by each router in the path.

All other fields, such as the source and destination addresses, the extension headers, the transport header and the packet data are protected by AH.

The protection is made by computing a cryptographic checksum over the protected fields. This checksum as well as the security association information is stored in the AH extension header, which is identified by the value 51 in the Next Header field.

Figure 13.6 shows a packet sent from N1 to N2 as in Figure 13.2 using the transport mode and AH. A security association is established between N1 and N2 prior to sending this packet. The AH extension header is inserted between the IP header and the transport (in this example: transport = 6 = TCP).

---

[1] Static keying can also be used, but has some problems and is not discussed here.

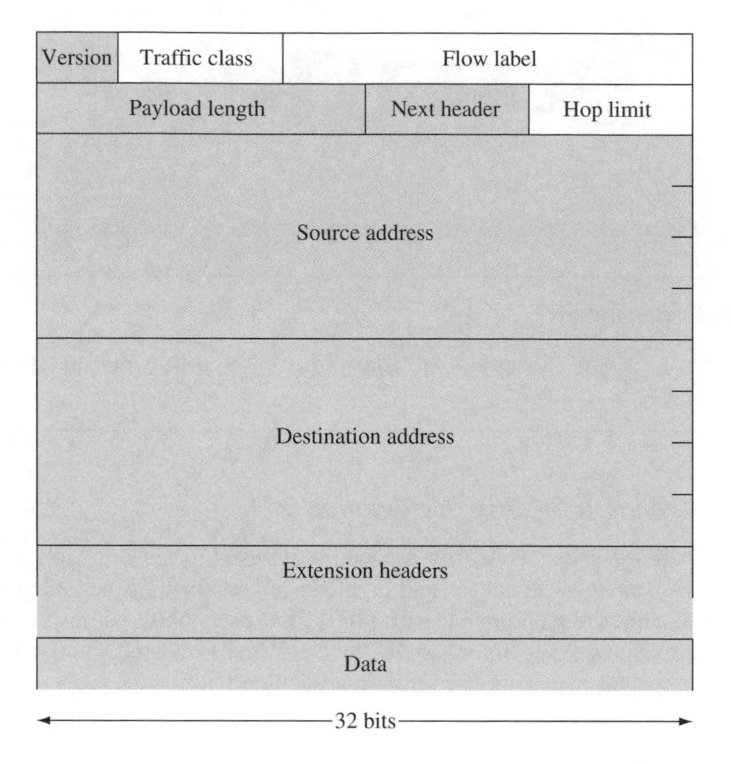

Figure 13.5 IPsec AH protected fields in the IP packet

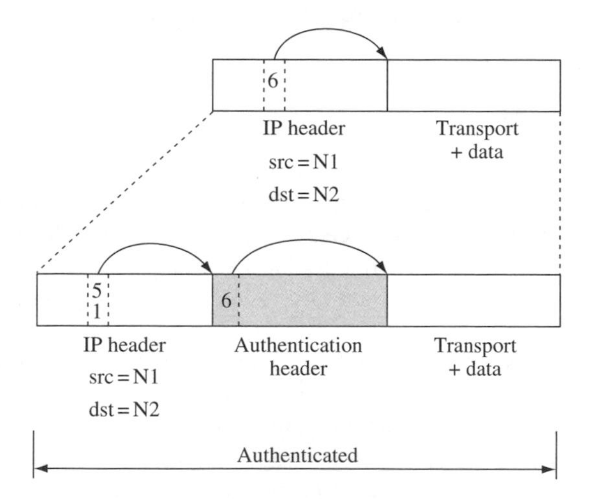

Figure 13.6 IPsec AH in transport mode

The whole new packet is protected by the AH cryptographic checksum: except for the traffic class, the flow label and the hop limit fields, a change in any other field, such as the source address, the transport port or the data is detected by the other peer (in this example, N2, the destination).

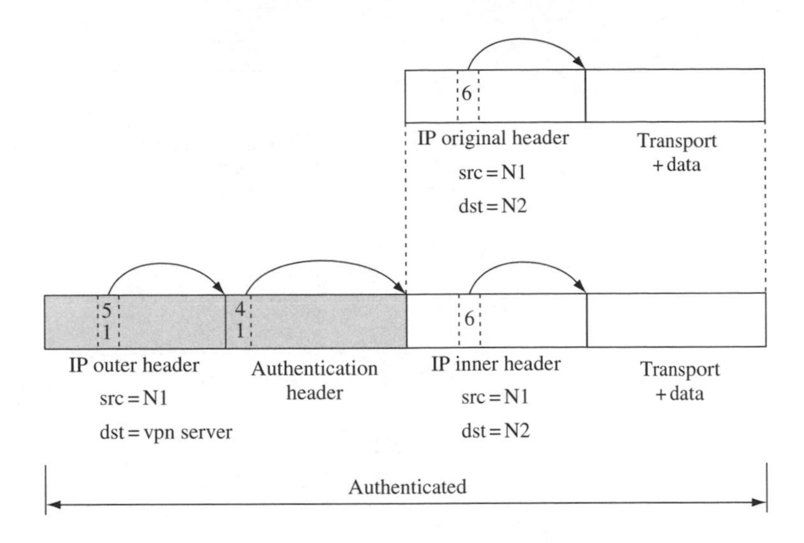

Figure 13.7 IPsec AH in tunnel mode

Figure 13.7 shows a packet sent from N1 to N2 as in Figure 13.3 using the tunnel mode to the VPN server with AH. A security association is established between N1 and the VPN server prior to sending this packet. The difference with the transport mode is that the original packet with the N2 destination is inserted completely in a new packet with the destination of the VPN server, together with the AH header.

The whole packet is protected. When the VPN server receives the packet, it verifies its integrity and then decapsulates the packet to forward the original packet, now unprotected, to N2.

Figure 13.8 shows a packet sent from N1 to N2 as in Figure 13.4 using the tunnel mode to the VPN server with AH, combined with the transport mode to N2 with AH. A security association is established between N1 and the VPN server, and between N1 and N2 prior to sending this packet. A AH header for the transport mode to N2 is first added to the original packet and then this modified packet is inserted in a new packet with the VPN server as destination and the AH header for the tunnel mode.

Figure 13.9 shows the AH header itself, which is inserted as an extension header in the previous examples.

The AH header contains the fields listed in Table 13.1.

In summary, IPsec AH provides integrity of the whole packet, authentication of the source and replay protection, using the AH extension header. It does not provide confidentiality. It can be used in the transport or in the tunnel mode.

Important to remember:

IPsec AH provides integrity of the whole packet, authentication of the source and replay protection. It does not provide confidentiality.

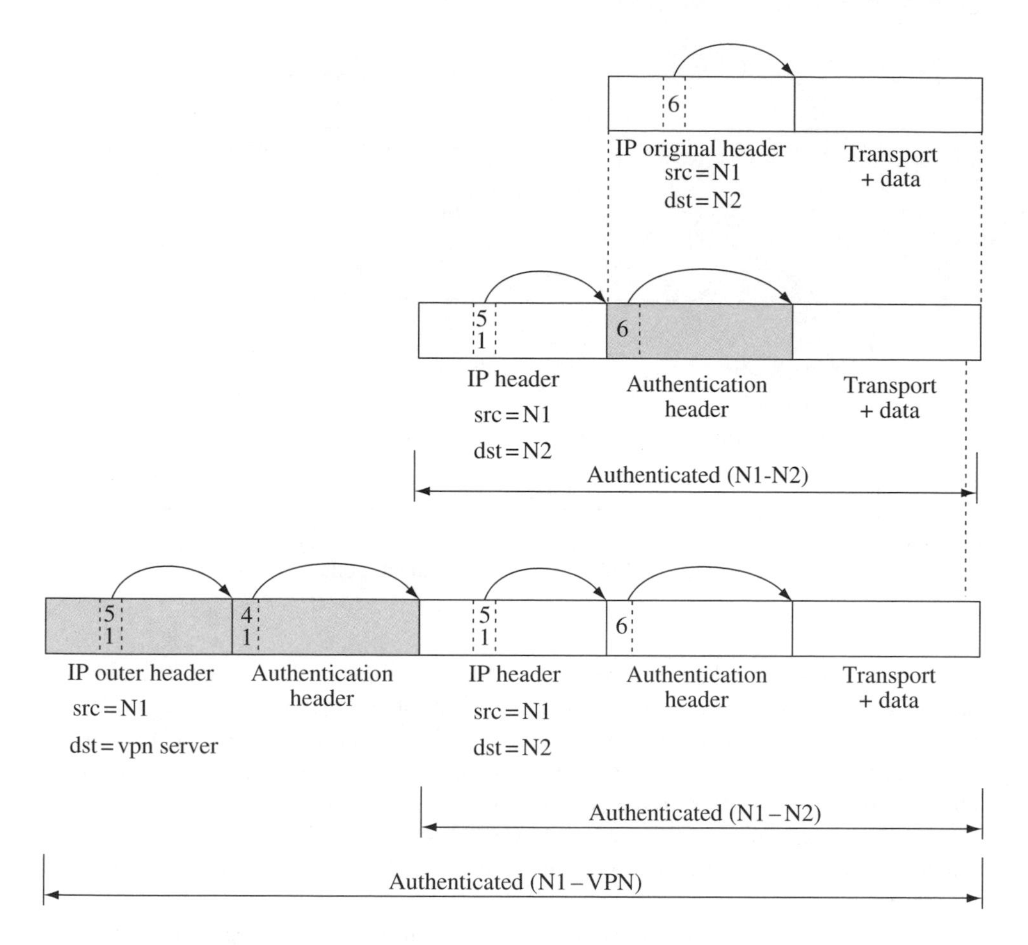

Figure 13.8 IPsec AH headers in tunnel and transport modes

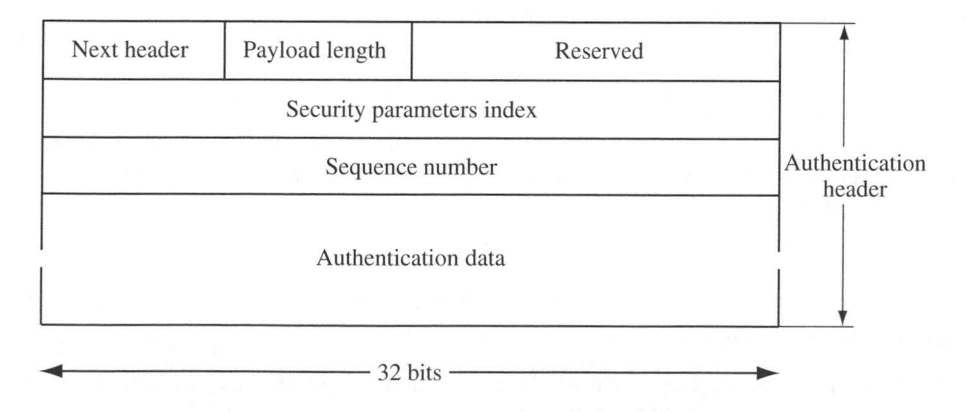

Figure 13.9 IPsec AH header

Table 13.1 IPsec AH header fields

AH field name	Description
Next Header	Identifies the payload following the AH header. If no other extension headers are present, in transport mode, it is usually the transport protocol number; in tunnel mode, it is usually the IP protocol number (41 for IPv6) since the payload is the original whole packet starting with the IP header.
Payload Length	Specifies the length of the AH header only.
Security Parameters Index (SPI)	With the destination address, the SPI uniquely identifies a security association with the other peer.
Sequence Number	Specifies a incremental counter to avoid the replay attacks. (A previously used number cannot be reused, so an attacker cannot reinsert an old packet.)
Authentication Data	The cryptographic checksum of the whole packet using a cryptographic algorithm identified by the security association.

13.1.4 ESP Header

IPsec Encapsulating Security Payload (ESP)[RFC2406] provides:

- confidentiality;
- integrity of the inner packet;
- authentication of the source;
- anti-replay protection.

Compared to AH, ESP adds confidentiality (encryption), but has a more limited integrity protection, covering only the payload.

An ESP extension header is identified by the value of 50 in the Next Header field.

Figure 13.10 shows the IPv6 packet, where the gray fields are protected by ESP and the white are not. None of the IPv6 header is protected. Only the content of the payload is protected, including the transport protocol header. If the ESP confidentiality service is used, then the payload is encrypted.

Figure 13.11 shows a packet sent from N1 to N2 as in Figure 13.2 using the transport mode and ESP. A security association is established between N1 and N2 prior to sending this packet. The ESP extension header is inserted between the IP header and the transport (in this example: transport = 6 = TCP).

The payload of the original packet, including the transport protocol header, is encrypted as shown by the crossed fields.

Figure 13.12 shows a packet sent from N1 to N2 as in Figure 13.3 using the tunnel mode to the VPN server with ESP. A security association is established between N1 and the VPN server prior to sending this packet. The difference with the transport mode is that the original packet with the N2 destination is inserted completely in a new packet with the destination of the VPN server, together with the ESP header.

The original packet is completely encrypted in the payload of the outer packet. But the outer header is not protected.

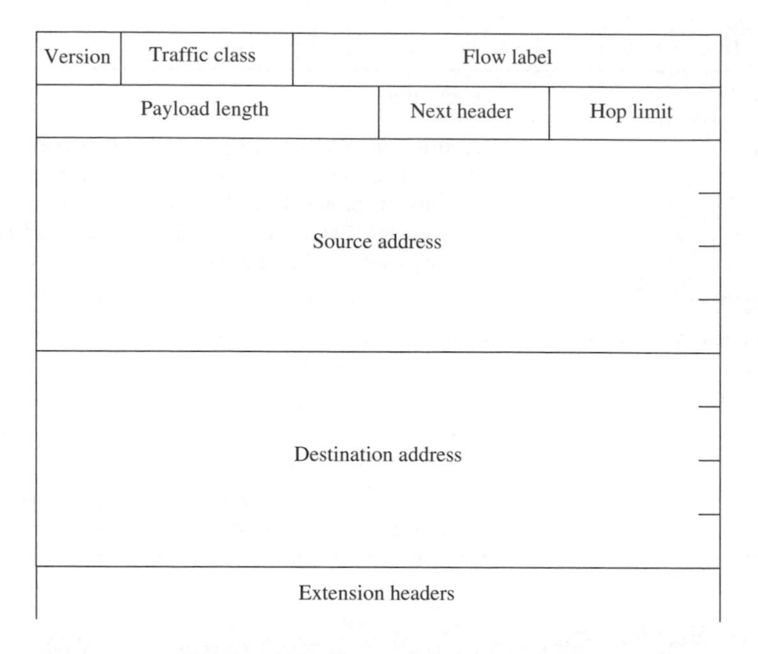

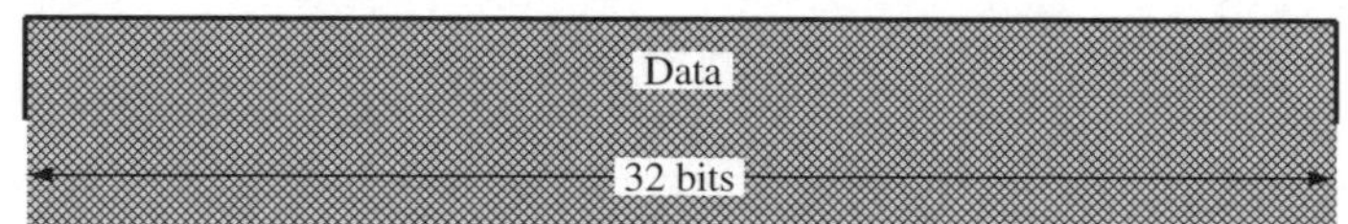

Figure 13.10 ESP payload protection

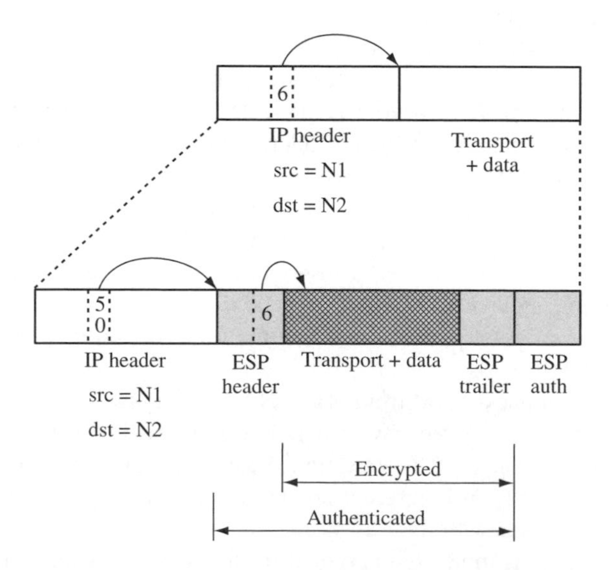

Figure 13.11 IPsec ESP header in transport mode

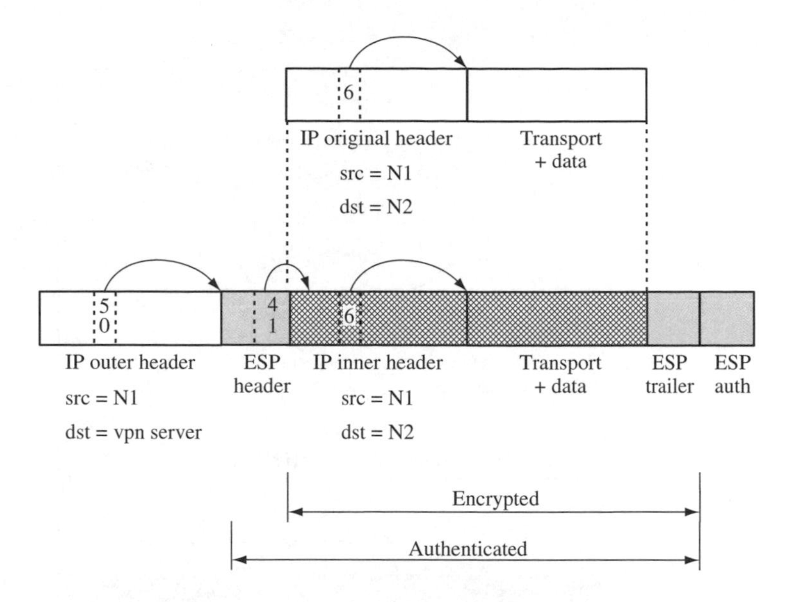

Figure 13.12 IPsec ESP header in tunnel mode

Figure 13.13 shows a packet sent from N1 to N2 as in Figure 13.4 using the tunnel mode to the VPN server with ESP, combined with the transport mode to N2 with AH. This shows the combination of modes (transport and tunnel) and security services (AH and ESP). A security association is established between N1 and the VPN server, and between N1 and N2 prior to sending this packet. A AH header for the transport mode to N2 is first added to the original packet and then this modified packet is inserted into a new packet with the VPN server as destination and the ESP header for the tunnel mode.

Figure 13.14 shows the ESP header itself, which is inserted as an extension header in the previous examples.

The ESP header contains the fields listed in Table 13.2.

In summary, IPsec ESP provides confidentiality and integrity of the inner packet, authentication of the source and replay protection, using the ESP extension header. It can be used in the transport or in the tunnel mode. ESP and AH can be combined together.

Important to remember:

IPsec ESP provides confidentiality and integrity of the inner packet, authentication of the source and replay protection. It does not protect the whole packet.

13.1.5 IPsec and IPv4 NAPT

IPsec is an example of a protocol which uses the IP address of the peers in its intrinsics. The IP addresses are protected by the AH header, as shown in Figure 13.5. IKE uses IP addresses to identify keys, as discussed in Section 13.1.2. When a NAPT device is in the path between two IPsec peers, IPsec does not work [RFC3715]. ESP does not work for TCP

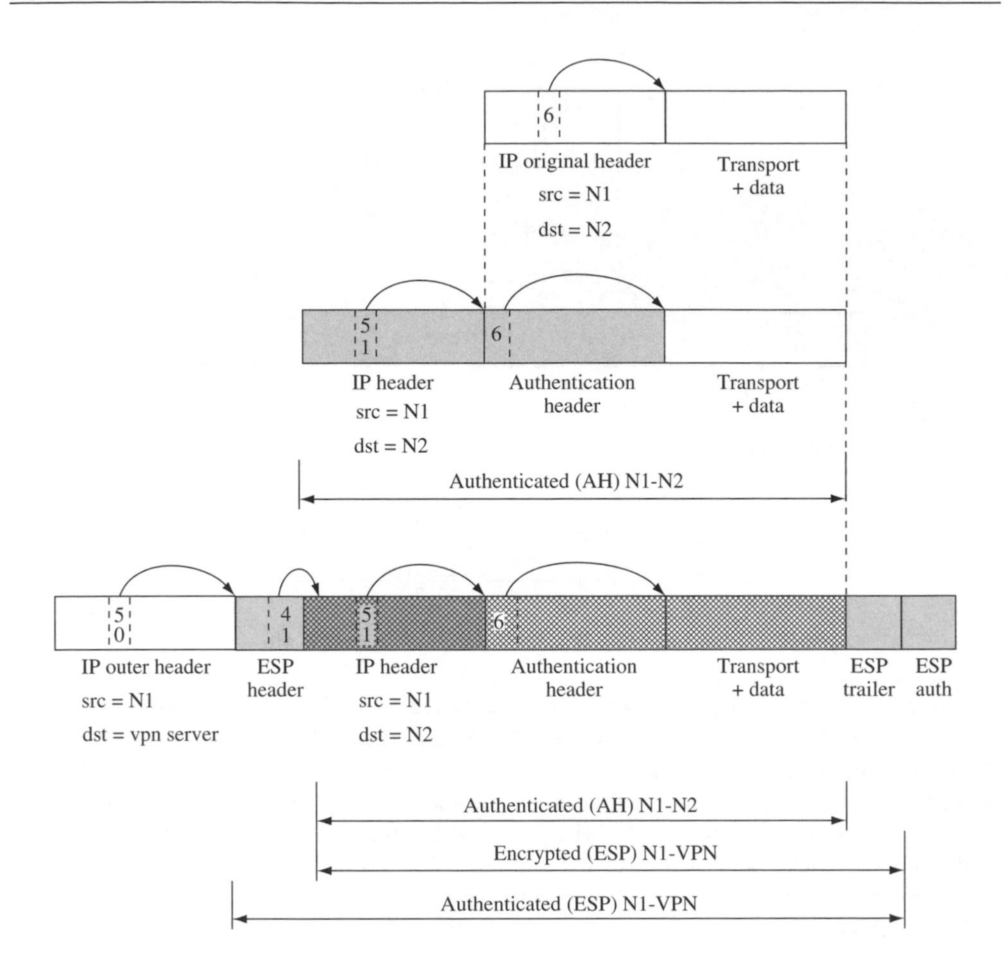

Figure 13.13 IPsec AH-transport and ESP tunnel combined modes

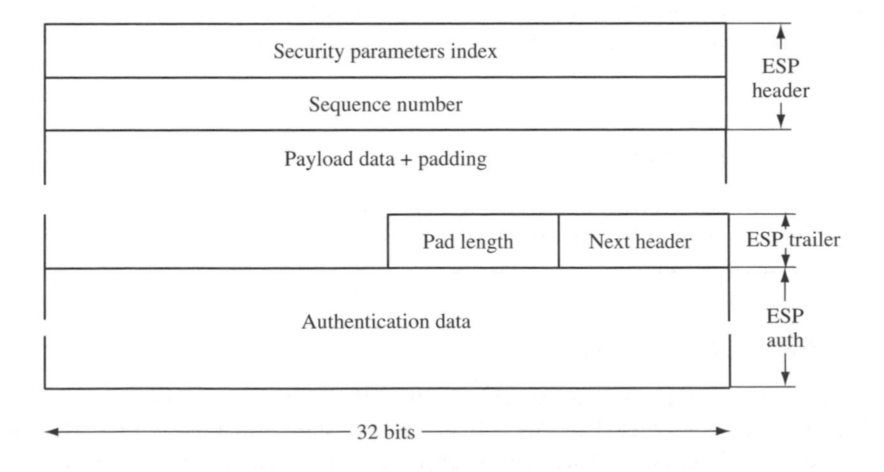

Figure 13.14 IPsec ESP header

Table 13.2 IPsec ESP header fields

ESP field name	Description
Security Parameters Index (SPI)	With the destination address, the SPI uniquely identifies a security association with the other peer.
Sequence Number	Specifies a incremental counter to avoid the replay attacks. (A previously used number cannot be reused, so an attacker cannot reinsert an old packet.)
Payload Data + Padding	Embed the encrypted payload of the original packet. Padding is used to synchronize with specific requested sizes of cryptoalgorithms, such as the block size of block ciphers.
Next Header	Identifies the payload following the ESP header. If no other extension headers are present, in transport mode, it is usually the transport protocol number; in tunnel mode, it is usually the IP protocol number (41 for IPv6) since the payload is the original whole packet starting with the IP header.
Authentication Data	The cryptographic checksum of the whole packet using a cryptographic algorithm identified by the security association.

and UDP transport because of a mismatch in the transport checksum. Since the addresses, port numbers and checksum are changed by the NAPT device in the path, the integrity check done by other peers upon receiving the modified packet, inevitably fails. The IETF is working on modifying IPsec to make it more NAPT friendly for IPv4, but not for all cases and at the important cost of decreasing the protection, which defeats the purpose of IPsec. Moreover, it will take years to upgrade NAPT devices and IPsec clients. Instead, IPv6 enables IPsec deployment without these problems.

13.1.6 IPsec and IPv6

Given the security threats on networks and the need for security at the IP layer, IPsec became popular. However, the IPv4 NAPT causes a lot of headaches for users and support staff, so IPsec is often not used because of the limitations of IPv4 NAPT. Moreover, either proprietary extensions or modifications to IPsec will take years to deploy and a lot of effort.

IPsec is mandatory for all IPv6 nodes and IPv6 does not have addressing limitations that required NAPT for IPv4. IPsec is easier to deploy over IPv6 and will help secure the Internet, the enterprise networks and the home. Moreover, IPsec, being mandatory, enables developers to design new applications and protocols using the IPsec infrastructure in place for IPv6. Therefore, the application and protocols will have more security embedded from the start.

13.2 Secure Shell (SSH)

Secure Shell (SSH)[Ylonen *et al.*, 2003] is a secure remote login and file transfer protocol and application, replacing telnet and ftp. It is widely used by network and system administrators to manage servers and network devices remotely. SSH is IPv6-ready. IPv6 configuration of SSH is discussed in Section 22.4.

13.3 Filtering and Firewalls

Security policies should be applied for IPv6 traffic for the same reasons as for IPv4. However, there are some specific considerations for IPv6. The book web site contains examples of security policies.

13.3.1 ICMP Filtering

IPv6 nodes rely on Path MTU discovery (see Section 3.6.2) to get the maximum performance for a connection. Path MTU discovery relies on the ICMP packet too big error messages (see Section 7.1.1.2) that are sent from routers on the network to the sending node. Some sites implement their security policies in IPv4 by preventing ICMP messages from passing through their firewall. For IPv6, specific ICMP messages, such as ICMP packet too big, should pass through to enable nodes to work properly.

13.3.2 MobileIPv6

The IPv4 source routing feature is known to be used for some attacks. For that reason, most IPv4 security policies block source routing in firewalls. IPv6 source routing is implemented by the routing header (see Section 9.2). Without further thinking, an IPv6 security policy would block the routing header too. However, the routing header is used by MobileIP (see Chapter 11); blocking the routing header at the border of a network would disable MobileIP from working properly with mobile nodes. Packet tunneling, other extension headers and specific ICMP messages are used by the MobileIPv6 protocol. An IPv6 security policy should take into account the use of MobileIP.

13.3.3 Network Address/Port Translation

NAPT hides the identity of the IPv4 nodes that are behind it. If the security policy is applied after the NAPT processing, there is no easy way to permit and deny traffic patterns from nodes behind the NAPT, since they are not recognizable from the IP source address and ports. Moreover, troubleshooting of NAPTed traffic is very difficult since a synchronization of the events (that is, the problem being troubleshot and the translation function) must be done. If there are multiple NAPTs in the path, troubleshooting becomes very difficult or impossible to do. Without NAPT in IPv6, applying the security policies and troubleshooting is much simpler, since the packet header fields are not changed en route.

13.4 Temporary Addresses

In IP, when a node moves to another network, it receives a new address. The change of address makes it difficult for some servers to track the node when it moves. This is why Web sites often use HTTP cookies to keep track of users, whatever IP address they might have.

When an IPv6 autoconfigured node moves, the prefix changes but the interface identifier remains the same, since the node uses the autoconfiguration mechanism on the new network and the identifier is based on the same hardware address. This raised some concerns on privacy, since the IPv6 node could be easily tracked based on the rightmost 64 bits of its address that do not change even when it moves.

The solution is to use temporary interface identifiers, changed 'often', to achieve the given goal of privacy. The interface identifier is no more based on the hardware address, but instead is randomized and changed periodically, implementing temporary addresses. The specification [RFC3041] defines a way to ensure the randomness of the changes to minimize the possibility of guessing the new addresses based on the older ones, which would defeat the purpose.

When a new temporary address is generated, the old one goes to the deprecated state (see Section 5.3), so it can still receive packets from already established connections but cannot be used for new connections.

However, temporary addresses make debugging of networking events more difficult since the address changes over time. It will be more difficult to correlate activities in log files, that are usually based on IP addresses.

Hardware cards are often designed with a few numbers of registers for multicast addresses and processing. Minimizing the number of multicast groups to listen to, helps multicast processing to be done directly by the card. The use of randomly generated temporary addresses breaks this optimization since the last 24 bits will be different so each temporary address will generate a new solicited-node multicast address to be listened to.

13.5 More Secure Protocols

With security in mind and the ubiquitousness of IPsec in IPv6 stacks, the IPv6 protocol suite is more secure and uses more security services. For example, some enhancements are:

- IPsec used in routing protocols (see chapter 9);
- IPsec used in MobileIPv6 (see section 11.8);
- Hop limit set to 255 (see section 7.3).

13.6 Securing IPv6 on the Link

When a rogue node has physical access to a link, as illustrated by N3 in Figure 13.15, many tools can be used to make attacks. Wireless networks make this more difficult since the rogue node just has to receive enough wireless signal, even if it is not in the building or room, where the wireless network is intended to service. Tools such as traffic tapping, specially formed packets or responding to local-link queries are security threats. These are similar for any layer 3 protocol, such as IPX, IPv4 or IPv6. IPv4 uses ARP, DHCP and ICMP for interaction on the local link. IPv6 uses Neighbor Discovery and ICMP.

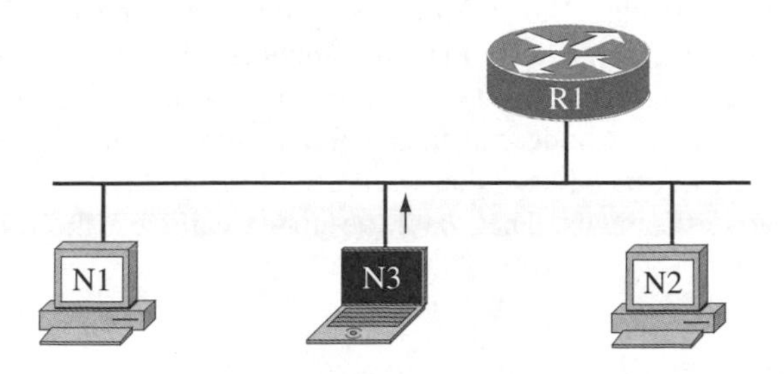

Figure 13.15 Link example with a rogue node

Table 13.3 Link access trust models

Trust model	Typical scenario
A node trusts all other nodes on the link.	Corporate internal network
A node only trusts the routers on the link.	Provider access network
A node does not trust any node on the link.	Ad-hoc network

13.6.1 Threats and Trust Models for IPv6 on the Link

Figure 13.15 shows a local link with nodes N1, N2, router R1 and a rogue node N3.

Inherent to a link access, a node may trust zero, some or all nodes on the link. Table 13.3 lists the trust models [RFC3756] and related typical scenarios.

Many threats have been identified [RFC2461, RFC2462, RFC3756, RFC3971] regarding the link-local interaction. Some are listed in Table 13.4, using Figure 13.15 for the examples.

These link-level threats are not so new in IPv6, since most are also possible in IPv4, using ARP, DHCP or ICMP messages.

IPsec was identified [RFC2461] as securing neighbor discovery but automatic keying (IKE) is based on an existing IPv6 address which is not yet configured at the time the neighbor discovery services are needed to secure the configuration messages. Manual keying is impracticable [Arkko, 2003, RFC3756] especially for mobile nodes and for the solicited-node multicast addresses. Instead, public-key based signatures added to the neighbor discovery messages and cryptographically generated addresses are used by the secure Neighbor Discovery protocol to remedy many of the listed threats for the trust models.

13.6.2 Secure Neighbor Discovery

Secure neighbor discovery (SEND) is defined in two parts: SEND itself [RFC3971] and cryptographically generated addresses (CGA) [RFC3972]. This section summaries the two mechanisms.

13.6.2.1 Finding a Trusted Router

Routers have to be trusted by the nodes. Prior to connecting, the node is pre-configured with a list of trust anchors, identified by public keys and a name, which are authoritative to issue certificates for routers [RFC3971]. When a node connects to a link, it sends a new solicitation message, named the certification path solicitation message, to all routers on the link asking for certificates related to the pre-configured trust anchors. Routers respond by sending certificates that were signed by the trust anchors specified in the solicitation.[2] The node then sends a router solitication and routers respond with a router advertisement signed by one of the certificates. The node can then validate the router advertisements since the host is pre-configured with trusted keys that were used to sign the received router certificates used to sign router advertisements. That way, the nodes can trust the router advertisements sent by trusted routers.

[2] These can be signed directly or through a certification path from the trust anchor. Direct certificates are discussed for simplicity although the specification allows certification paths.

Table 13.4 Threats on link-local interaction

#	Threat	Description	Message
1	Neighbor solicitation(NS) and advertisement(NA) spoofing	N3 sends a Neighbor solicitation (NS) or a Neighbor advertisement (NA) message with N1, N2 or R1 IP addresses and N3 link-layer address. Traffic goes to N3 instead of valid nodes.	NS, NA
2	Neighbor Unreachability Detection (NUD) failure	N3 sends NA responding to NUD NS messages of others. Nodes think another node is still reachable, while it is not.	NS, NA
3	Duplicate Address Detection (DAD) failure	N3 responds to each DAD NS message claiming it owns the address. Nodes do not configure an address.	NS, NA
4	Announcing as default router	N3 sends router advertisements (RA) announcing itself as a default router. Traffic goes to N3 instead of R1.	RA, RS
5	Disabling a default router	N3 sends RA as if it comes from R1, with a router lifetime of zero. R1 is no more used to forward traffic.	RA, RS
6	Redirecting traffic	N3 sends ICMP redirect messages as if it comes from R1. Traffic goes to N3 instead of R1.	ICMP redirect
7	Invalid on-link prefix	N3 sends RA with a non-valid prefix identified on-link. Off-link traffic to this prefix is denied or sent to N3 (by adding threat #1).	RA
8	Flooding prefixes	N3 sends RAs with so many prefixes that the nodes drop the valid prefixes.	RA
9	Invalid autoconfig prefix	N3 sends RA with a non-valid prefix identified for autoconfiguration. Nodes autoconfigure and send traffic with source addresses including the non-valid prefix; response traffic never comes back.	RA
10	Invalid parameters in RA	N3 sends RA with invalid parameters such as Hop limit too small or with the flags to specify a DHCPv6 configuration, which would be done by N3.	RA
11	Keep routers busy	N3 sends packets with random addresses, causing R1 to send NS messages.	NS

13.6.2.2 Verifying the Address Ownership of a Neighbor

Each node on the link involved in SEND generates and uses a cryptographically generated address (CGA) [RFC3972]. This is composed of the link prefix (/64) and the interface identifier generated from the hash[3] of the public-key of the address owner (the node who owns the address).

The CGA and public-key are sent inside new options of neighbor solicitation and neighbor advertisement messages. The whole neighbor discovery message is signed by the private key corresponding to the public key used to generate the CGA.

[3] Actual format is more complicated than just the hash: the interested reader should read the specification [RFC3972].

With this information, the receiving node can verify the neighbor discovery messages by verifying the signatures and by recomputing the hash of the public key that generated the CGA. This way, the receiving node is sure that the message comes from the node that owns that address, and not by a rogue node.

13.7 Is IPv6 More Secure?

IPv6 is a new protocol and is implemented with new code. New code means new security flaws. However, IPv6 as a protocol fixes many issues regarding IPv4 security. It is hoped that the IP stack developers learned while working on the IPv4 issues and code IPv6 more securely. Some new attacks were discovered and some old attacks are now difficult to achieve. For example, the typical network scan of addresses on a subnet is now much more difficult from outside, because there are 2^{64} possibilities to try; sure, there are some optimizations to decrease this number, but it is still very long by any measure. This attack is even more difficult if nodes use temporary addresses, which are ephemeral compared to the scan time.

From the framework perspective, IPv6 does not have NAPT. The absence of NAPT helps designing security policies, troubleshooting and deploying security services such as IPsec.

Protocols in the IPv6 protocol suite are more secured, for example by using IPsec for their security services. New security protocols such as SEND are only available in IPv6 and add new security features (securing link access for SEND) not available in IPv4.

After an initial period of fixing bugs in the new implementation and flaws in the new protocol, IPv6 is becoming more secure than IPv4. The community should have learned to make protocols and implementations more secure.

13.8 Configuring Security on Hosts and Routers

This section describes how to configure temporary addresses, filtering, IPsec and SSH on various implementations.

13.8.1 FreeBSD

FreeBSD supports temporary addresses, filtering through various tools, IPsec and SSH.

13.8.1.1 Temporary Addresses

To enable temporary addresses, the `sysctl` variable `net.inet6.ip6.use_tempaddr` is used.

```
# sysctl net.inet6.ip6.use_tempaddr = 1
```

13.8.1.2 Filtering

There are various filtering/firewall softwares available on FreeBSD, such as ipfw and ipfilter. IPfilter is discussed here.

IPfilter [IPFILTER] has supported IPv6 since version 3.4. The most noticeable impact is when an address is expected in the config file, it could be either IPv4 or IPv6. The example below shows the blocking of any traffic coming from a 2002:: address.

```
# more /etc/ipf.rules
block in on fxp0 from 2002::/16 to any
```

As with IPfilter 4.0beta, it also supports the keywords shown in Table 13.5.

13.8.1.3 IPsec

IPsec is available for IPv4 and IPv6 in FreeBSD. The code is from the Kame stack. To enable IPsec, the kernel must be compiled with the `IPSEC` and `IPSEC_ESP` options, as shown below.

```
# more /usr/src/sys/i386/conf/MYKERNEL
options IPSEC
options IPSEC_ESP
```

The `setkey` command is used to manipulate the security association (SA) and security policy database(SPD) of IPsec in the kernel. The default configuration file read by `setkey` at boot time is `/etc/ipsec.conf`. To enable IPsec at boot time, set the `ipsec_enable` to '`YES`' in the `/etc/rc.conf` file.

```
# more /etc/rc.conf
ipsec_enable="YES"
```

The `/etc/ipsec.conf` configuration file contains the `setkey` operations to make an IPsec tunnel. Manual keying is all done through `setkey`, while automatic keying (IKE) is done by the `racoon` daemon, available in the `ports/security`.

The following example shows a static IPsec tunnel between two IPv6 hosts, using the `3des-cbc` crypto algorithm and the '`thisisaverysecretkey`' secret key. The first host (`3ffe:b00:1:1::1`) has the following configuration:

```
# more /etc/ipsec.conf
add 3ffe:b00:1:1::1 3ffe:b00:1:2::2 esp 9998 -E 3des-cbc "
thisisaverysecretkey" ;
add 3ffe:b00:1:2::2 3ffe:b00:1:1::1 esp 9999 -E 3des-cbc "
thisisaverysecretkey" ;
spadd 3ffe:b00:1:1::1 3ffe:b00:1:2::2 any -P out ipsec
esp/transport//use;
```

Table 13.5 IPfilter IPv6 keywords

Keyword	Description
dstopts	Destination Options extension headers
hopopts	Hop-by-hop options extension headers
v6hdrs	IPv6 header

The add commands create the security associations for both sides. Each security association has a different number, called the security parameter index, which is arbitrarily chosen and must be unique within a node and synchronized with the other peer. The spadd command creates the security policy to initiate the IPsec tunneling between the two peers.

The other peer (3ffe:b00:1:2::2) has the following configuration:

```
# more /etc/ipsec.conf
add 3ffe:b00:1:1::1 3ffe:b00:1:2::2 esp 9998 -E 3des-cbc "
thisisaverysecretkey" ;
add 3ffe:b00:1:2::2 3ffe:b00:1:1::1 esp 9999 -E 3des-cbc "
thisisaverysecretkey" ;
spadd 3ffe:b00:1:2::2 3ffe:b00:1:1::1 any -P out ipsec
esp/transport//use;
```

As soon as the config file is executed by setkey, the kernel is configured for that security association, and any traffic going to the other peer will use the IPsec tunnel.

13.8.1.4 SSH

SSH supports IPv6 by default. If the SSH server is known in the DNS with a AAAA record, then the SSH client tries to connect to that address. The –6 and –4 options in the command line forces the use of IPv6 or IPv4 respectively.

13.8.2 Windows

Windows XP supports temporary addresses. Service Pack 2 (SP2) includes an IPv6 firewall.

13.8.2.1 Privacy/Temporary Addresses

Temporary addresses can be enabled by setting the ‘privacy state’ variable.

```
C> netsh interface ipv6 set privacy state=enabled
```

13.8.3 Cisco

Cisco IOS supports IPv6 filtering and SSH.

13.8.3.1 Filtering

IOS can filter the following fields in the IPv6 packet: source and destination address, transport protocol source and destination port, diffserv codepoints, ICMP types, flow label and presence of the routing or fragments header.

The following example configures the ACL1 IPv6 access-list to accept any ICMP packet, traffic to port 80, but denies anything else.

```
configure terminal
ipv6 access-list ACL1
 permit icmp any any
 permit tcp any any eq 80
 deny ipv6 any any
```

13.8.3.2 SSH

The SSH client on Cisco IOS supports IPv6 transport.

```
ssh 3ffe:b00:0:1::2
```

13.8.4 Juniper

Juniper supports IPv6 filtering as shown below.

13.8.4.1 Filtering

Table 13.6 lists the fields of an IPv6 packet that can be filtered by JunOS.

An IPv6 filter is defined using the 'family inet6' statement under 'firewall'.

The following example shows a filter F1 to discard IPv6 packets traversing the router and going to node 3ffe:b00:0:4::b with a routing header (next header code = 43) as the first next header in the packet. This filter is applied on input of the IPv6 traffic on the fe-0/0/1 interface.

```
firewall {
 family inet6 {
 filter F1 {
  term F1-T1 {
   from {
    destination-address 3ffe:b00:0:4::b/128;
```

Table 13.6 JunOS filtering of IPv6 packet fields

IPv6 packet field	JunOS match statement
Source and destination address	`address, source-address, destination-address`
Transport protocol source and destination port	`port, source-port, destination-port`
Next header (the first one in the basic header)	`next-header`
Traffic class	`traffic-class`
ICMP type and code	`icmp-type, icmp-code`
Packet length	`packet-length`

```
      next-header 43;
     }
     then {
      discard;
       }
      }
     }
   }
 }
 interfaces {
   fe-0/0/1 {
     unit 0 {
       family inet6 {
         filter {
           input F1;
         }
       }
     }
   }
 }
```

Filtering IPv6 in IPv4 encapsulated packets is done on the IPv4 traffic by matching the IP protocol field with the '`protocol 41`' statement. The following example shows a filter F2 to discard all IPv6 in IPv4 packets on input of interface fe-0/0/1.

```
firewall {
  family inet {
    filter F2 {
      term F2-T1 {
        from {
          protocol 41;
        }
        then {
          discard;
        }
      }
    }
  }
}
interfaces {
  fe-0/0/1 {
    unit 0 {
      family inet {
        filter {
          input F2;
        }
      }
    }
  }
}
```

13.9 Summary

IPv6 helps the deployment and management of IP security by mandating IPsec and through the no-NAPT feature making it easier to deploy IPsec. IPv6 also has a temporary address feature enabling anonymity when necessary. Threats on the local link are identified, a secure enhancement of the neighbor discovery protocol is available making the IPv6 link more secure than a IPv4 link. The IPv6 protocol suite uses more IPsec services, making the suite more secure. IPv6 should be more secure than IPv4, however, there are advantages and disadvantages and only time will tell if IPv6 is indeed more secure.

13.10 References

Arkko, J., 'Manual Configuration of Security Associations for IPv6 Neighbor Discovery', draft-arkko-manual-icmpv6-sas-02 (work in progress), March 2003.

[IPFILTER] IPfilter Web site, http://www.ipfilter.org, August 2003.

[RFC1752] Bradner, S. and Mankin, A., 'The Recommendation for the IP Next Generation Protocol', RFC 1752, January 1995.

[RFC2401] Kent, S. and Atkinson, R., 'Security Architecture for the Internet Protocol', IETF RFC 2401, November 1998.

[RFC2402] Kent, S. and Atkinson, R., 'IP Authentication Header', RFC 2402, November 1998.

[RFC2406] Kent, S. and Atkinson, R., 'IP Encapsulating Security Payload (ESP)', RFC 2406, November 1998.

[RFC2409] Harkins, D. and Carrel, D., 'The Internet Key Exchange (IKE)', RFC 2409, November 1998.

[RFC2461] Narten, T., Nordmark, E. and Simpson, W., 'Neighbor Discovery for IP Version 6 (IPv6)', RFC 2461, December 1998.

[RFC2462] Thomson, S. and Narten, T., 'IPv6 Stateless Address Autoconfiguration', RFC 2462, December 1998.

[RFC3041] Narten, T. and Draves, R., 'Privacy Extensions for Stateless Address Autoconfiguration in IPv6', RFC 3041, January 2001.

[RFC3715] Aboba, B. and W. Dixon, 'IPsec-Network Address Translation (NAT) Compatibility Requirements', RFC 3715, March 2004.

[RFC3756] Nikander, P., Kempf, J. and Nordmark, E., 'IPv6 Neighbor Discovery (ND) Trust Models and Threats', RFC 3756, May 2004.

[RFC3971] Arkko, J., Kempf, J., Zill, B. and Nikander, P., 'Secure Neighbor Discovery (SEND)', RFC 3971, March 2005.

[RFC3972] Aura, T., 'Cryptographically Generated Addresses (CGA)', RFC 3972, March 2005.

Ylonen, T., Kivinen, T., Saarinen, M., Rinne, T. and Lehtinen, S., 'SSH Protocol Architecture', draft-ietf-secsh-architecture-14 (work in progress), July 2003.

14

Quality of Service

The 'best effort' delivery of packets is fundamental in the design of IP [Cerf and Kahn, 1974], which means variable delays and congestion losses on busy IP networks. Real-time applications and guaranteed services require an extension of this best effort design to support quality of service (QoS). QoS is often an overloaded keyword, but in this context can be defined as: 'any mechanism that provides distinction of traffic types, which can be classified and administered differently throughout the network' [Ferguson and Huston, 1998].

At the IP layer, three QoS frameworks are discussed in this chapter: IPv5, discussed briefly for historical reasons, Differentiated Services and Integrated Services. IPv4 and IPv6 use the last two frameworks in the same way. Differences between IPv4 and IPv6 such as flow label, network address translation and hardware processing are discussed.

This chapter is far from a comprehensive overview of QoS. The goal is to show the major differences between the two IP version regarding QoS. For an excellent book on QoS, read *Quality of Service* [Ferguson and Huston, 1998].

14.1 IPv5: Streaming Protocol

One QoS approach uses a different layer 3 protocol than IP, such as the Stream Protocol (ST-2) [RFC1190]. This approach is a circuit-based protocol requiring a new layer 3 network which would not be integrated with IP. This results in a new and parallel layer 3 network. ST-2 uses the same IP framing so it uses the number 5 in the IP version, as shown in Section 3.4.1, and is known as IPv5. IPv5 or ST is experimental and is not implemented nor deployed.

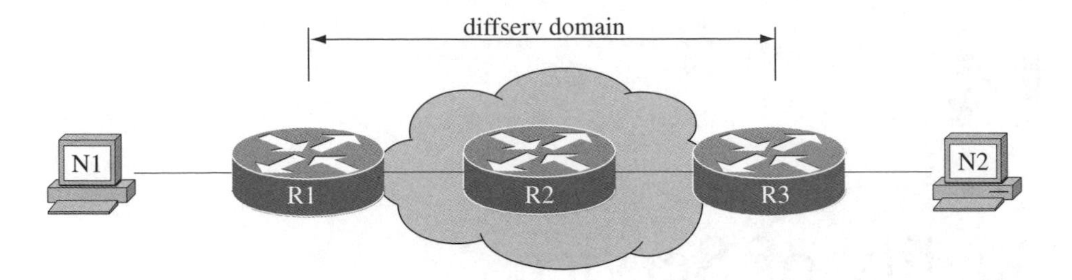

Figure 14.1 Diffserv domain example

14.2 Diffserv

Differentiated services (diffserv) [RFC2475] mark packets and have the routers in the network make differences in their processing of these marked packets, known as per-hop behavior (PHB). Diffserv packets are marked by setting some bits, called DS bits, in the TOS field of the IPv4 header or in the Traffic Class field of the IPv6 header. Diffserv codepoints (DSCP) are standardized [RFC2474] or of local domain specific definition.

The source node or a router at a network boundary marks the packets using a DSCP based on a predefined QoS policy. An example of such a policy could be to mark the voice-over-IP packets for high priority. The routers in the network, seeing this DSCP in the packets, process the marked packets with higher priority.

As an example, in Figure 14.1, node N1 sends real-time traffic such as video to N2. Router R1 marks the packets from N1 to N2 by putting a DSCP in the TOS field of the IPv4 header if the packets are IPv4, or in the Traffic Class field of the IPv6 header if the packets are IPv6. This DSCP maps to a per-hop behavior for high priority processing. R2 receives these marked packets and processes them with high priority compared to non-marked packets. R3 proceeds similarly, but might clear the DSCP bits. N2 receives these marked packets before any best-effort packets. This processing is identical for IPv4 or IPv6.

Important to remember:

Diffserv is identical in IPv4 or IPv6. Diffserv code points (DSCP) are put in the TOS field of the IPv4 header or in the Traffic Class field of the IPv6 header.

Compared to integrated services described in the next section, diffserv does not require any signaling or flow management. Diffserv is generally used more in the networks, because it is easier to deploy than integrated services.

14.3 Integrated Services

The integrated services (intserv) model [RFC1633] is based on the use of resource reservation and admission control to achieve QoS, in a guaranteed and predictive manner. The resource reservation protocol (RSVP) [RFC2210] is used to signal the QoS for the flow from the source to the destination. Routers in the network maintain a per-flow state to achieve the reserved resource properly. Admission control authenticates packets before receiving special router treatment.

A flow identifies the packets to apply QoS processing. A flow is defined as a 'distinguishable stream of related datagrams that results from a single user activity and requires the same QoS' [RFC1633].

Integrated services is implemented by having the source node start a RSVP reservation through the path to the destination node. Along the path, each router processes this reservation. When reaching the destination, an acknowledgment is sent back on the reverse path to the source node to confirm the reservation. This reservation specifies the flow to be processed for some specified QoS level, such as high priority. When the packets of this specified flow are sent from the source node, they are processed by each router in the same path with the specified QoS level.

14.3.1 RSVP

The resource reservation protocol (RSVP) [RFC2205] is used to reserve the resource along the path from the source to the destination of a flow that requires specific QoS.

Figure 14.2 shows an example where N1 is the receiver of the flow and N2 the transmitter.

The receiver N1 sends an RSVP request for QoS resource (the resv message) towards N2. The packet includes a hop-by-hop extension header called Router Alert [RFC2711], as discussed in Section 3.5.1.1. Each router in the path processes the RSVP request and reserves the appropriate resource. N2 receives the packet and sends back a RSVP acknowledgment (the path message) to N1. Then N2 can send traffic to N1 and the network (routers R1, R2, R3 in the figure) has reserved the appropriate QoS, such as dedicated bandwidth.

An RSVP packet sent by N1 to N2 is illustrated in Table 14.1. The Next Header field in the IPv6 basic header indicates, with a value of zero, a hop-by-hop option extension header following the basic header. In this hop-by-hop option extension header, the Next Header field indicates, with a value of 46, an RSVP message following the extension header. The Option-Type field indicates a Router Alert. Router Alert means that every router receiving this packet should look inside it.

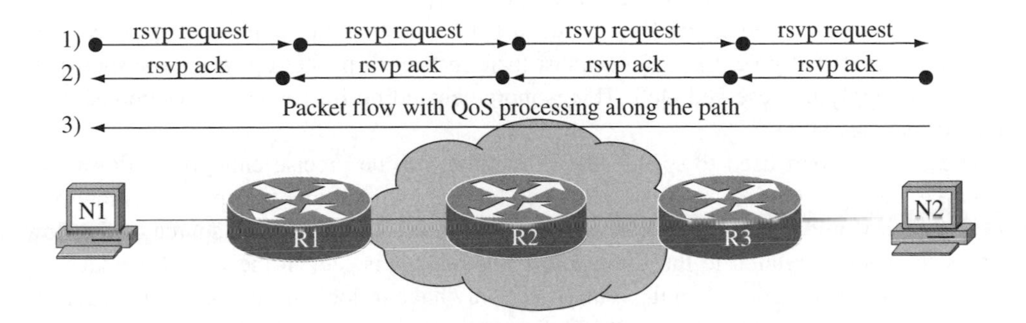

Figure 14.2 RSVP example

Table 14.1 RSVP IPv6 packet

Name	Value	Description
	IPv6 Basic Header	
Version	6	= 6 for IPv6
Traffic Class	0	Diffserv and Explicit Congestion Notification bits
Flow Label	0	Per-flow identification
Payload Length		Size of the inner datagram (after the basic header)
Next Header	0	Specifies a Hop-by-Hop Option Extension Header to follow the IPv6 basic header
Hop Limit		Maximum number of hops
Source Address	<N1>	N1 Source address
Destination Address	<N2>	N2 Destination address
	Hop-by-Hop Option Extension Header	
Next Header	46	RSVP message is following this extension header
Extension Header Length	4	4 octets are included in this extension header
Option Type	5	5 = Router Alert. The first three bits of this octet designate the type of option: all zeros means if a router does not understand the option, it must skip to the next header and keep the option in the packet when forwarding it.
Option Length	2	2 octets are following
Option Value	1	1 means this is a Router Alert for an RSVP message
	RSVP message following	

While the actual reserved values are not really useful (unless you are a developer or are sniffing the traffic with a raw sniffer), this packet structure shows the user many features of the IPv6 header, such as chaining of extension headers, use of Next Header to designate the next payload type, hop-by-hop options and options type.

In the reservation setup, the flow is identified by destination address, IP protocol and destination transport port. Each router must then open each packet up to the transport port in order to apply the reserved QoS. This is more demanding for routers to accomplish than just normal forwarding.

The transport port used to specify the flow might not be precise enough for flows. For example, during a multimedia session between N1 and N2, multiple transport ports might be used, which would each require a new reservation setup. Instead, if the source of the flow marks the packets related to the flow, and if this marker is sent in the RSVP messages to the routers in the network, then the routers will only have to look at this marker to apply the reserved QoS on the flow, making the QoS processing efficient. This marker, named flow label, only exists in IPv6.

14.3.2 *Flow Label*

The Integrated Services (IS) model classifies the packets by looking at addresses and port numbers in the packets. Often, the routers have to look inside the application protocol payload to identify the flows. This makes the router implementation more complex, thus potentially slower. Moreover, the port numbers might be hidden because of the use of encryption or fragmentation.

If the source node identifies the flow by some label before it reaches the network, then the routers do not need to open the packet to find the flow. Hence, it eases the classification and processing of flows in the routers. The IPv6 flow label is designed for that purpose and was suggested as part of the IS architecture.

The flow label field in the IPv6 header is 20 bits long, as described in Section 3.4.3 and shown gray in Figure 14.3.

The value of all zeros does not identify a flow [RFC3697]. Any non-zero value is not currently assigned: all non-zero values have no specific meaning. The current default behavior for source nodes is to assign a flow label not already in use by itself. If using integrated services, then the source node also signals this flow label to the network using some reservation protocol such as RSVP. Then the source sends all packets related to this flow with the flow label in the IPv6 header set to the previously assigned flow label.

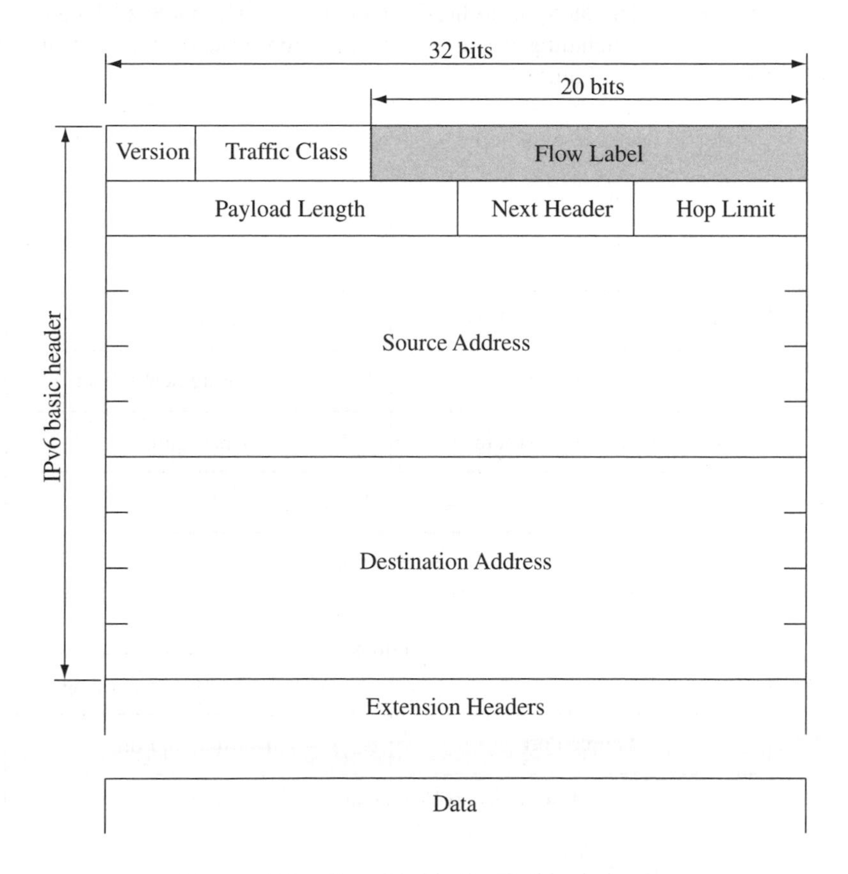

Figure 14.3 Flow label in the IPv6 basic header

14.4 Network Address Translation

As discussed in this chapter, QoS is applied using the IP addresses and port numbers, unless the IPv6 flow label is used. IPv4 network address translation (NAT) changes the source address and the source port of the packets. Therefore, NAT creates an artificial boundary by disabling the use of QoS across NAT. Since IPv6 has no NAT, thase artificial boundaries do not exist and QoS can be applied to larger domains than in IPv4.

14.5 Hardware processing

To apply QoS in traditional IPv4 techniques, a router has to look at the transport port number. As shown in Figure 14.4, the IPv4 header contains the header length (HL) which includes the options in the calculation [RFC791], and the transport protocol (Protocol).

The header length defines the start of the transport protocol header and the transport protocol (TCP or UDP) number locates the transport protocol number inside the transport header. These two values enable a fast lookup of the transport port number in the transport header. UDP and TCP both have the source and destination port at the beginning of the transport header. With or without options, the lookup of the transport port numbers can be done efficiently in hardware.

In the IPv6 packet header, there is no header length field. The payload length gives the length of the total packet, including basic header, extension headers, transport header, and application data, as shown in Figure 14.5.

The transport protocol number is in the next header field in the basic IPv6 header if no extension header is used. If extension headers are used, as shown in Figure 14.5, the

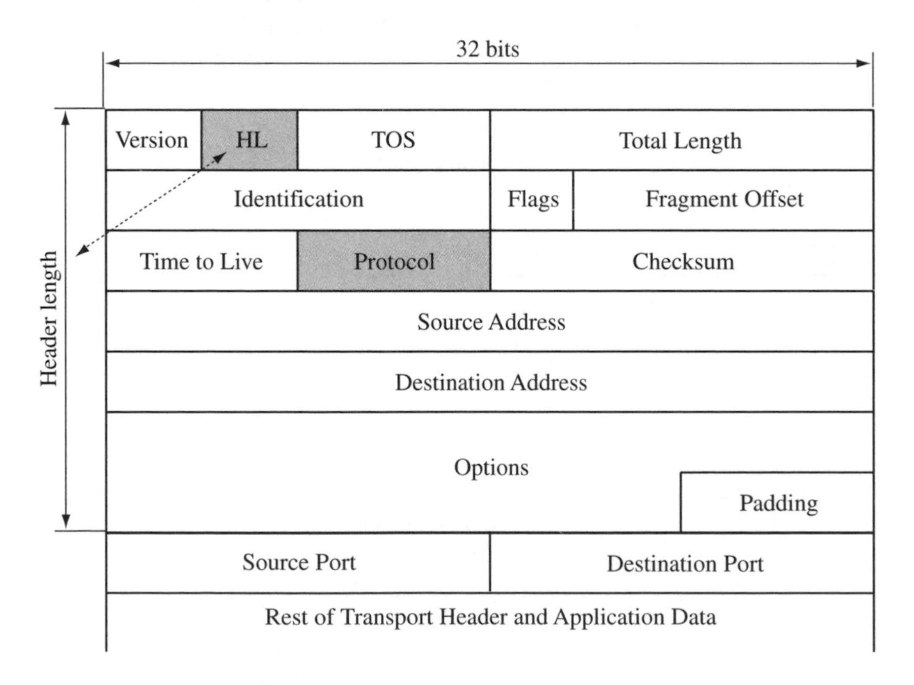

Figure 14.4 IPv4 header with transport ports

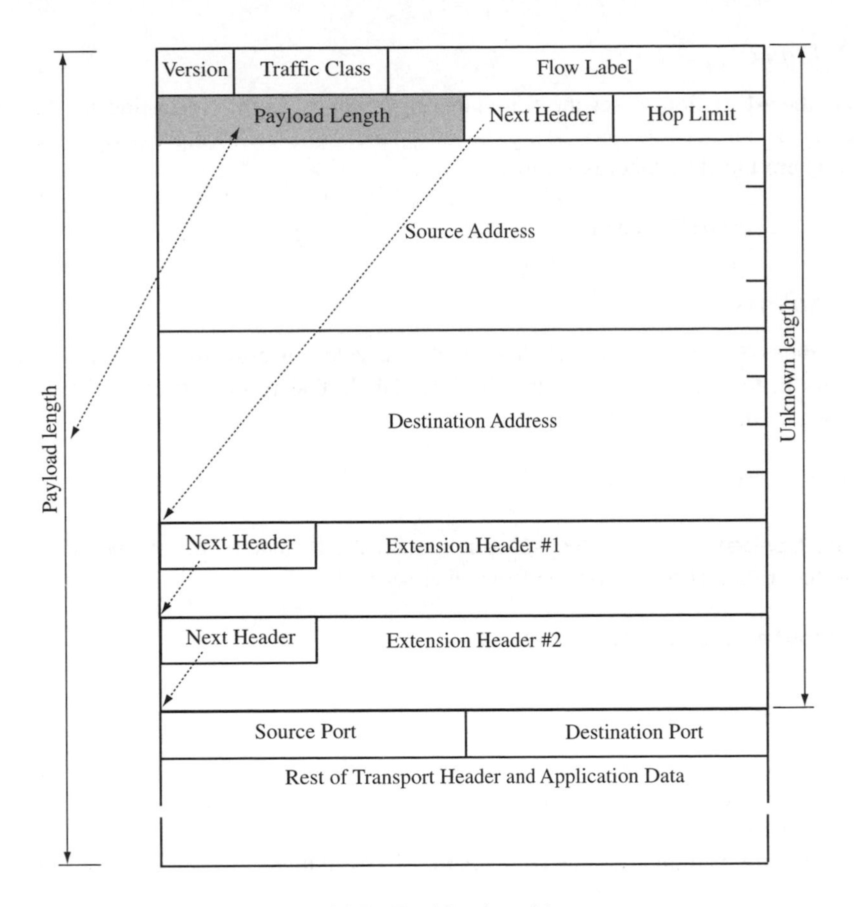

Figure 14.5 IPv6 header with ports

transport protocol number is in the next header field of the last extension header of the daisy-chain. If no extension headers are used, the lookup is as efficient as in IPv4. However, if extension headers are used, the router has to parse the daisy-chain of extension headers to find the start of the transport protocol header. This additional processing is less efficient and may impact QoS performance based on transport ports when extension headers are used. Extension headers, however, are seldom used. To avoid this impact, the use of the IPv6 flow label should be considered.

14.6 Configuring QoS on Hosts and Routers

This section describes the specific commands for the IPv6 flow label.

14.6.1 FreeBSD

The Kame stack automatically assigns a new flow label for every new flow.

14.6.2 Linux

Ping has the `-F` option to set the flow label to a value, useful for testing the behavior of the routers in the network processing the flow label. The following example sends ICMP packets where the flow label is set to 1.

```
# ping6 -F 1 3ffe:b00:0:1::2
```

14.6.3 Solaris

Ping has the `-F` option to set the flow label to a value, useful for testing the behavior of the routers in the network processing the flow label. The following example sends ICMP packets where the flow label is set to 1.

```
# ping -F 1 3ffe:b00:0:1::2
```

Similarly, `traceroute` has the `-L` option to set the flow label to a value. The following example has `traceroute` use the flow label set to 1.

```
# traceroute -L 1 3ffe:b00:0:1::2
```

14.6.4 Cisco

Cisco IOS implementation enables the use of access-lists to be triggered by a specific flow label. Any IOS service using an access-list for filtering can apply specific processing for IPv6 flow labels.

The following example defines an `access-list` named `flow1` triggered by the flow label value of 1 of any IPv6 packets.

```
configure terminal
ipv6 access-list flow1
  permit ipv6 any any flow-label 1
```

14.7 Summary

Quality of service is defined as 'any mechanism that provides distinction of traffic types, which can be classified and administered differently throughout the network' [Ferguson and Huston, 1998]. Three frameworks are presented: IPv5, diffserv and intserv. While diffserv and intserv are done similarly in IPv4 and IPv6, a few differences exist: the use of the IPv6 flow label, network address translation and hardware processing issues. The IPv6 flow label promises better flow processing where implemented.

14.8 References

Cerf, V., and Kahn, R., 'A Protocol for Packet Network Intercommunication', *IEEE Trans on Comm.*, Vol. Com-22, No. 5, May 1974.

Ferguson, P. and Huston, G., *Quality of Service*, Wiley, 1998.

[RFC791] Postel, J., 'Internet Protocol', STD 5, RFC 791, September 1981.

[RFC1190] Casner, S., Lynn, C., Park, P., Schroder, K. and Topolcic, C., 'Experimental Internet Stream Protocol: Version 2 (ST-II)', RFC 1190, October 1990.
[RFC1633] Braden, B., Clark, D. and Shenker, S., 'Integrated Services in the Internet Architecture: an Overview', RFC 1633, June 1994.
[RFC2205] Braden, B., Zhang, L., Berson, S., Herzog, S. and Jamin, S., 'Resource ReSerVation Protocol (RSVP) – Version 1 Functional Specification', RFC 2205, September 1997.
[RFC2210] Wroclawski, J., 'The Use of RSVP with IETF Integrated Services', RFC 2210, September 1997.
[RFC2474] Nichols, K., Blake, S., Baker, F. and Black, D., 'Definition of the Differentiated Services Field (DS Field) in the IPv4 and IPv6 Headers', RFC 2474, December 1998.
[RFC2475] Blake, S., Black, D., Carlson, M., Davies, E., Wang, Z. and Weiss, W., 'An Architecture for Differentiated Services', RFC 2475, December 1998.
[RFC2711] Partridge, C. and Jackson, A., 'IPv6 Router Alert Option', RFC 2711, October 1999.
[RFC3697] Rajahalme, J., Conta, A., Carpenter, B. and Deering, S., 'IPv6 Flow Label Specification', RFC 3697, March 2004.

15

Multicast and Anycast

Multicast is a special addressing scheme with signaling protocols that enables efficient transmission of packets to groups of nodes. This chapter describes the basics of multicast, the multicast listener discovery protocol, the mechanisms to join and leave a multicast group and multicast routing, within the context of IPv6. After that, anycast is described.

As an aside, the inventor of Multicast, Steve Deering, is also the co-inventor of IPv6.

15.1 Multicast Basics

IPv6 Multicast is first used for efficiency of basic IPv6 functions on the local link, such as neighbor discovery. This is accomplished by the senders sending to a multicast destination address and the receivers listening to that address. This works without any other special mechanism, such as multicast routing.

However, multicast is also used across links where a source of traffic sends a datagram to an arbitrary number of nodes in the most efficient way: the multicast routing takes care of sending only one datagram on each link that contains at least one listener for that multicast group. A group is identified by a unique multicast address.

Figure 15.1 shows an example. The node S is the source of the traffic. Nodes A, B, C and D are listeners. Using unicast, S must send four copies of the each datagram, one addressed to each node.

If the routers in the path are multicast-enabled, then the source sends one copy of the datagrams to a multicast address, the nodes listen to that address and the routers forward only one copy of the datagram to the links where there is at least one listener. If a link contains no listener, no traffic is forwarded, as shown in Figure 15.2.

However, the multicast-enabled routers, called simply routers for the rest of this chapter, have to find if there is at least one listener on each of their connected links. For each multicast

Migrating to IPv6: A Practical Guide to Implementing IPv6 in Mobile and Fixed Networks Marc Blanchet

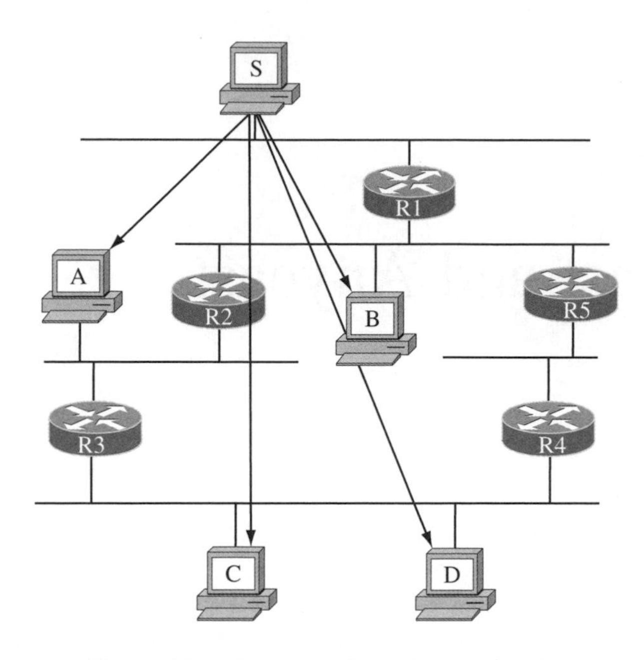

Figure 15.1 Source sending unicast to 4 nodes

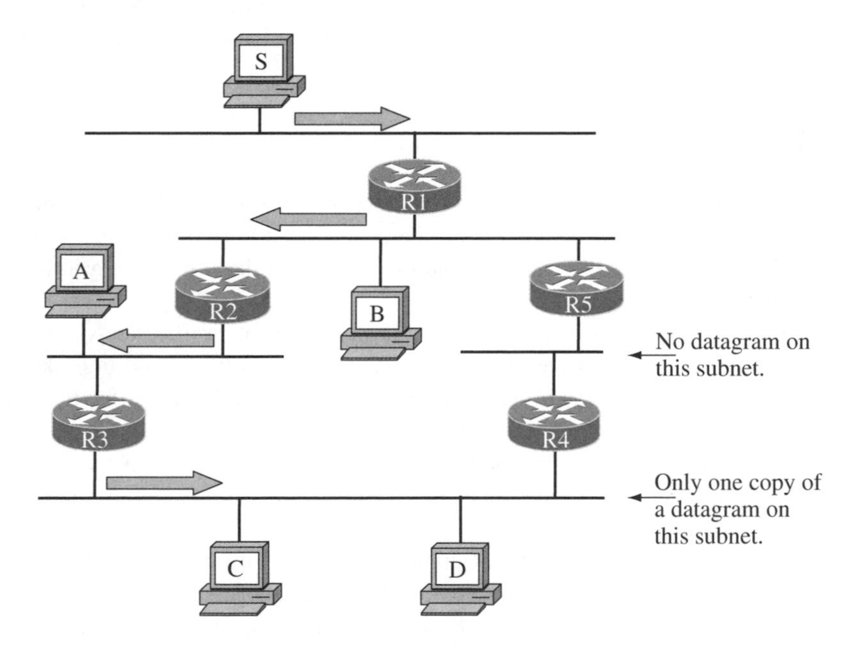

Figure 15.2 Source sending multicast

group, they do not need to know how many listeners or who they are: they only need to know that at least one is listening. IPv4 nodes and routers use Internet Group Management Protocol (IGMP) to register the presence of multicast listeners on a link. IPv6 uses Multicast Listener Discovery (MLD) for the same purposes.

Routing multicast datagrams between routers is handled by specific multicast routing protocols, discussed in Section 15.3.

15.2 Multicast Listener Discovery

Multicast Listener Discovery (MLD) [RFC2710] is used by nodes and routers to inform routers of the existence of at least one listener for each multicast group.

MLD is very similar to IGMPv2 [RFC2236]. The important difference is that MLD uses ICMP as the transport protocol, so MLD messages are ICMP datagrams. The source address of MLD messages is always a link-local address.

MLD datagrams have a special use of two IP header fields: the Hop Limit and Router Alert extension header. Since MLD interaction is only between the nodes and their directly connected routers, all MLD datagrams have the IP header Hop Limit field equal to 1, so these datagrams will not be forwarded by routers.

A router manages all possible multicast groups. Instead of listening to all possible groups by default, all MLD messages use the Router Alert hop-by-hop extension header, which tells routers specifically to look at the datagram, as described in Section 3.5.1.

15.2.1 Node Joining a Multicast Group

When a node wants to join a multicast group, it sends a message to the routers on its link reporting that it is joining a specific multicast group. The message is a MLD Report message containing the address of the multicast group and sent to that address of the multicast group, as described in Table 15.1. Although, routers do not necessarily listen to all possible multicast addresses, they do process the datagram because they are configured to listen to all multicast link-layer addresses and because the IP header has the Router Alert option.

If the router does not have that multicast group in its multicast forwarding table, then it pushes the request to the multicast routing engine in order to receive and forward the datagrams for that group. If the router already forwards the group traffic, then it does nothing except resetting an internal timer. All routers on the link do the same processing for the MLD report message.

The node resends the report periodically, unless it receives the same report from another node. Since the destination address of the MLD report is the address of the multicast group, then all nodes listening to that group receive that message.

Table 15.1 Node joining a group

From → To	Type of MLD datagram	Source and destination addresses	Description and content
1 Node → Routers	Report	Src: node link-local Dst: the joining multicast group address Hop limit = 1 Router Alert option	The node tells the routers that it is joining a multicast group. The address of the group is contained in the MLD datagram.

MLD is designed to send messages efficiently, only when necessary. Recall that the only purpose of MLD is to inform routers of the existence of at least one listener for each multicast group. If a report was recently sent to a multicast group, then there is no need to resend the same report, even from a different node, since the router knows that at least one node is listening. So the other nodes will not resend the same report.

15.2.2 Node Leaving a Multicast Group

When a node wants to leave a multicast group, it sends a MLD Done message to the link-local all-routers multicast address, as described in Table 15.2. Since the routers need to find if this node is the last listener leaving, one of the routers, the elected router (Section 15.2.4), will send a MLD query message to all listening nodes to that multicast address on the link, asking them to send a report. If none sends a report, then the Done message was from the last listener and the routers will cease forwarding the traffic for that multicast group.

The usefulness of the Done message is that the router stops forwarding the datagrams for that group. This happens only when the last member leaves the group. If a leaving node recently received a Report message for the same multicast group it wants to leave, then it may not send Done message, since it is obviously not the last member of that group on the link.

15.2.3 Router Verifying Group Membership

Routers have to make sure that there is at least one listener for each multicast group to which they are forwarding the traffic. It is possible that listeners do not send Done messages because of accidental reboots, link failures or something else. As described in Table 15.3, the routers periodically verify the group membership by sending a MLD General Query message asking all nodes to report all multicast groups to which they are listening.

Table 15.2 Node leaving a group

From → To	Type of MLD datagram	Source and destination addresses	Description and content
1 Node → Routers	Done	Src: node link-local Dst: link-local all-routers multicast (FF02::2) Hop limit = 1 Router Alert option	The node tells the routers that it is leaving a specific multicast group. The address of the group is contained in the MLD datagram.
2 Elected Router → all nodes listening to the multicast group	Multicast-Address-Specific Query	Src: elected router link-local Dst: the leaving multicast group address Hop limit = 1	Router verifies that this node is the last one by requesting a report to all nodes listening to that multicast group. If no report is received, then all routers delete this multicast group from their multicast forwarding table.

Table 15.3 Periodic verification of group membership

From → To	Type of MLD datagram	Source and destination addresses	Description and content
1 Elected router → All nodes	General Query	Src: client link-local Dst: all-nodes link-local: FF02::1 Hop limit = 1 Router Alert option	The elected router is asking all nodes to report their memberships for all multicast groups.
2 Every node for every joined multicast group → All routers	Report	Src: node link-local Dst: the joining multicast group address Hop limit = 1	Each node sends a specific report datagram for every joined multicast group. A randomized interval is used between reports of a node.

However, this query could result in flooding the network with administrative messages of multiple queries and reports. To avoid multiple queries from all routers on the link, only the elected router, described in Section 15.2.4, sends that General Query message. To avoid the same reports for the same multicast group from all listeners, reports are sent to the multicast group address, so all other listeners receive it and decide not to send the same report since it was just sent.

When a router boots, it starts as an elected router and sends a General Query message so it can quickly collect the membership of all groups.

15.2.4 Electing a Router

The only purpose of an elected router is to minimize the number of queries sent on the link. The elected router is the only multicast router to send the queries. Recall that all MLD messages are received and processed by all routers, both non-elected and elected.

The election mechanism is simple: when the current elected router receives a query from another router whose IPv6 link-local address is lower, then the elected router become a non-elected router and the other router becomes the elected router. The decision of using the lower address is essentially arbitrary: the only important thing is to have a decision criteria that all routers can do locally based on the information they have.

The process of electing a router should be fault-tolerant so that if the elected router disappears suddenly, another router takes the task. An internal timer in each router triggers a self-election if no elected router has been heard after some delay. When a router boots, it starts in the elected mode, so any link has one elected router.

15.2.5 Multicast Listener Discovery Version 2

A new version of MLD, called MLDv2 [RFC3810] is synched with the new version of IPv4 IGMP, IGMPv3 [RFC3376]. This adds the ability for listeners to request multicast group traffic only from some specific source addresses. By default, a listener to a multicast group will receive the traffic from all sources for that group.

15.3 Multicast Routing

Many multicast routing protocols have been defined: Distance Vector Multicast Routing Protocol (DVMRP), Multicast OSPF (MOSPF), Protocol Independent Multicast (PIM) and Multicast BGP, among others.

DVMRP [RFC1075] does not support IPv6. The original MOSPF [RFC1584] is based on OSPFv2 [RFC2328] which only supports IPv4. However, OSPFv3 [RFC2740], as discussed in Section 9.6, supports IPv6 and multicast extensions for IPv6. PIM-SM [RFC2362] supports IPv6 since it is protocol-independent.

These multicast routing protocols are used between routers to control the forwarding of the multicast traffic only where there are listeners. For example, OSPFv3 maintains a link state database for normal IP routing. The multicast members are pinpointed in the database. A tree is constructed using the shortest path for any multicast group. All links without listeners for a particular group will be pruned from the tree. In Figure 15.2, all links except the one between R4 and R5 will be in the tree. Note that a tree is specific to, and constructed for, a multicast group. The MLD membership messages, Report and Done, are used to tell the multicast routing infrastructure the location of the listeners so the routing infrastructure constructs the trees and forwards the groups traffic to the links where listeners are present.

15.4 Multicast Address Allocation

Section 4.4 describes the basic multicast address structure, as shown in Figure 15.3.

In this basic address structure, the multicast group identifier is 112 bits long. However, for allocation of multicast groups, only the last 32 bits are used [RFC3513], as shown in Figure 15.4.

The last 32 bit group identifier maps directly to the 32 bit IEEE 802 link layer multicast addresses, as discussed in Section 6.6.2.

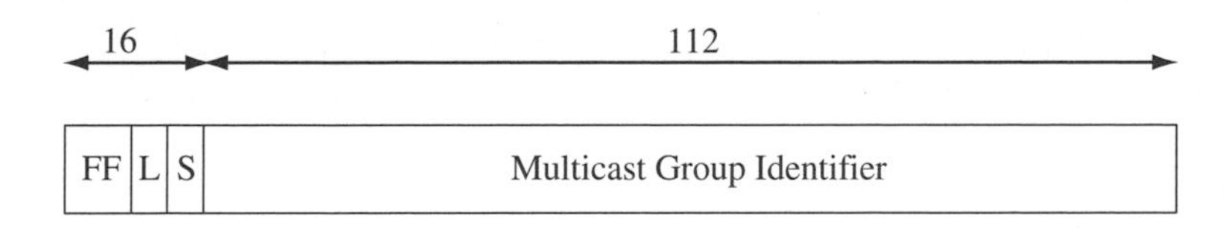

Figure 15.3 Multicast address structure

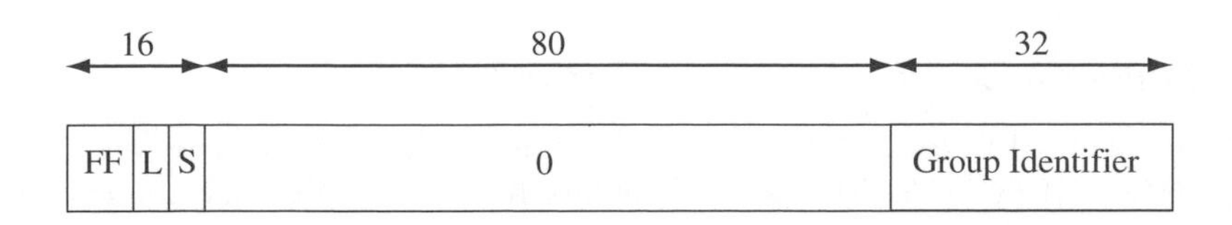

Figure 15.4 Allocation of group addresses

15.5 Unicast based Multicast Addressing

Allocating multicast groups to connected domains needs synchronization: a group number used in one domain cannot be used in another connected domain. In IPv4, multicast address allocation protocols [RFC2909] are designed to manage this task.

Given the large address space in IPv6, the allocation [RFC3306] is based on the unicast prefix owned by a domain, which enables unique multicast address allocation. The intention is to localize the allocation of multicast addresses inside a network, without having to take care of the outside allocation. The prefix of the network is embedded inside the multicast address structure in the 80 bits left. This makes the last 32 bit group identifier local to the prefix being used. The prefix may be of any length (for a network, which means length ≤ 64) so the multicast group may be local to a subnet, to a site, or even to a large provider network.

The new structure of the multicast address is shown in Figure 15.5. The prefix length (PL) field is 8 bits and contains the length of the prefix embedded in the next 64 bits.

To distinguish between these multicast addresses and the usual ones, the third bit in the L field is used. The value of one (1) means that the multicast address contains a prefix-based structure. The new possible values for the L field bits are described in Table 15.4 (an extension of Table 4.6).

Note that value 2 is prohibited, since a group using the prefix-based structure must be temporary.

15.6 Allocation of Multicast Addresses

Allocation guidelines [RFC3307] are available to help network managers allocate multicast group addresses. The use of random group identifiers for temporary groups is recommended to decrease the chance of choosing the same group identifier for different groups.

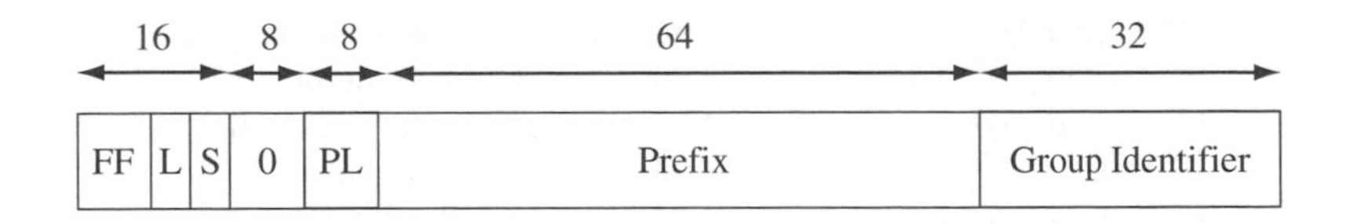

Figure 15.5 Allocation of group addresses based on prefix

Table 15.4 L bits in a multicast address

Value of L (4 bits binary)	Value of L (4 bits hex)	Description
0000	0	Permanent group
0001	1	Temporary group not based on network prefix
0011	3	Temporary group based on network prefix

Table 15.5 Decomposition of a prefix-based group address

Value	Description
ff	Multicast prefix
3	Temporary group based on network prefix (see Table 4.6)
5:	Site scope
00	Reserved field to 0
30:	Prefix length (/48) of the site, expressed in hexadecimal
3ffe:b00:0001:0000:	Site prefix padded with 0 up to 64 bits
6f3b:1148	Group identifier randomly generated

If one manages a site network with the 3ffe:b00:1::/48 prefix and wants to setup an internal (site-scoped) multicast channel for broadcasting a videoconference, then one could use the prefix-based structure of multicast addresses, as shown in Section 15.5. If the 32 bit group identifier is randomly chosen as 6f3b1148, then the full multicast address of this group would be: ff35:0030:3ffe:b00:0001:0000:6f3b:1148, as decomposed in Table 15.5.

15.7 Multicast Reserved Addresses

Table 15.6 lists some multicast reserved addresses, extracted from the specification [RFC2375]. IANA maintains a list of reserved multicast addresses [IANA, 2004].

Table 15.6 Multicast Reserved Addresses

Address	Scope	Description
FF01:0:0:0:0:0:0:1	Interface	All interfaces on the node
FF01:0:0:0:0:0:0:2	Interface	All routers on the node
FF02:0:0:0:0:0:0:1	Link	All nodes on the link
FF02:0:0:0:0:0:0:2	Link	All routers on the link
FF02:0:0:0:0:0:0:4	Link	DVMRP routers
FF02:0:0:0:0:0:0:5	Link	OSFP routers
FF02:0:0:0:0:0:0:6	Link	OSPF designated routers
FF02:0:0:0:0:0:0:9	Link	RIP routers
FF02:0:0:0:0:0:0:B	Link	MobileIP agents
FF02:0:0:0:0:0:0:D	Link	PIM routers
FF02:0:0:0:0:0:1:2	Link	DHCP agents
FF02:0:0:0:0:1:FFXX:XXXX	Link	Solicited-node
FF05:0:0:0:0:0:0:2	Site	Routers on the site
FF05:0:0:0:0:0:1:3	Site	DHCP servers on the site
FF05:0:0:0:0:0:1:4	Site	DHCP agents on the site
FF05:0:0:0:0:0:1:1000-13FF	Site	Service location
FF0X:0:0:0:0:0:0:101	Any	Network Time Protocol

15.8 Anycast

Anycast enables sending a datagram to one node belonging to a group of nodes. The routing identifies the nearest node in the group that will receive the datagram, based on the routing metrics. Anycast has been defined for IPv4 [RFC1546] but its use is limited.

An anycast address is the same unicast address configured on multiple nodes. The address is also configured on the routers as a host route (a full /128 IPv6 route) so the routers on the network will deliver the datagram to the nearest node that is member of the anycast group. The choice of the anycast address is arbitrary.

An anycast address can be used for any size of networks: a site, a link, an organization, etc. The host route must be injected in the network supporting the anycast address.

Figure 15.6 shows an example of a network with two DNS servers, S1 and S2. They are configured with the same anycast address (3ffe:b00:1::2)[1] and both answer DNS queries sent to that IP address. These servers are located in different locations in order to provide redundancy and best performance.

The clients, A and B, send their queries to the anycast address, 3ffe:b00:1::2, and the border routers of the site, R1 and R2, have in their routing table a /128 route for 3ffe:b00:1::2 which points to S1 and S2. The datagram always reaches the nearest server. However, if the S1 server is down, as shown in Figure 15.7, then the routing entry for that server is withdrawn and the routers, R1 and R2, forward to the other server (S2).

A mechanism has to be used between the host S1 and its routers to register and withdraw the host route for S1. A proposal is based on an extension of the Multicast Listener Discovery protocol [Haberman and Thaler, 2002] where the host sends an MLD Report message to the routers on the link informing that it is subscribing to the anycast group.

The anycast mechanism is used for the discovery of services on a network. MobileIPv6, described in Section 11.10.2, defines a reserved anycast group address [RFC2526] identifying

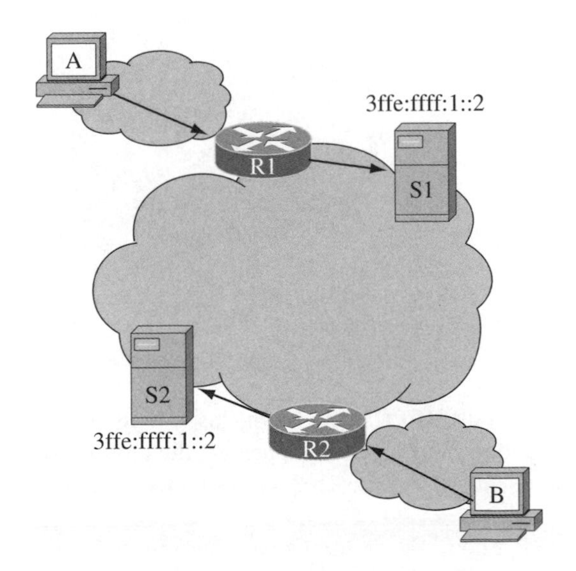

Figure 15.6 Servers with an anycast address

[1] The anycast address is in addition to the host's primary address.

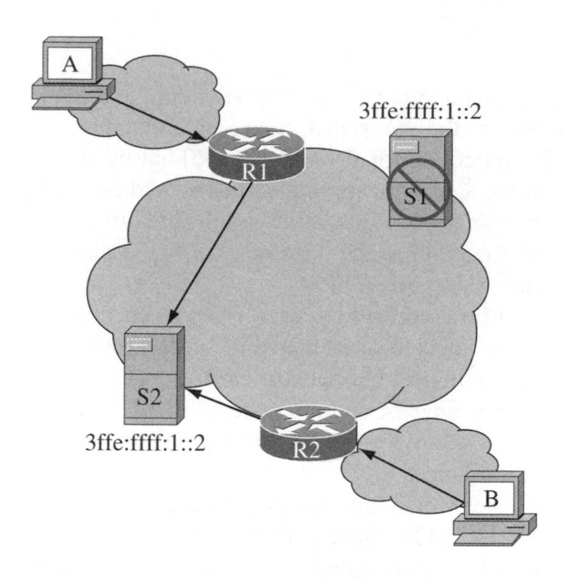

Figure 15.7 Redundancy with an anycast address

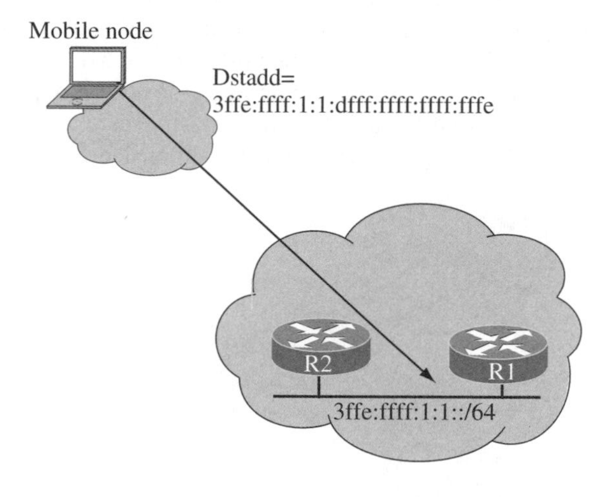

Figure 15.8 MobileIP home agent discovery anycast address

home-agents on a subnet, as dfff:ffff:ffff:fffe for the last 64 bits. When a mobile node is away from its home network and its home agent disappeared, the mobile node has to discover other home agents on its home subnet. Figure 15.8 shows an example with the 3ffe:b00:1:1::/64 home prefix.

The mobile node sends a home agent discovery request message to the home agent anycast group address of its home subnet, 3ffe:b00:1:1:dfff:ffff:ffff:fffe. One of the home agents will reply to the mobile node with a list of available home agents on the link.

Anycast needs further work [Hagino and Ettikan, 2001] and is currently restricted in its use [RFC3513]. A list of reserved anycast addresses is available [IANA, 1999].

15.9 Configuring Anycast and Multicast on Hosts and Routers

This section describes the configuration of anycast and multicast on some host and router implementations.

15.9.1 FreeBSD

To configure the 3ffe:b00:1:1::1 anycast address on an interface, use the anycast parameter on ifconfig.

```
# ifconfig fxp0 3ffe:b00:1:1::1 anycast
```

To see the multicast groups to which a FreeBSD node is subscribed, use the `ifmcstat` command, `netstat -ain` or `netstat -g`.

```
# ifmcstat
# netstat -ian
# netstat -g
```

15.9.2 Linux

Use the `-g` flag in `netstat` to see the multicast groups in Linux.

```
# netstat -g
```

15.9.3 Solaris

Use the `-g` flag in `netstat` to see the multicast groups in Solaris.

```
# netstat -g
```

15.10 Summary

Multicast is more efficient in IPv6, with larger allocation space, better handling of scope and local allocation of group addresses. Anycast is used to enable load balancing, fault tolerance and performance. It is achieved as a host route in the routing realm.

15.11 References

Hagino, J. and Ettikan, K., 'An Analysis of IPv6 Anycast', draft-itojun-ipv6-anycast-analysis-02 (work in progress), March 2001.

Haberman, B. and Thaler, d., 'Host-based Anycast using MLD', Internet-Draft draft-haberman-ipngwg-host-anycast-01, May 2002.

IANA, 'Internet Protocol Version 6 Anycast Addresses', http://www.iana.org/assignments/ipv6-anycast-addresses, August 1999.

IANA, 'Internet Protocol Version 6 Multicast Addresses', http://www.iana.org/assignments/ipv6-multicast-addresses, 2004.

[RFC1075] Waitzman, D., Partridge, C. and Deering, S., 'Distance Vector Multicast Routing Protocol', RFC 1075, November 1988.

[RFC1546] Partridge, C., Mendez, T. and Milliken, W., 'Host Anycasting Service', RFC 1546, November 1993.

[RFC1584] Moy, J., 'Multicast Extensions to OSPF', RFC 1584, March 1994.

[RFC2236] Fenner, W., 'Internet Group Management Protocol, Version 2', RFC 2236, November 1997.

[RFC2328] Moy, J., 'OSPF Version 2', STD 54, RFC 2328, April 1998.

[RFC2362] Estrin, D., Farinacci, D., Helmy, A., Thaler, D., Deering, S., Handley, M. and Jacobson, V., 'Protocol Independent Multicast-Sparse Mode (PIM-SM): Protocol Specification', RFC 2362, June 1998.

[RFC2375] Hinden, R. and Deering, S., 'IPv6 Multicast Address Assignments', RFC 2375, July 1998.

[RFC2526] Johnson, D. and Deering S., 'Reserved IPv6 Subnet Anycast Addresses', RFC 2526, March 1999.

[RFC2710] Deering, S., Fenner, W. and Haberman B., 'Multicast Listener Discovery (MLD) for IPv6', RFC 2710, October 1999.

[RFC2740] Coltun, R., Ferguson, D. and Moy, J., 'OSPF for IPv6', RFC 2740, December 1999.

[RFC2909] Radoslavov, P., Estrin, D., Govindan, R., Handley, M., Kumar S. and Thaler, D., 'The Multicast Address-Set Claim (MASC) Protocol', RFC 2909, September 2000.

[RFC3306] Haberman, B. and Thaler D., 'Unicast-Prefix-based IPv6 Multicast Addresses', RFC 3306, August 2002.

[RFC3307] Haberman, B., 'Allocation Guidelines for IPv6 Multicast Addresses', RFC 3307, August 2002.

[RFC3376] Cain, B., Deering, S., Kouvelas, I., Fenner, B. and Thyagarajan A., 'Internet Group Management Protocol, Version 3', RFC 3376, October 2002.

[RFC3513] Hinden, R. and Deering, S., 'Internet Protocol Version 6 (IPv6) Addressing Architecture', RFC 3513, April 2003.

[RFC3810] Vida, R. and L. Costa, 'Multicast Listener Discovery Version 2 (MLDv2) for IPv6', RFC 3810, June 2004.

16

Deploying IPv6 in IPv4 Dominant Networks

Current networks are dominated by the IPv4 protocol. IPv6 can be deployed:

1. together with the IPv4 protocol;
2. tunneled in IPv4;
3. in parallel with or separated from the IPv4 network;
4. or as a new network without IPv4.

The third case is discussed in Chapter 17. Combining IPv4 and IPv6 in a single network is discussed in Section 16.1, while the rest of this chapter discusses the various approaches available for tunneling IPv6 in IPv4.

This chapter describes many techniques in detail, but most sections are independent and links between them are provided. The reader can skip the sections that are of less interest.

16.1 Combined IPv4 and IPv6 Network

The primary migration strategy [RFC2893] is to have all nodes as dual-stack, which means they support IPv4 and IPv6 at the same time. The applications are supporting both protocols and the network is routing both protocols.

When starting a new connection, an application on a dual-stack node chooses the IP protocol based on the responses from the DNS. When an application is resolving a name to an IP address, the following happens:

- If the DNS response contains only an IPv4 address, then the source node will use IPv4.
- If the DNS response contains only an IPv6 address, then the source node will use IPv6.
- If the DNS response contains both IPv4 and IPv6 addresses, then the source node will use one or the other depending on its configuration, as discussed in Section 8.6.

Migrating to IPv6: A Practical Guide to Implementing IPv6 in Mobile and Fixed Networks Marc Blanchet

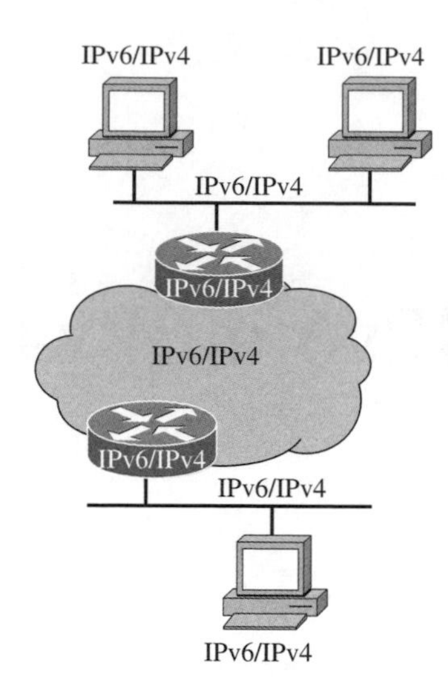

Figure 16.1 Combined IPv6 and IPv4 network

Not only are the nodes dual-stack, but there should be some IPv6 network between any two IPv6 nodes. The IPv6 network could be run together with IPv4, meaning that the routers are dual-stack, they forward both IPv4 and IPv6 datagrams on their interfaces and they run IPv4 and IPv6 routing protocols, as shown in Figure 16.1.

This is similar to the days when corporate networks were running the Novell IPX protocol together with the IPv4 protocol. Some care should be taken with routing protocols, as discussed in Chapter 9. Deployment scenarios of combined IP protocols are discussed in Chapter 23.

16.2 Tunneling IPv6 in IPv4

It is sometimes difficult to deploy a fully combined IPv6/IPv4 network, for example, because of old routers that do not have enough memory to run the latest releases of their operating system which includes IPv6. In that context, part of a network might be IPv6/IPv4 and other parts might be IPv4-only.

The tunneling technique is often used to overlay a new protocol over a legacy one. IPX datagrams were often encapsulated in IPv4 datagrams over IPv4-only backbones. The same technique is used to encapsulate IPv6 datagrams over an IPv4 network.

16.2.1 Encapsulation

The encapsulation of IPv6 datagrams over an IPv4 network [RFC2473] uses the IPv4 protocol number 41. The encapsulation node can be either a host or a router and the decapsulation node can be either a host or a router as well.

The IPv6 datagram is put inside the payload of an IPv4 datagram, as shown in Figure 16.2.

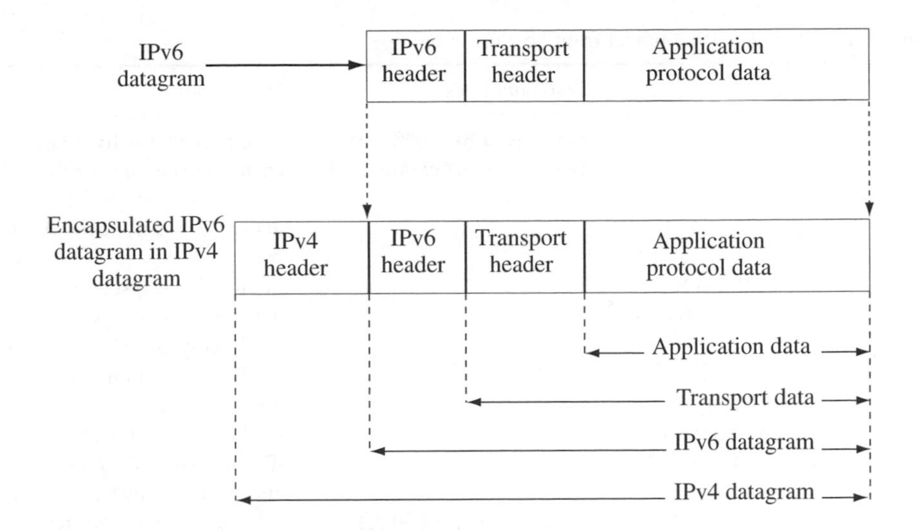

Figure 16.2 Encapsulation of an IPv6 datagram in an IPv4 datagram

The IPv4 source and destination addresses of the IPv4 datagram are the addresses of the encapsulation and decapsulation node, which may or may not be the source and destination of the IPv6 datagram.

16.2.2 Host to Router Encapsulation

Figure 16.3 shows an example where node A is dual-stack and wants to send an IPv6 datagram to an IPv6-only node B. Node A is manually configured to send all IPv6 datagrams within an IPv6-in-IPv4 tunnel to the R1 router.

As described in Table 16.1, source node A internally makes an IPv6 datagram for the destination node B (step 1 in Table 16.1). The source address of the IPv6 datagram is A's address (3ffe:b00:a:1::1) and the destination address is B's address (3ffe:b00:a:3::2). This IPv6 datagram is not sent on the wire, but instead it is encapsulated inside an IPv4 datagram by A (step 2), since A is configured to send all IPv6 traffic encapsulated to the R1 router.

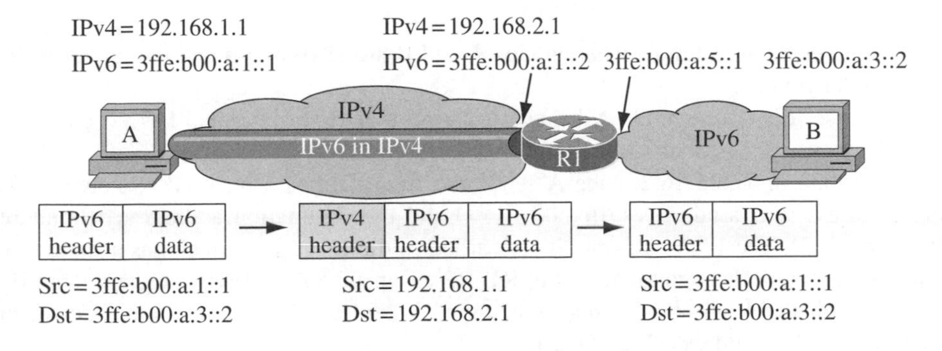

Figure 16.3 IPv6 over IPv4 tunnel from host to router

Table 16.1 IPv6 over IPv4 tunnel from host to router

Step	IPv4 addresses	IPv6 addresses	Description
1	None	Src = A = 3ffe:b00:a:1::1 Dst = B = 3ffe:b00:a:3::2	Source A internally makes an IPv6 datagram for the destination node B. This datagram is still inside the A IP stack.
2	Src = A = 192.168.1.1 Dst = R1 = 192.168.2.1		Source A encapsulates the IPv6 datagram into the payload of the IPv4 datagram. Destination of the IPv4 datagram is the end of the tunnel, R1.
3	Src = A = 192.168.1.1 Dst = R1 = 192.168.2.1		IPv4 datagram travels on the IPv4 network. R1 receives the encapsulated IPv4 datagram.
4		Src = A = 3ffe:b00:a:1::1 Dst = B = 3ffe:b00:a:3::2	R1 removes the IPv4 header and sends the IPv6 original datagram on its IPv6 network. Destination of the IPv6 datagram is B.
5		Src = A = 3ffe:b00:a:1::1 Dst = B = 3ffe:b00:a:3::2	Destination B receives the datagram.

The source of the IPv4 datagram is the A IPv4 address (192.168.1.1) and the destination address is the R1 IPv4 address (192.168.2.1). The payload of the IPv4 datagram contains the IPv6 datagram as illustrated in Figure 16.2.

R1 receives the IPv4 datagram (step 3), removes the IPv4 header and pulls the original IPv6 datagram. R1 forwards the IPv6 datagram (step 4) on its IPv6 network to the destination B, since the destination address of the decapsulated IPv6 datagram is B's address (3ffe:b00:a:3::2). Destination B receives the IPv6 datagram (step 5).

B has no clue that the datagram was encapsulated over some parts of the path.

16.2.3 Router to Router Encapsulation

Figure 16.4 shows an example where nodes A and B are IPv6 nodes and the tunnel is done by the border routers.

In this case, the source node A behaves as a normal IPv6 node and has no configured tunnel. However, R2 is dual-stack and has a configured tunnel with R1.

As described in Table 16.2, node A sends the IPv6 datagram on its IPv6 network. The source address is A's address (3ffe:b00:a:1::1) and the destination address is B's address (3ffe:b00:a:3::2). The IPv6 datagram is forwarded to the R2 router which has a route to the destination network through the tunnel to R1. R2 encapsulates the IPv6 datagram in an IPv4 datagram with the IPv4 source address being R2's address (192.168.1.1) and destination address being R1's address (192.168.2.1).

A and B have no clue that the datagram was encapsulated over some parts of the path.

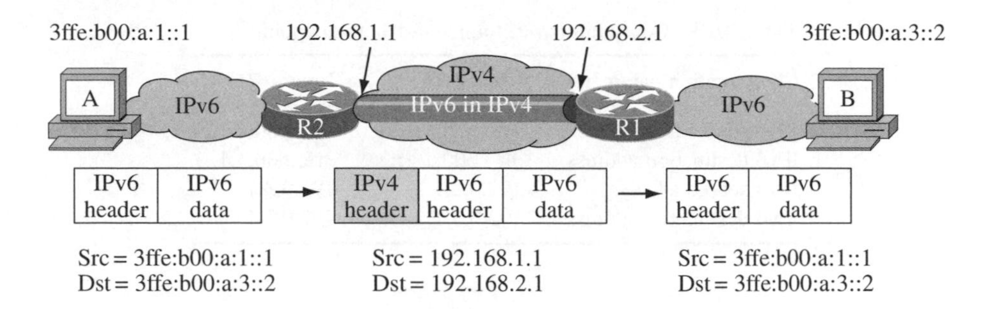

Figure 16.4 IPv6 over IPv4 tunnel between two routers

Table 16.2 IPv6 over IPv4 tunnel between two routers

Step	IPv4 addresses	IPv6 addresses	Description
1	None	Src = A = 3ffe:b00:a:1::1 Dst = B = 3ffe:b00:a:3::2	Source A makes an IPv6 datagram for the destination node B.
2	None	Src = A = 3ffe:b00:a:1::1 Dst = B = 3ffe:b00:a:3::2	IPv6 datagram travels on the IPv6 network. Router R2 receives the IPv6 datagram from A.
3	Src = R2 = 192.168.1.1 Dst = R1 = 192.168.2.1		R2 encapsulates the IPv6 datagram into the payload of the IPv4 datagram. Destination of the IPv4 datagram is the end of the tunnel, R1.
4	Src = R2 = 192.168.1.1 Dst = R1 = 192.168.2.1		IPv4 datagram travels on the IPv4 network. R1 receives the encapsulated IPv4 datagram.
5		Src = A = 3ffe:b00:a:1::1 Dst = B = 3ffe:b00:a:3::2	R1 removes the IPv4 header and sends the IPv6 original datagram on its IPv6 network. Destination of the IPv6 datagram is B.
6		Src = A = 3ffe:b00:a:1::1 Dst = B = 3ffe:b00:a:3::2	Destination B receives the IPv6 datagram.

16.2.4 Static Tunneling

Manual configuration of the source and destination IPv4 and IPv6 addresses on both endpoints of the tunnel is required to make the tunnel work. Both endpoints have to be dual-stack.

In the second column of Table 16.3, the configuration data needed for router R2 in Figure 16.5 are shown. The third column is for router R1.

Some implementations do not need the IPv6 destination address in the configuration, since a tunnel is a point-to-point link and point-to-point links do not have to know the other endpoint address. However, the IPv4 destination address is essential since it is the primary information to forward the encapsulated datagram.

Table 16.3 Configuration of tunnels for both endpoints

Configuration parameter	Router R2	Router R1
IPv6 source address	3ffe:b00:1:1::1	3ffe:b00:1:1::2
IPv6 destination address	3ffe:b00:1:1::2	3ffe:b00:1:1::1
IPv4 source address	192.0.2.1	192.0.3.1
IPv4 destination address	192.0.3.1	192.0.2.1

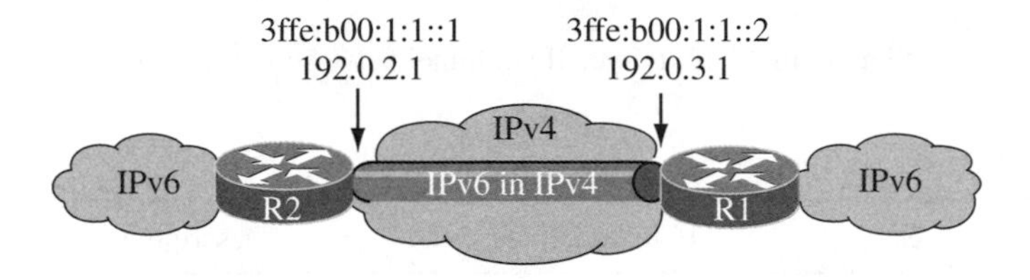

Figure 16.5 Static tunnel between two routers

16.2.4.1 Requirements for Static Tunneling

The following requirements have to be met in this router to router configuration:

- Routers R1 and R2 are dual-stack.
- Router R1 has a reachable IPv4 address from R2, and vice-versa.

After configuring the two endpoints of the tunnel, routing, either static or dynamic, has to be configured so the IPv6 traffic will go through the tunnel. Using Figure 16.5 as an example, if R1 is connected to the IPv6 Internet, then R2 would have a static default IPv6 route pointing to the tunnel interface.

16.2.4.2 Limitations

Manual configuration of the tunnel endpoints is tedious

- if many tunnels have to be done;
- if the IP addresses are changing for many tunnels.

Static tunnels does not traverse IPv4 NAT, as discussed in Section 16.2.7.8.

IPv6 in IPv4 packets using protocol 41 might be filtered by security gateways: in this case, the tunnel does not work, as discussed in Section 16.2.7.7.

16.2.4.3 Applicability

IPv6 in IPv4 static tunnels are applicable when a very small number of tunnels are needed and no IPv4 NAT is present in the path.

6to4, ISATAP, TSP tunnel broker and Teredo are mechanisms automating the creation of tunnels. They are described in the next sections.

16.2.5 6to4

The 6to4 mechanism [RFC3056] builds IPv6 in IPv4 tunnels on demand and allocates IPv6 address space.

16.2.5.1 Addressing

The 6to4 IPv6 address space is built by the 2002::/16 prefix reserved for the 6to4 mechanism, followed by the 32 bits of the IPv4 external address of the border router of the site, giving the site a /48 prefix, as shown in Figure 16.6.

Figure 16.7 shows an example of an IPv6 site using the 6to4 mechanism.

The border router has an external IPv4 address (192.0.2.1). The IPv6 site behind the border router uses 2002:c000:0201::/48 to number its whole network. The address space is based on 2002:<ipv4 external address in hex>::/48, where the IPv4 address is the border router external IPv4 address (192.0.2.1), represented in hexadecimal as c000:0201.

The 6to4 mechanism needs to be only implemented in border routers. Hosts inside the IPv6 site do not need to support 6to4.

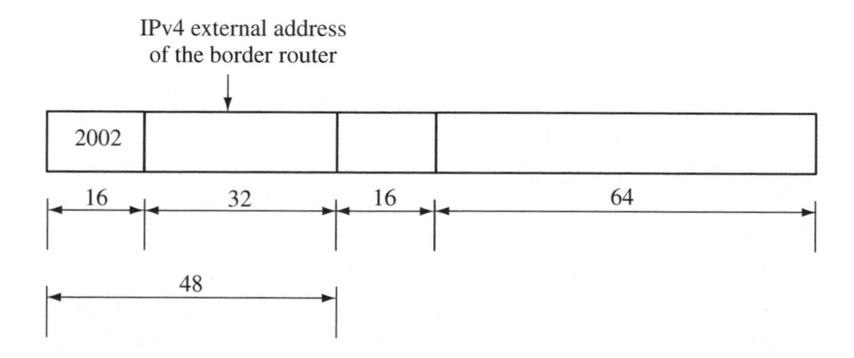

Figure 16.6 6to4 address structure

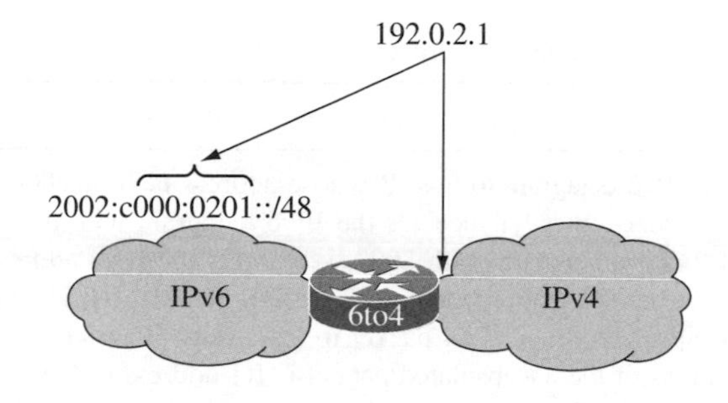

Figure 16.7 6to4 site address space based on the border router IPv4 address

16.2.5.2 Router to Router Tunneling

Figure 16.8 shows an example of the 6to4 automatic tunneling process. Host A (2002:c000:201:1::1) sends a packet to host B (2002:c000:301:2::2). The process is explained in Table 16.4.

B's reply packet will show the same behavior but in the reverse path.

16.2.5.3 Host to Router Tunneling

Similarly, a single host can use its IPv4 address and create the 6to4 /48 prefix for its own use. Figure 16.9 shows the same process as above, but the host is 6to4 capable and encapsulates the IPv6 traffic in the IPv4 packet to the router destination. The process is explained in Table 16.5.

B's reply packet will show the same behavior but in the reverse path.

16.2.5.4 Host to Host Tunneling

A host to host scenario works the same way, where the two hosts are 6to4 capable, as shown in Figure 16.10. Both hosts encapsulate and decapsulate the packets before sending on the wire.

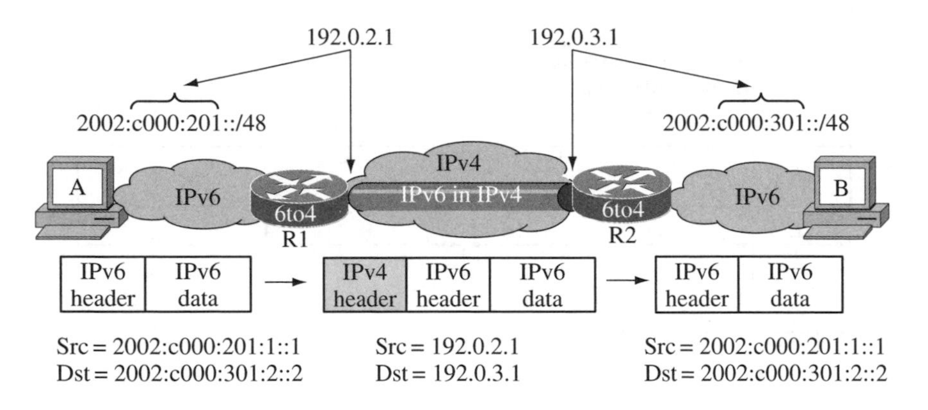

Figure 16.8 6to4 router to router tunneling

Table 16.4 6to4 router to router tunneling process

Step	Description
1	Host A sends an IPv6 datagram to host B whose address is 2002:c000:301:2::2.
2	When the 6to4 border router R1 receives the IPv6 datagram with the destination address 2002:c000:301:2::2, which starts with 2002, it extracts the IPv4 address in the 32 bits following 2002 in the destination address(c000:0301 =>192.0.3.1), and then forwards the IPv6 packet encapsulated in an IPv4 packet to R2 whose IPv4 address is 192.0.3.1. The IPv4 source address of the encapsulated packet is R1 address (192.0.2.1).
3	R2 receives the IPv6 in IPv4 encapsulated packet, decapsulates it and forwards it to B.

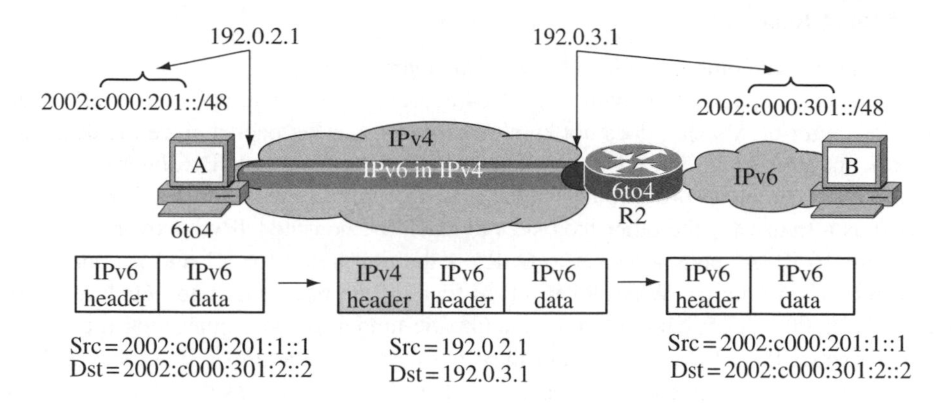

Figure 16.9 6to4 host to router tunneling

Table 16.5 6to4 host to router tunneling process

Step	Description
1	Host A sends an IPv6 datagram to host B whose address is 2002:c000:301:2::2.
2	Host A is 6to4 enabled. When the internal stack processes the IPv6 datagram with the destination address 2002:c000:301:2::2, which starts with 2002, it extracts the IPv4 address in the 32 bits following 2002 in the destination address (C000:0301 => 192.0.3.1), and then forwards the IPv6 packet encapsulated in an IPv4 packet to R2 whose IPv4 address is 192.0.3.1. The IPv4 source address of the encapsulated packet is A address (192.0.2.1).
3	R2 receives the encapsulated packet and then decapsulates it and forwards it to B.

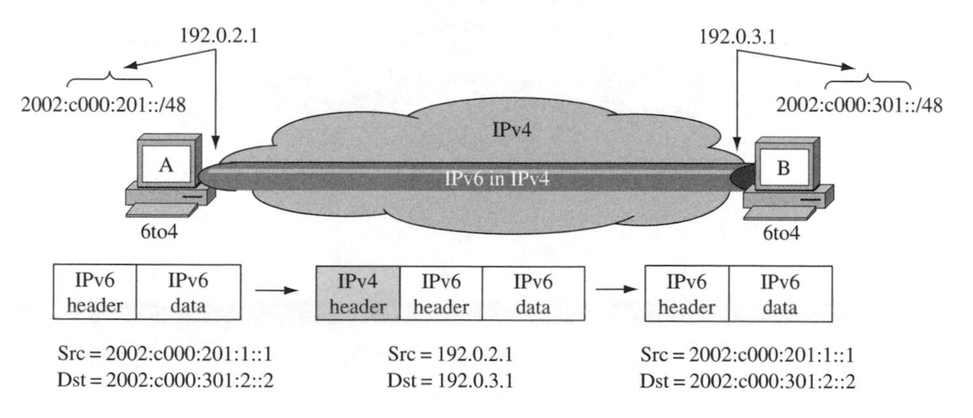

Figure 16.10 6to4 Host to host tunneling

16.2.5.5 6to4 Relay

Figure 16.11 shows communications between 6to4 sites.

If host A sends a packet to C with a non-6to4 address such as 3ffe:b00:1::1, the R1 router, the border router of A's site, does not know where to route the packet since the destination address is not a 6to4 address. R1 needs a 6to4 relay to the non-6to4 IPv6 Internet.

A 6to4 relay is a 6to4 border router that has connectivity to the rest of the IPv6 networks. It is used as a transit for the other 6to4 sites to reach the non-6to4 IPv6 networks.

In Figure 16.12, A sends a packet to C. R3 is the 6to4 relay for R1 and is connected to the non-6to4 IPv6 Internet. When R1 wants to forward the packet to a non-6to4 destination address (3ffe:b00:1::1), it cannot build an automatic tunnel to some other 6to4 router, since it cannot extract the IPv4 address from the IPv6 destination address. If R1 has a default route to go through R3 via the 6to4 mechanism, then R1 encapsulates the IPv6 packet to R3 and R3 decapsulates it and forwards it to the IPv6 network. R3 and the links between R3 and the non-6to4 IPv6 Internet are used as a transit network. In this context, there should be some agreement between the 6to4 sites to use R3 as transit. As a counterpart, any 6to4 border router should be tightly configured with ingress filtering to avoid being used as a transit for other 6to4 sites unless this is intended.

To enable the 6to4 relay, the 6to4 relay router is a 6to4 router with a default route to the IPv6 Internet. The 6to4 relay should contain some ingress filtering. A 6to4 site that is using a 6to4 relay installs in the 6to4 border router an IPv6 default route pointing to the 6to4 address of the relay. 6to4 relay routers do not require specific features to act as 6to4 relays, just a static route entry.

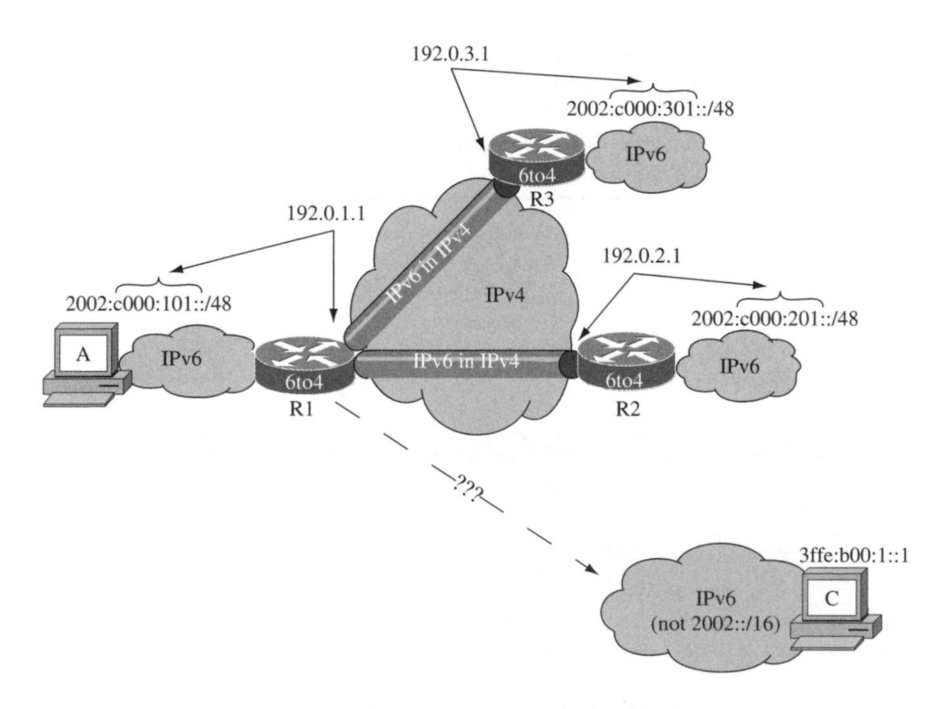

Figure 16.11 6to4 sites without relay

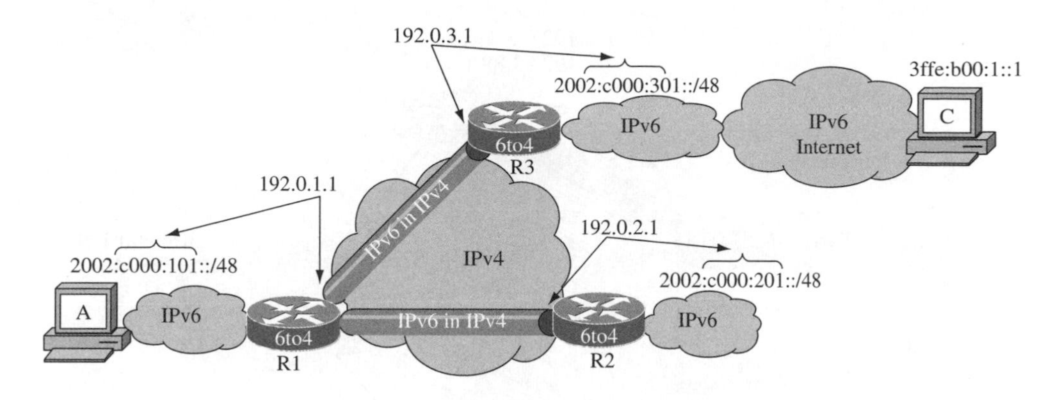

Figure 16.12 6to4 relay

16.2.5.6 Discovery of 6to4 Relays

In the previous example, R1 has to be manually configured to have a default route pointing to the R3 address. The 6to4 mechanism by itself does not depend on static information, but the static route entry for the relay is static.

Anycast, discussed in Section 15.8., is used in a 6to4 additional mechanism [RFC3068] to discover the 6to4 relay automatically. An anycast IPv4 address is reserved: 192.88.99.1. A 6to4 relay is configured with this address as a secondary address and the IPv4 routing infrastructure in the network is configured to route packets to that IPv4 special address.

In Figure 16.13, R1 is not configured with the address of a specific 6to4 relay but with the 6to4 reserved anycast address for its IPv6 default route. R1 forwards the encapsulated packets for the non-6to4 IPv6 Internet to the IPv4 anycast reserved address. The IPv4 routing sends it to the nearest relay (R3). The 6to4 relay (R3) is configured with the anycast address 192.99.88.1 as a secondary address. It also injects the 192.99.88.0/24 prefix in the IPv4 routing table.

Not only does the anycast address bring automatic discovery of the 6to4 relay, but it also provides redundancy and optimal network reachability. In Figure 16.14, R3 and R4 are 6to4 relays that are configured with the anycast address 192.99.88.1 as a secondary address.

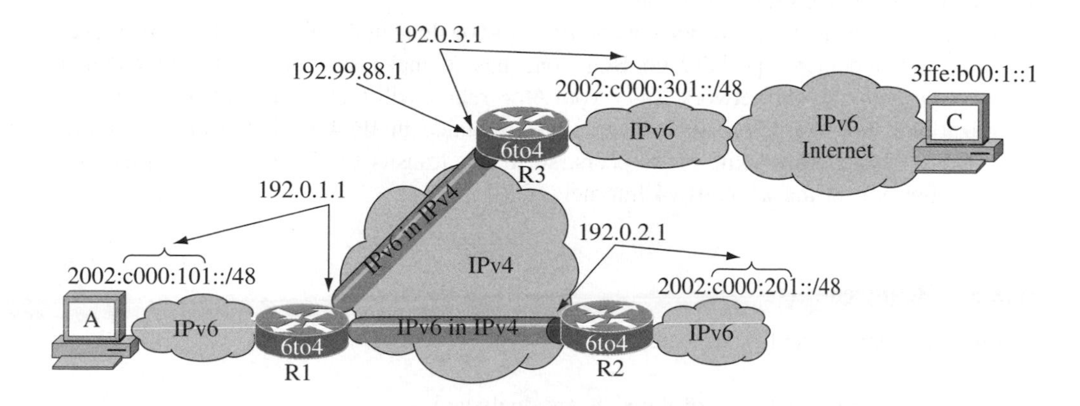

Figure 16.13 6to4 relay discovery

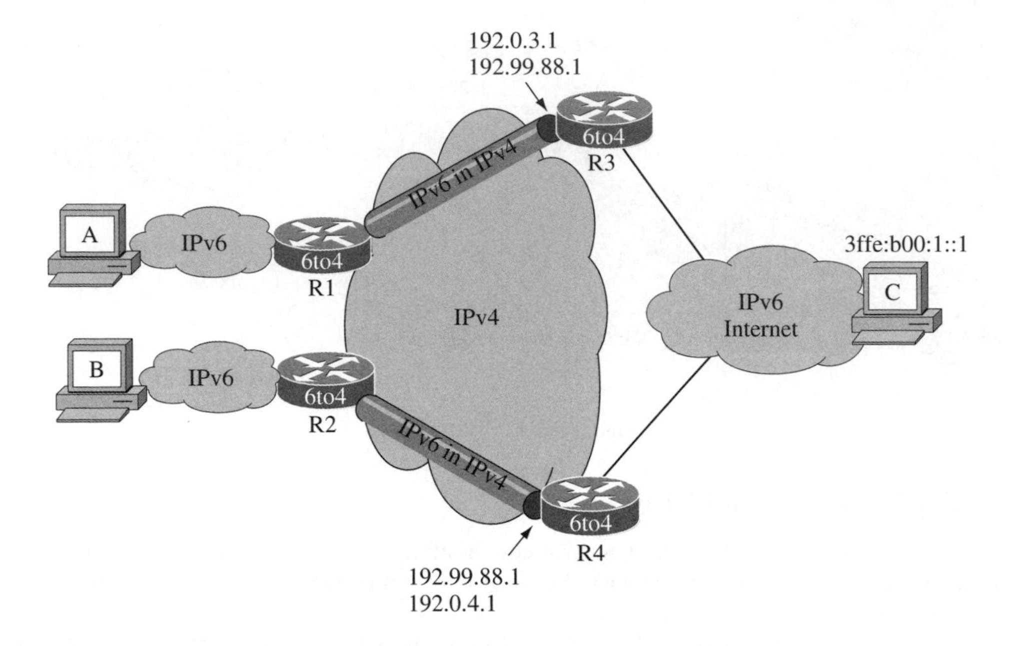

Figure 16.14 Multiple 6to4 anycast relays

When host A sends a packet to C, it has to go through a 6to4 relay since C's address is not a 6to4 address. R1 receives the packet and forwards it – encapsulated – to the IPv4 6to4 anycast address (192.99.88.1). The routing inside the IPv4 network forwards the packet to the nearest router that has this address configured, in this case R3. The same process happens for the packet from B where R2 forwards the packet encapsulated to the IPv4 6to4 anycast address. The IPv4 routing forwards to the nearest router: R4.

When R3 is down, the routing tables in the IPv4 network withdraw the route to R3 for the anycast address. When A sends a new packet, R1 again encapsulates it and forwards it to the IPv4 6to4 anycast address, but the IPv4 routing forwards the encapsulated packet to R4, since R3 no longer advertises the 6to4 anycast address. Anycast then gives automatic redundancy of the 6to4 relay function.

The advertisement of the 6to4 anycast prefix could be done in the IGP or in BGP. However, in each case and more specially on BGP, one has to make sure that this advertisement does not go beyond your network, since your 6to4 relays will then transit traffic from other organizations. For example, a provider could announce in IPv4 BGP the anycast prefix (192.99.88.0/24) to only some of its peers, so that it transits traffic for its customers and partners, but not for the whole IPv4 Internet.

16.2.5.7 Requirements

The requirements to deploy 6to4 are:

- The border routers (R1, R2) of the sites are dual-stack.
- The border routers (R1, R2) of the sites support 6to4.

- All 6to4 sites are reachable through their IPv4 border router address.
- All 6to4 sites have a 6to4 relay, statically configured on the site border router, to transit the non-6to4 IPv6 traffic.
- Hosts do not need to support or know about 6to4.

Important to remember:

The 6to4 mechanism uses the 2002::/16 prefix. Any 6to4 site has the following address space: 2002:<ipv4 external address in hex>::/48. Only the border router has to support 6to4.

16.2.5.8 Limitations

6to4 embeds the IPv4 address inside the IPv6 prefix. So all nodes of a 6to4 site have the IPv4 address of the site border router embedded in their IPv6 address, in the prefix part.

Embedding of the IPv4 address inside the IPv6 address introduces some issues, such as:

- IPv4 address binding;
- redundancy;
- scalability;
- routing priority.

The border router external IPv4 address is used to define the /48 of the site. In the event that the border router changes its IPv4 address, then the whole site has to renumber: all nodes within the site have to change their IPv6 address. This also means DNS changes, configuration changes, network management setup changes, etc. Changing IP addresses of all nodes within a site is a painful experience. Even if IPv6 has an IPv6 renumbering protocol (see Section 9.10), it is still costly and inconvenient to renumber a site.

The border router external IPv4 address is used inside the IPv6 addresses. All the traffic going to the 6to4 site must go through the router that has its IPv4 external address embedded. Even if the site has multiple connections to the IPv4 Internet, all IPv6 encapsulated traffic is always going through the same border router, reachable and known by its IPv4 address. If the router is dead, then the full site is dead, since there is no possible redundant path because the IPv6 address is bound to the external IPv4 address.

If all users of the current IPv4 Internet use 6to4 for IPv6 connectivity, then the full IPv4 address space (2^{32} addresses) will be inserted into the IPv6 site routing table, which has scaling issues.

When a dual-stack node in a combined IPv4/IPv6 network has 6to4 support, if the destination is a 6to4 address, does it send the packet to the 6to4 router or does it make the automatic 6to4 tunnel itself? By default, the node tries to do the 6to4 automatic tunneling, which results inside the site that this packet will be routed through the IPv4 routing instead of the IPv6 routing.

Security considerations for 6to4 have been documented [RFC 3964].

Finally, 6to4 cannot traverse NATs.

16.2.5.9 Applicability

All these above issues make the use of 6to4 relevant in small/home networks where there is no NAT in the path, but typically not in enterprise or large networks.

16.2.6 ISATAP

Intra-Site Automatic Tunnel Addressing Protocol (ISATAP) [Templin *et al.*, 2003] is a mechanism to automate the creation of tunnels from nodes to a router and from nodes to nodes inside a site.

16.2.6.1 Addressing

This mechanism embeds the IPv4 address of the node in the last 32 bits of the interface identifier part of its IPv6 address. Figure 16.15 shows the format of an ISATAP address. The first 32 bits of the interface identifier are '00:00:5E:FE', reserved by IANA for ISATAP, and define the ISATAP interface identifier.

16.2.6.2 Tunneling

ISATAP creates one virtual link over a full IPv4 site. A /64 from the /48 prefix of the site is dedicated for the ISATAP link. This will be used by the ISATAP nodes as their /64 prefix, as shown in Figure 16.15. Link-local (fe80::/64) is also used on the virtual ISATAP link.

Figure 16.16 shows an example of an ISATAP network. Host A, host B and router R1 are all dual-stack and have an enabled implementation of ISATAP. Host C is IPv6 without ISATAP. The network manager uses the 3ffe:b00:1::/48 for its site. It assigns 3ffe:b00:1:2::/64 to the ISATAP virtual link over the IPv4 network.

Host A has the 192.0.2.1 IPv4 address and creates its link-local address based on the ISATAP format: fe80::5efe:c000:0201, computed using the ISATAP 32 bit interface identifier (0000:5efe) and the hexadecimal representation (c000:0201) of its IPv4 address (192.0.2.1). Host B and router R1 do the same respectively, which builds a virtual link as shown in Figure 16.17.

Host A sends a router solicitation to the ISATAP address of the ISATAP router (R1) statically configured in the node. Receiving a router advertisement from the ISATAP router R1 over the ISATAP virtual link, host A configures its global address as 3ffe:b00:1:2:5efe:c000:0201, since 3ffe:b00:1:2::/64 is the prefix of the virtual link received

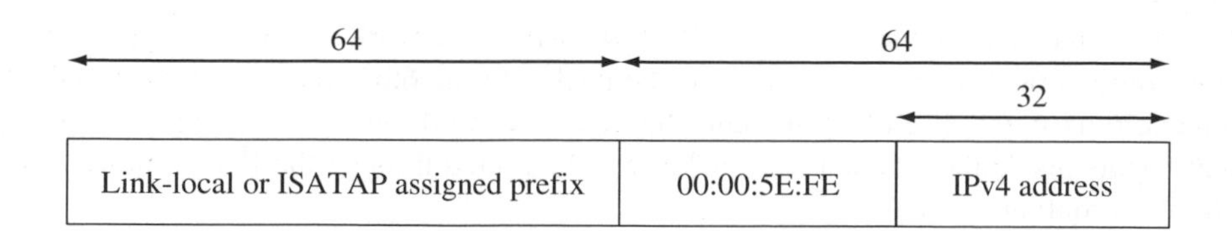

Figure 16.15 ISATAP address format

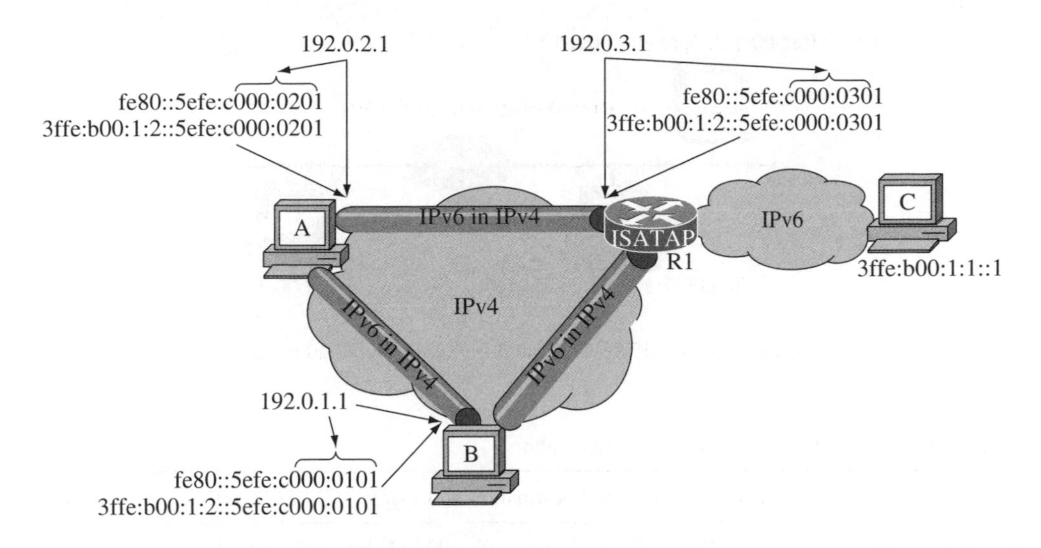

Figure 16.16 ISATAP overlay network

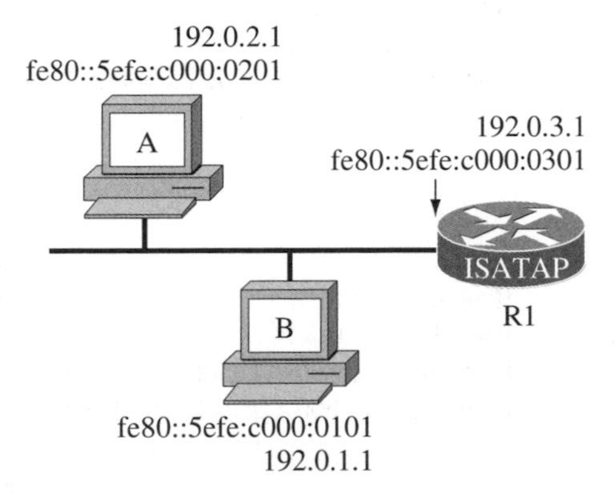

Figure 16.17 ISATAP virtual link with link-local addresses

from the router. The router advertisements also include that R1 is identifying itself as a default router, advertising its ISATAP address as the next hop. Host B does the same respectively, which enhances the virtual link with global addresses and a default router to reach the IPv6 network inside the site, as shown in Figure 16.18.

When A sends a packet to B, it sees B as on the same link, so A sends a neighbor solicitation to the B address. B responds with a neighbor advertisement. These IPv6 messages are encapsulated inside unicast IPv4 packets.

Table 16.6 describes the process when the ISATAP node A sends the first IPv6 packet to ISATAP node B, using Figure 16.16 as the example network.

Table 16.7 describes the process when the ISATAP node A sends the first IPv6 packet to non-ISATAP node C, using Figure 16.16 as the example network.

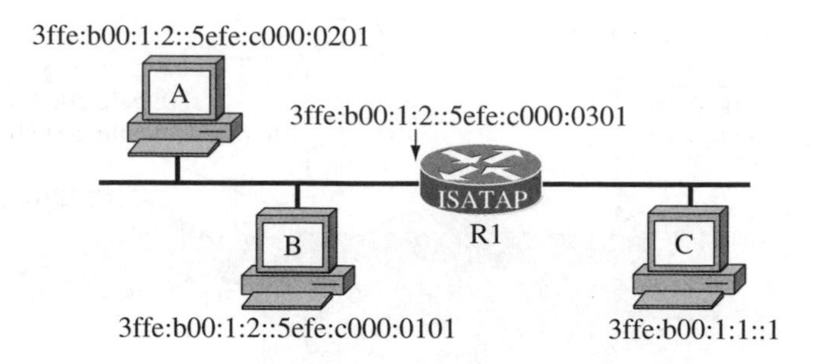

Figure 16.18 ISATAP virtual link with global addresses

Table 16.6 ISATAP node to node communication

From → To	Type of packet	Source and destination addresses	Description and content
0 A wants to send a packet to B. B's IPv6 address (3ffe:b00:1:2::5efe:c000:201) has the same link prefix (3ffe:b00:1:2::/64) as A (3ffe:b00:1:2::5efe:c000:101) on the ISATAP virtual interface of A. Therefore, A does a neighbor solicitation on the ISATAP virtual interface for B.			
1 A → B	IPv6 neighbor solicitation over IPv4	IPv4 src: 192.0.2.1 IPv4 dst: 192.0.1.1 IPv6 src: 3ffe:b00:1:2::5efe:c000:201 IPv6 dst: ff02::1:ff00:101	The neighbor solicitation is encapsulated in IPv4. The IPv4 destination address is taken from the IPv6 destination address. The IPv6 destination address is the solicited-node multicast address (see Section 6.1).
2 B → A	IPv6 neighbor advertisement over IPv4	IPv4 src: 192.0.1.1 IPv4 dst: 192.0.2.1 IPv6 src: 3ffe:b00:1:2::5efe:c000:101 IPv6 dst: 3ffe:b00:1:2::5efe:c000:201	The neighbor advertisement is encapsulated in IPv4. Returns confirmation of 3ffe:b00:1:2::5efe:c000:0101 with link-layer address of 192.0.1.1.
3 A → B	IPv6 packet over IPv4	IPv4 src: 192.0.2.1 IPv4 dst: 192.0.1.1 IPv6 src: 3ffe:b00:1:2::5efe:c000:201 IPv6 dst: 3ffe:b00:1:2::5efe:c000:101	IPv6 packet sent encapsulated.
4 B → A	IPv6 packet over IPv4	IPv4 src: 192.0.1.1 IPv4 dst: 192.0.2.1 IPv6 src: 3ffe:b00:1:2::5efe:c000:101 IPv6 dst: 3ffe:b00:1:2::5efe:c000:201	Reply IPv6 packet.

16.2.6.3 Requirements

The requirements to deploy ISATAP are:

- IPv6 nodes in the site are dual-stack.
- IPv6 nodes implement ISATAP.

Table 16.7 ISATAP node to non-isatap node communication

	From → To	Type of packet	Source and destination addresses	Description and content
0	A wants to send a packet to C. C's IPv6 address (3ffe:b00:1:1::1) does not have the same link prefix (3ffe:b00:1:2::/64) as A (3ffe:b00:1:2::5efe:c000:101) on the ISATAP virtual interface of A. Therefore, A needs to send the packet to C through the default router R1 (statically configured in A). A does a neighbor solicitation on the ISATAP virtual interface for R1.			
1	A → R1	IPv6 neighbor solicitation over IPv4	IPv4 src: 192.0.2.1 IPv4 dst: 192.0.3.1 IPv6 src: fe80::5efe:c000:201 IPv6 dst: ff02::1:ff00:301	The neighbor solicitation is encapsulated in IPv4. The IPv4 destination address is R1, the default ISATAP router, whose address is statically configured in A.
2	R1 → A	IPv6 neighbor advertisement over IPv4	IPv4 src: 192.0.3.1 IPv4 dst: 192.0.2.1 IPv6 src: fe80::5efe:c000:301 IPv6 dst: fe80::5efe:c000:201	The neighbor advertisement is encapsulated in IPv4. Returns confirmation of fe80::5efe:c000:0301 with link-layer address of 192.0.3.1.
3	A → C through R1	IPv6 packet over IPv4	IPv4 src: 192.0.2.1 IPv4 dst: 192.0.3.1 IPv6 src: 3ffe:b00:1:2::5efe:c000:201 IPv6 dst: 3ffe:b00:1:1::1	IPv6 packet sent encapsulated in IPv4 to R1, while C is the final IPv6 destination.
4	R1 → C	IPv6 native packet	IPv6 src: 3ffe:b00:1:2::5efe:c000:201 IPv6 dst: 3ffe:b00:1:1::1	R1 receives the IPv6 in IPv4 encapsulated packet, decapsulates it and send it natively to C.
5	C → A	IPv6 reply packet	IPv6 src: 3ffe:b00:1:1::1 IPv6 dst: 3ffe:b00:1:2::5efe:c000:201	

- A dual-stack router with ISATAP is available on the IPv4 network to reach the rest of IPv6 in the site.
- The ISATAP default router is statically configured in all ISATAP nodes.
- ISATAP is used within a site, or within one administrative domain.

Important to remember

All nodes must support ISATAP in order to communicate. An ISATAP router is needed and must be statically configured in ISATAP nodes.

16.2.6.4 Limitations

ISATAP creates a virtual link over a potentially wide IPv4 network. Any broadcast-like packet, such as one sent to ff01::1, on the link will create a potentially large number of packets on the network. So the more ISATAP nodes that are deployed, the less it is scalable.

ISATAP does not traverse NAT.

16.2.6.5 Applicability

ISATAP is suited for enterprise deployment, but not for providers, and not for home networks. In the enterprise context, no NAT should be in the path between any two ISATAP nodes.

16.2.7 IPv6 in IPv4 Tunneling Considerations

Static tunnels, 6to4 and ISATAP are all using IPv6 in IPv4 encapsulation using IPv4 protocol 41.

Encapsulation of one protocol datagram inside another protocol datagram hides the underlying infrastructure from the encapsulated protocol. This makes routing less efficient, troubleshooting more difficult, tricks the Hop limit processing and requires special processing of MTU discovery and ICMP messages and requires traversing firewalls and NATs.

16.2.7.1 Routing Inefficiency

When overlaying one IP protocol over the other, the routing realms of the two IP protocols are unrelated which sometimes results in less optimal routing decisions of the encapsulated protocol.

Figure 16.19 shows an example of IPv6 tunnels over an IPv4 network. Router R1 has a tunnel to R4 which goes through R2 and R3 in the IPv4 network. R1 also has another tunnel to R5. When the host H1 sends an IPv6 datagram to H2, it sends it to R1. R1 encapsulates the IPv6 datagram in an IPv4 datagram and forwards it to R4, which decapsulates it and sends the IPv6 datagram to H2. R1 may also forward the encapsulated datagram to R5, which decapsulates it and sends the IPv6 datagram to R6 which forwards it to H2.

Figure 16.20 shows the IPv6 routing view of the Figure 16.19 network, without the underlying (IPv4) infrastructure.

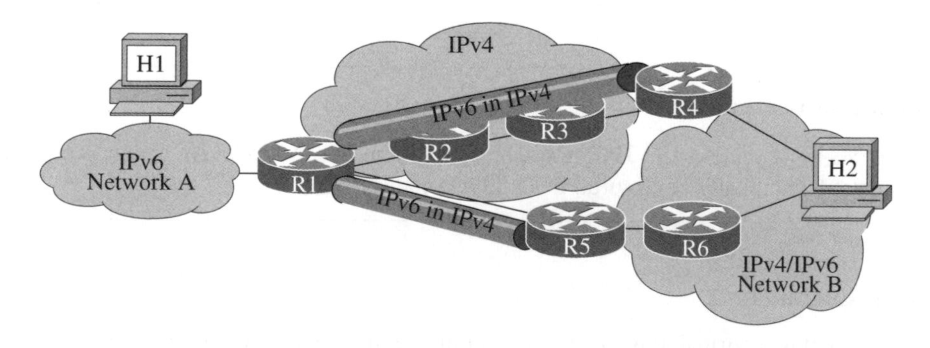

Figure 16.19 IPv4/IPv6 routing realms example

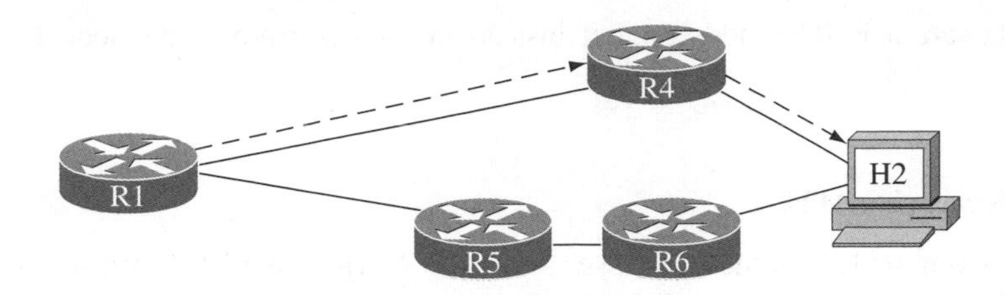

Figure 16.20 IPv6 routing realm

In default distance-vector routing, R1 chooses R4 as the next hop since the path to H2 is shorter in the number of hops: 1 hop (R4) instead of 2 hops (R5–R6). So IPv6 datagrams are forwarded through the R1–R4 tunnel.

However, in the underlying IPv4 infrastructure, the reality is different. Figure 16.21 shows the underlying (IPv4) routing infrastructure of the Figure 16.19 network.

In a default distance-vector routing, an IPv4 datagram sent to H2 is forwarded to R1 and R5 since the path is shorter: 2 hops (R5–R6) instead of 3 hops (R2–R3–R4). Since the underlying IPv4 infrastructure corresponds to the real routing infrastructure, the chosen IPv6 forwarding path (through R4), as shown in Figure 16.20, is not the optimal one (through R5).

While this is a generic issue, it does not apply with the same level of concern in all cases. To minimize the effect, make the tunneled virtual network topology as close as possible to the underlying network topology. In other words, keep the underlying hop count as small as possible and as linear as possible.

16.2.7.2 Troubleshooting is more difficult

Since the encapsulated IPv6 datagrams are not directly seen in the underlying IPv4 infrastructure, the troubleshooting is more difficult. For example, the management tools of the IPv4 network may show some congestion between two routers. Without encapsulation, it is easy to find the source and destination of the datagrams that generate heavy traffic and then fix the problem, by shutting down the source node for example. However, with encapsulated datagrams, the real source of the datagrams is not shown directly but is encapsulated in the packets. For example, the traffic generating the tunneled datagram traveling between R1 and R4 in Figure 16.19 does not originate from R1, even if the source address of

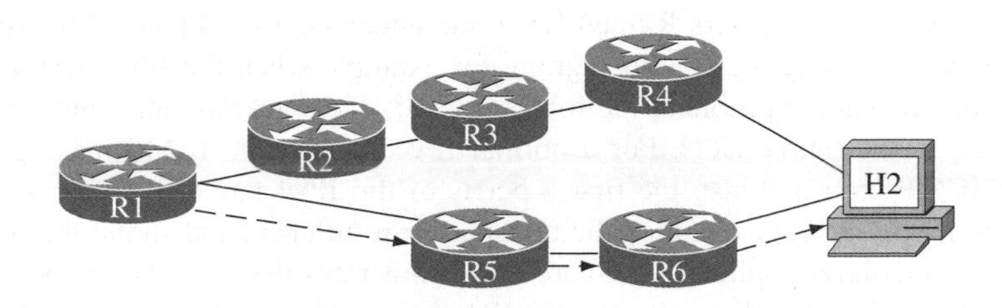

Figure 16.21 IPv4 routing realm

the IPv4 datagram is R1's address, but instead originates from some node in IPv6 network A.

16.2.7.3 Decreasing MTU

An encapsulating node sees the IPv4 layer as a layer 2, where the IPv4 infrastructure is seen as a virtual link with an 'unlimited' MTU. Since the maximum IPv6 MTU is 65536 octets, then the maximum theoretical MTU over the IPv4 tunnel is $65536 - 20\text{octets} = 65516$, since 20 octets are used for the IPv4 header.

If the IPv4 path MTU, for example 1400 octets, between the two encapsulation points (R1 and R4 in Figure 16.19) is smaller than the MTU used at the IPv6 layer, for example 1500 octets, then the IPv6 source node sends larger datagrams, for example 1450 octets, than the IPv4 path can support, and fragmentation happens at the IPv4 layer. Fragmentation impacts on the performance of routers as well as on destination nodes. It also increases the end-to-end delay since the decapsulating router or node has to wait for all fragments before forwarding.

Since the decapsulating router is the IPv4 destination, for example R4 in Figure 16.19, it has to store the IPv4 fragments as they arrive and wait until all IPv4 fragments arrive to reassemble the IPv4 datagram and then decapsulate in order to forward the IPv6 datagram. This causes additional memory storage and processing time on the decapsulating router and delivery delays.

To avoid this situation, the encapsulating router should do an IPv4 path MTU discovery. Then the IPv6 MTU on the tunnel is the IPv4 path MTU found minus the size of IPv4 header (20 octets). The encapsulating router should also set the 'Don't fragment' bit in the IPv4 header to avoid fragmentation. However, recall that the guaranteed minimum IPv6 MTU is 1280 octets, as discussed in Section 3.6.2. If the calculated IPv6 MTU on the tunnel, based on the discovered IPv4 path MTU, is less than the minimum IPv6 MTU (1280 octets), then whenever the source node sends a datagram larger than the IPv6 MTU on the tunnel, IPv4 fragmentation must be used, since that would be the only way to succeed sending in the datagram to the destination node. If IPv4 fragmentation is not used in this special case, then all datagrams larger than the IPv6 MTU on the tunnel will be discarded and an ICMP 'Packet too big' error message sent back to the source.

Most current link-layers have a MTU of at least 1500 octets, so these issues may not happen very often.

16.2.7.4 ICMP Processing

In Figure 16.19, the IPv4 routers R2 and R3 in the underlying IPv4 path of the tunnel may have problems forwarding the IPv4 datagram, for example when the IPv4 destination (R4) is unreachable. Seeing the problem, the router discards the datagram and sends an ICMPv4 message back to the source (R1). For a normal IPv4 packet, the ICMP message contains the type of ICMP error and also the first 8 octets of the IPv4 payload, so the source node receiving the ICMP message can read the transport port numbers and signal the ICMP error to the proper upper layer application. Figure 16.22 illustrates the extraction done by a router of the IP header and the first 8 octets after the IP header, which contains the transport header. This extraction is sent in the payload of the ICMP message.

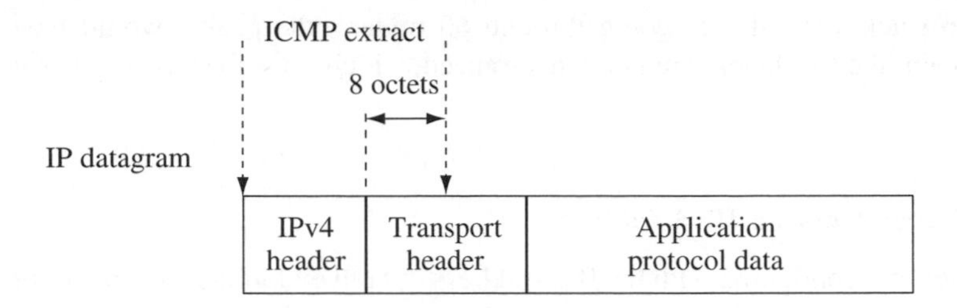

Figure 16.22 ICMP extraction

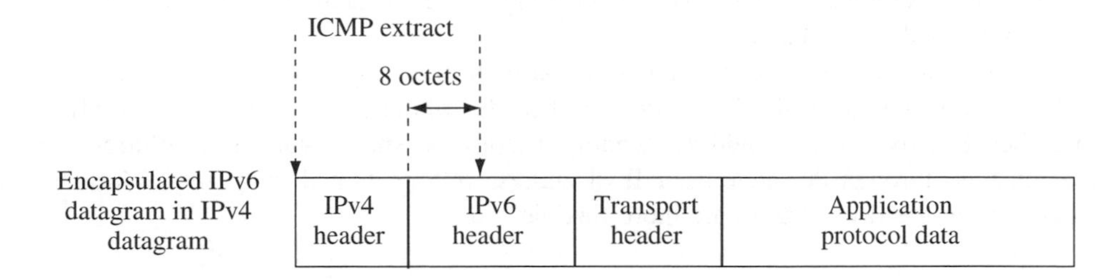

Figure 16.23 ICMP extraction of an encapsulated datagram

However, if the datagram is an encapsulated datagram as in transit between R1 and R4 in Figure 16.19, then the first 8 octets of the IPv4 payload is the first 8 octets of the inner IPv6 header, as illustrated in Figure 16.23. Obviously, this does not include the transport port numbers as in Figure 16.22, but instead contains the version, traffic class, flow label, payload length, next header and hop limit fields. No IPv6 source and destination address and no transport information is available in the first 8 octets of the IPv4 payload of an IPv6 encapsulated packet.

The source of the IPv4 datagram, R1 in Figure 16.19, receiving the ICMP message, does not have any idea how to find the IPv6 source of the inner datagram, since the ICMP data are not large enough.

Even worse, the error should be notified to the IPv6 source (H1) of the original packet, but the router R1 does not receive the IPv6 source address and the port numbers inside the ICMP payload.

To circumvent this problem, encapsulating routers maintain a soft-state database of the tunnels. In the case of an ICMP IPv4 error message, the router is then able to reconstruct from its soft-state table the IPv6 information and create an IPv6 ICMP error message to be sent to the IPv6 source node of the original packet.

16.2.7.5 Hop Limit

The hop limit in IPv6 or TTL in IPv4 is decremented each time a router forwards the packet, as discussed in Section 3.4.5. For an IPv6 in IPv4 tunnel, the IPv4 packet might go through 60 routers, where the IPv6 packet might still be in the same tunnel. The IPv6 traffic is not aware of this behavior. If the IPv6 source stack sets the hop limit to 32 hops while the IPv4 stack TTL is set to 64, then the packet will not be discarded, but one can guess that if the IPv6 stack knew it was going through 60 routers, the packet would have been discarded. Again, the two layers are not synchronized, which, in some cases, gives non-optimal behaviors.

16.2.7.6 Dependency on IPv4 Address

Static tunnels are configured with the IPv4 address of both endpoints. Whenever one endpoint changes its IPv4 address, the tunnel needs to be reconfigured to work again. If static tunnels are used by nodes that have a probability of changing IPv4 address, such as DHCP clients, then they are not the best tool.

6to4 and ISATAP are mechanisms embedding the IPv4 address of the node inside the specially formed IPv6 address.

If the node moves, then the IPv6 address is also changed. Moreover, if a gateway receives an IPv6 prefix based on the IPv4 address, then the whole network using this prefix will renumber. Finally, the IPv4 address binding disables possible redundancy solutions, since one must go through the advertised IPv4 address for the tunnel endpoint. Solutions not dependent on the IPv4 address are more flexible.

16.2.7.7 Filtering IPv4 Protocol 41

An IPv6 packet encapsulated in an IPv4 packet is identified with 41 in the protocol field of an IPv4 header. Since security gateways have the typical rule of denying everything unless specifically permitted, protocol 41 packets are often blocked by security gateways. IPv6 in IPv4 tunnels usually do not work across a security gateway unless the manager of the gateway specifically permits them. Static tunnels, 6to4 and ISATAP do not work if such a security gateway policy is applied in the path of the tunnel.

16.2.7.8 IPv4 Network Address Translation

An IPv6 in IPv4 tunnel requires that both tunnel endpoints are reachable by their IPv4 address. If one or many NATs are in the path between the two endpoints, then the tunnel does not work. This is because the IP address of the endpoint is translated by the NAT, as shown in Figure 16.24.

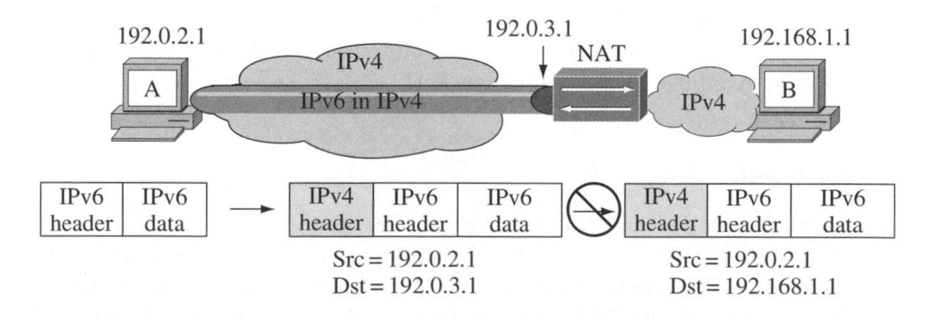

Figure 16.24 Tunnels do not traverse NAT

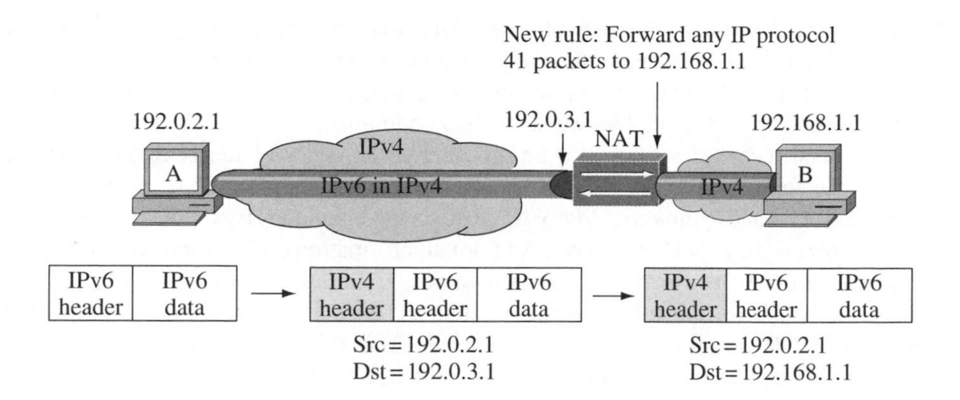

Figure 16.25 NAT forwarding IP protocol 41

In Figure 16.24, node A knows B by the external address of the NAT (192.0.3.1). When node A sends an encapsulated packet to B, it reaches the NAT but it is not forwarded to B, nor is the destination address changed.

Workarounds are possible to enable the tunnel when only one NAT is in the path and when this NAT is under the control of one of the tunnel endpoint administrators.

The first workaround is to configure the NAT statically to forward any IPv4 protocol 41 packet to the specific IPv4 internal address of the tunnel endpoint, as shown in Figure 16.25.

In Figure 16.25, a rule is added to the NAT where any IP protocol 41 packets (containing IPv6 packets) are forwarded to 192.168.1.1. Only one node behind the NAT can receive the packets, limiting this workaround to one node per NAT. If multiple NATs are in the path, it is nearly operationally impossible to add a rule to each of them to let the traffic through. Moreover, many NAT implementations do not support specific processing and forwarding of IPv4 protocol 41.

The other workaround is to assign a secondary address on the external interface of the NAT to be the mapped IP address of the internal endpoint. Any traffic going to the secondary external address is forwarded to the internal endpoint as shown in Figure 16.26. However, secondary addresses are often not available.

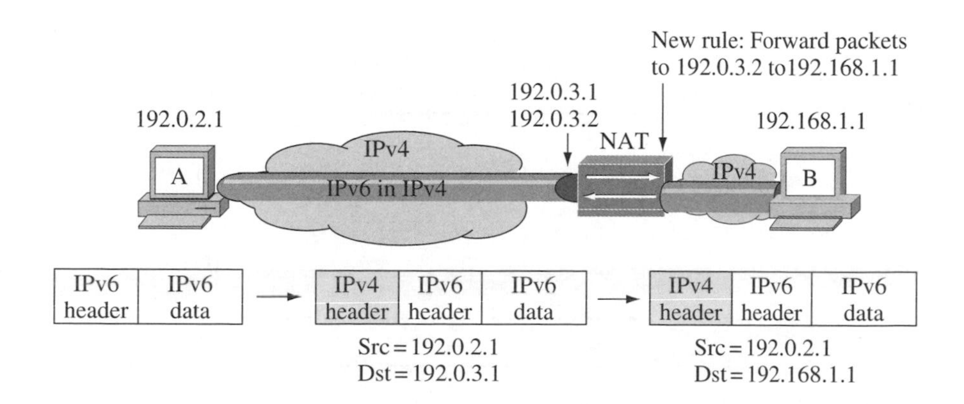

Figure 16.26 NAT mapping to a secondary address

These two workarounds obviously do not scale: only one tunnel is possible per NAT external IPv4 address. Moreover, they require administrative control of the NAT, static configuration of the NAT and support of this forwarding capability, and they will not work if multiple NATs are in the path between the two endpoints.

NAT traversal techniques are based on the internal behavior of the NAT port mapping [RFC2663]. Different categories [RFC3489] of NAT are defined: Full-cone, restricted cone, port restricted cone and symmetric. Many NAT traversal techniques try to identify the kind of NAT to optimize the traversal of packets. Additional information on NAT traversal techniques and categories of NAT are available on the book Web site (http://www.ipv6book.ca).

To traverse NAT, most of the solutions, such as STUN [RFC3489] for IPv4 applications, are based on UDP encapsulation of the IP packet. The same technique is used for IPv6 tunneling techniques over IPv4 with the presence of NATs in the path.

16.2.8 Encapsulating IPv6 in UDP IPv4

As discussed in section 16.2.7, IPv4 address translation (NAT) blocks IPv6 in IPv4 tunnels, since the node behind the NAT cannot be reached by the other party. One solution is to move up in the networking stack by encapsulating IPv6 packets in UDP transport in IPv4 packets, as shown in Figure 16.27.

The IP protocol is connectionless as UDP, which makes UDP a better candidate for transport encapsulation than TCP. TCP connection algorithm and header length are too important overhead for layer 3 encapsulation. Stream Control Transport Protocol (SCTP) [RFC2969] is a possible alternative to UDP but its lack of support by NATs and firewall implementations makes it useless to traverse deployed NATs in the field.

UDP being connectionless, NAT implementations cannot manage, as in TCP, an outgoing connection and close the mapped port when the connection is terminated. In UDP, no such

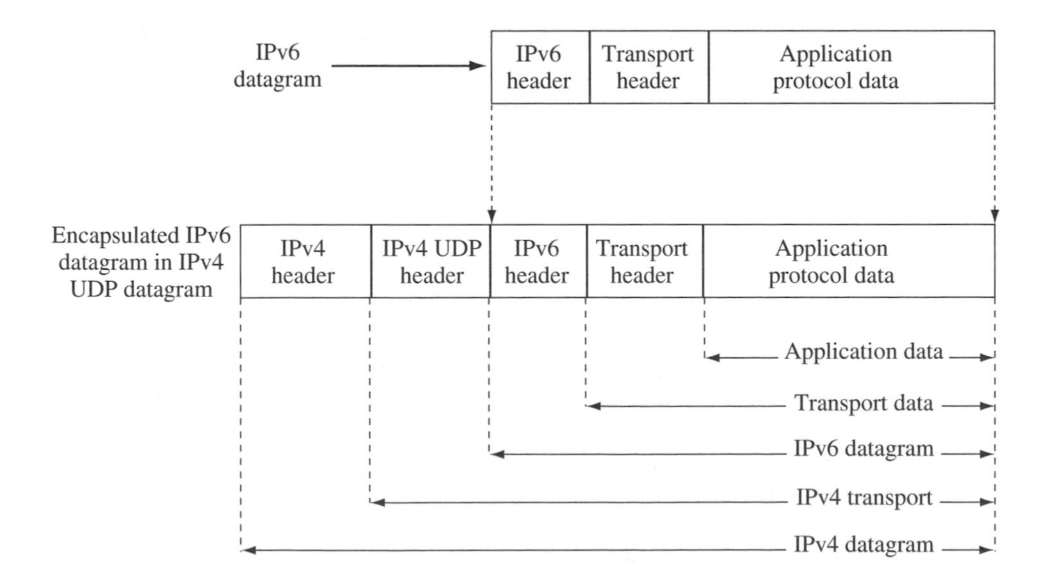

Figure 16.27 Encapsulation of an IPv6 datagram in a UDP-IPv4 datagram

termination happens. NATs manage the UDP connections by maintaining a timer of the absence of traffic on that connection before closing the mapped port. The timeout is usually between 30 seconds and a few minutes [Huitema, 2005; RFC3489].

16.2.9 Tunnel Setup Protocol (TSP) Tunnel Broker

The Tunnel Setup Protocol (TSP) [Blanchet and Parent, 2005] is designed to automate the process of establishing IPv6 in IPv4 tunnels in order to provide ubiquitous and transparent access to IPv6. It is a signaling protocol between the client and the broker, where the client requests a tunnel and the broker sends the information about the assigned tunnel. Both parties then configure their respective tunnel endpoints and the tunnel is established. TSP is an enhanced version of the tunnel broker model [RFC3053].

16.2.9.1 Basic Architecture

Figure 16.28 shows the basic components of the TSP tunnel broker architecture.

A node N1 has the TSP client software and connects to the Tunnel broker TB using TCP or UDP on the TSP well known port (3653). After authentication[1], the client requests the tunnel and the broker assigns and configures the new tunnel to the tunnel server TS1. The broker replies to the tunnel request from N1 by sending the tunnel info. The TSP client on node N1 configures its tunnel endpoint as a static tunnel (discussed in Section 16.2.4), but managed by TSP. The tunnel is established and IPv6 packets are encapsulated in IPv4 from the node N1 to the tunnel server TS1.

16.2.9.2 Combining Tunnel Broker and Server

The tunnel broker (TB) and server (TS) may also be combined on the same hardware, as shown in Figure 16.29.

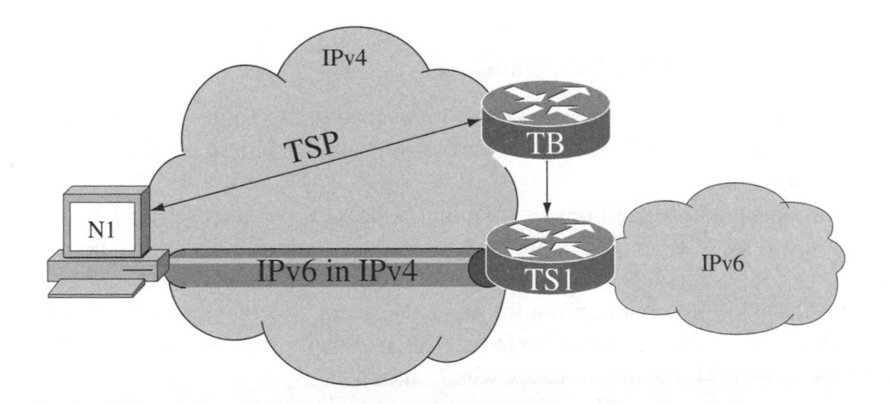

Figure 16.28 TSP tunnel broker

[1] Authentication is optional.

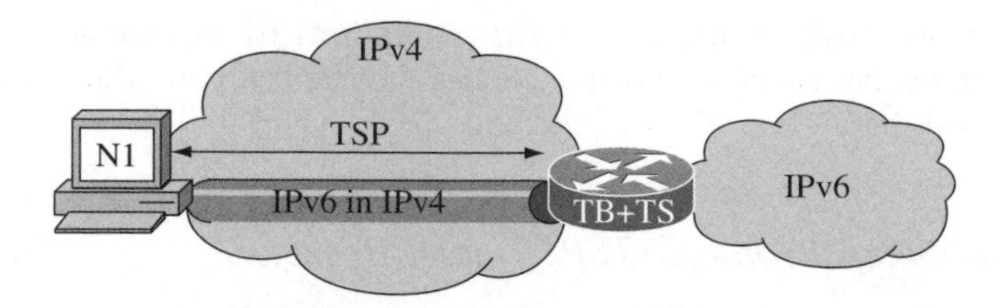

Figure 16.29 TSP tunnel broker and server combined

For the rest of the section, the two components are discussed as if they were separate. An implementation can decide to combine the two in the same hardware. However, to support the NAT traversal feature for all kinds of NAT, the tunnel broker and the tunnel server must be on the same interface sharing the same IPv4 address. Since most deployments over IPv4 networks require NAT traversal, then most implementations are combining the tunnel broker and the tunnel server together.

16.2.9.3 TSP Flow

The tunnel is setup between the two endpoints with the TSP exchange. When a tunnel client wants to establish a tunnel with the tunnel broker, the TSP client connects to the TSP server over TCP on port 3653 on the tunnel broker, authenticates itself to the broker, requests the tunnel and gets the tunnel information response from the broker. Both parties end the TSP connection and then configure on their side the tunnel endpoint. The tunnel is then established. Figure 16.30 shows the process of a TSP client requesting a tunnel.

Table 16.8 describes each step of Figure 16.30 in detail.

Although described in 12 steps, the setup consists of only a few round trip packets, and therefore is a pretty lightweight signaling protocol.

16.2.9.4 TSP Tunnel Broker Capabilities

In step 2 of Figure 16.30, the broker sends all its capabilities: Table 16.9 shows the list of some defined capabilities. A broker can advertise some or all of these capabilities, depending on its configuration.

Tunnel encapsulation capabilities of TSP tunnel brokers, such as IPv6 in IPv4, IPv6 in UDP-IPv4 and IPv4 in IPv6, enable a client to establish connectivity in all these contexts, making the solution versatile. For example, the client might not know if it is behind a NAT. With the TSP protocol capabilities and subsequent processing discussed below, the broker is able to find if the client is behind a NAT and then proposes an IPv6 in UDP-IPv4 tunnel to the client. Moreover, in cases where the client is on an IPv6-only network and needs IPv4 connectivity, the TSP negotiation manages this situation automatically. The TSP negotiation permits the adaptation of the context to establish the connectivity for the other IP protocol to which the client is not yet connected. When the TSP client is a mobile node, the service enables the client to be connected on both IP networks, whether its current point of attachment has one or the other IP protocol available.

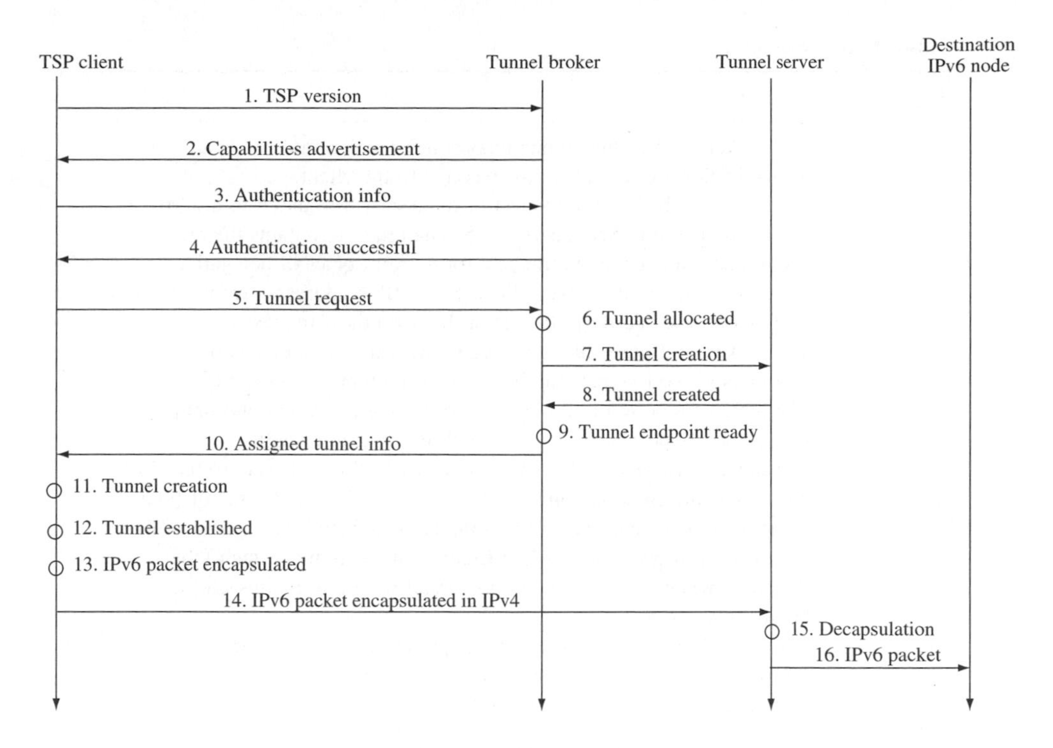

Figure 16.30 TSP flow

Table 16.8 TSP flow detailed

Step	Description
1	The TSP client connects to the TSP server port (3653) using TCP over IPv4 of the tunnel broker. The IPv4 address of the tunnel broker is configured in the TSP client. When the TCP connection is established, the client sends the version of the TSP protocol it is using to the broker. The version declaration enables the TSP protocol to evolve while keeping the TSP client simple and putting the burden of managing the multiple versions on the broker.
2	The tunnel broker verifies that it supports the received version of the TSP protocol and then sends a list of its capabilities, based on its configuration. For example, the broker advertises to the client the supported authentication mechanisms, tunneling encapsulation modes, prefix delegation and DNS delegation modes.
3	Based on the received capabilities and the TSP client configuration, the TSP client sends its authentication information, using the SASL [RFC2222] format. For example, the client sends its username and password.

(*continued overleaf*)

Table 16.8 (*continued*)

Step	Description
4	The broker verifies the authentication information and sends an authentication successful status message to the client.
5	The client sends a tunnel request to the server, formatted in a XML object, as shown in Section 16.2.9.5. The request contains the client IPv4 address and may contain a request for a prefix or DNS delegation.
6	For a new tunnel, the broker allocates the IPv6 addresses from a pool of IPv6 addresses, and may allocate an IPv6 prefix if requested.
7	The broker configures the allocated tunnel on its tunnel server.
8	The tunnel server signals the broker that the tunnel is created.
9	The tunnel endpoint on the broker/server side is created and tunnel information is permanently saved with the user information.
10	The tunnel information, formatted in a XML object, is sent to the TSP client. The information contains the IPv4 and IPv6 addresses of both tunnel endpoints and may contain the allocated IPv6 prefix and DNS information. It also contains the lifetime of the tunnel which tells the client when it should renew its tunnel before it expires on the broker/server side.
11	The TSP client creates its tunnel endpoint based on the tunnel information received from the broker in the previous step.
12	The tunnel is established between the two endpoints.
13	The client can then start sending IPv6 packets over the IPv6-over-IPv4 tunnel. For example, an IPv6 packet is encapsulated into an IPv4 packet with the destination address to the tunnel server.
14	The IPv6 in IPv4 packet is sent to the tunnel server.
15	The tunnel server receives the packet and decapsulates it to forward the IPv6 packet natively on the IPv6 network to the IPv6 destination node.
16	The destination node receives the IPv6 packet.

Table 16.9 TSP tunnel broker capabilities

Capability keyword	Description
TUNNEL = V6V4	This broker offers IPv6 in IPv4 tunnels.
TUNNEL = V6UDPV4	This broker offers IPv6 in UDP in IPv4 tunnels.
TUNNEL = V4V6	This broker offers IPv4 in IPv6 tunnels.
AUTH = ANONYMOUS	This broker supports anonymous authentication [RFC2245].
AUTH = PLAIN	This broker supports plain text authentication [RFC2222], where the username and passwords are transfered in the clear, without any encryption.
AUTH = DIGEST-MD5	This broker supports the digest-md5 authentication mechanism [RFC2831], where the password is not sent in the clear.

16.2.9.5 TSP Tunnel Request XML Messaging

In step 5 of Figure 16.30, the client sends its tunnel request in XML format. A request for an IPv6 in IPv4 tunnel is formatted as follows:

```
<tunnel action="create" type="v6v4">
 <client>
  <address type="ipv4">192.0.2.1</address>
 </client>
</tunnel>
```

The client requests a tunnel (`<tunnel>`) to be created (`action="create"`) using IPv6 in IPv4 encapsulation (`type="v6v4"`), advertising its IPv4 address (`<address type="ipv4"> 192.0.2.1</address>`).

The broker responds with the tunnel information (step 10 in figure 16.30), confirming the tunnel creation, formatted in XML as in the following:

```
<tunnel action="info" type="v6v4" lifetime="1440">
 <server>
  <address type="ipv4">192.0.1.1</address>
  <address type="ipv6"> 3ffe:b00:0:1::</address>
 </server>
 <client>
  <address type="ipv4">192.0.2.1</address>
  <address type="ipv6"> 3ffe:b00:0:1::1</address>
 </client>
</tunnel>
```

The IPv6 addresses of both endpoints and the IPv4 address of the tunnel server endpoint (192.0.1.1) are sent to the client. The client has all the information needed to configure its own tunnel endpoint.

16.2.9.6 Lifetime of the Tunnel

In the previous example, the broker also includes in its response the lifetime of the tunnel (1440 seconds) to tell the client when the tunnel will be expired. It is the responsibility of the client to reconnect to the broker before the lifetime expires to keep the tunnel up. A typical TSP client implementation usually sleeps for the lifetime period and then wakes up before the lifetime expires to reconnect to the broker. The renewal process uses the same messages: the client requests the creation of a tunnel and the broker sends the same tunnel information as before. If the tunnel is still up on both sides, then the broker and the client do not create or change the tunnel configuration on their respective sides.

16.2.9.7 Complete TSP Transaction

Table 16.10 shows an example of a simple TSP connection with anonymous authentication. The first column contains the step number. The second column identifies if the client (C) or the server (S) sends the message described in the third column.

Table 16.10 Anonymous TSP connection

		TSP message
0		TCP connection established on port 3653
1	C	VERSION=1.0
2	S	CAPABILITY TUNNEL=V6V4 AUTH=DIGEST-MD5 AUTH=ANONYMOUS
3	C	AUTHENTICATE ANONYMOUS
4	S	200 Authentication successful
5	C	Content-length: 123 <tunnel action="create" type="v6v4"> <client> <address type="ipv4">192.0.2.1</address> </client> </tunnel>
6	S	Content-length: 234 200 OK <tunnel action="info" type="v6v4" lifetime="1440"> <server> <address type="ipv4">192.0.1.1</address> <address type="ipv6">3ffe:b00:0:1::</address> </server> <client> <address type="ipv4">192.0.2.1</address> <address type="ipv6">3ffe:b00:0:1::1</address> </client> </tunnel>
7	C	Close the TCP connection.

Step 0 is the TCP connection established between the client and the broker over IPv4 on port 3653.

Step 1 shows the initial TCP connection of the client to the broker advertising the highest version number (VERSION=1.0) of the TSP protocol supported by the client. This enables further versions of the protocol to be deployed gracefully. The broker has the burden to support multiple versions of the protocol.

In step 2, if the broker supports the version advertised by the client, the broker advertises its capabilities to the client. In this example, the broker supports IPv6 over IPv4 tunnels (TUNNEL=V6V4) and authentication using digest-md5 (AUTH=DIGEST-MD5) or anonymous (AUTH=ANONYMOUS). This advertisement enables the client to know the broker capabilities and chooses among them.

Step 3 shows the client choosing the anonymous authentication (AUTHENTICATE ANONYMOUS), where no username or password is required.

In step 4, the broker verifies the authentication, in this case nothing, and reports successful authentication (200 Authentication successful). If the client chooses the digest-md5 authentication, then it sends the appropriate userid and the digest-md5 hash of its password to the broker, and the broker verifies the user credentials.

In step 5, after successful authentication, the client sends its tunnel creation request (action="create") for an IPv6-in-IPv4 tunnel (type="v6v4"), using its IPv4 address (<address type="ipv4">192.0.2.1</address>) as its endpoint of the tunnel.

In step 6, the broker processes the answer, allocates the IPv6 addresses, configures the tunnel server for the tunnel and then responds positively to the client request (200 OK) and sends to the client the necessary information for the client to configure its tunnel endpoint: the server's and the client's IPv4 and IPv6 addresses. It also defines a lifetime (lifetime="1440") for the tunnel.

Then the client closes the TSP TCP connection and configures its endpoint of the tunnel. The tunnel is then active and IPv6 traffic can go through the tunnel.

16.2.9.8 TSP Client Behind a NAT

As discussed in Section 16.2.7.8, an IPv6 in IPv4 tunnel does not work if a NAT is in the path between the two endpoints. To traverse NATs, the TSP client and broker handle the TSP session over UDP IPv4, as shown in Figure 16.31.

When the TSP session is terminated, the tunnel is established over the same UDP channel, ensuring that the NAT port mapping is preserved. The IPv6 traffic is encapsulated in UDP IPv4 and traverses the NAT through the same port mapping.

A TSP broker discovers when a client is behind a NAT because the embedded IPv4 address inside the TSP XML tunnel request is different from the source IPv4 address of the TSP packet, as shown in Figure 16.32.

When the TSP client N1 assembles its TSP packet, its IPv4 address 10.1.1.1 is put as the source address of the packet and inside the TSP XML tunnel request. The NAT changes the source address to 192.0.2.1 but does not change the address in the TSP XML tunnel request. The TSP broker compares the two addresses received and detects the presence of a NAT.

When a NAT is detected, the broker offers to the client an IPv6 in UDP IPv4 tunnel, using the 'v6udpv4' keyword, as shown below.

```
<tunnel action="create" type="v6v4">
 <client>
  <address type="ipv4">10.1.1.1</address>
 </client>
</tunnel>
```

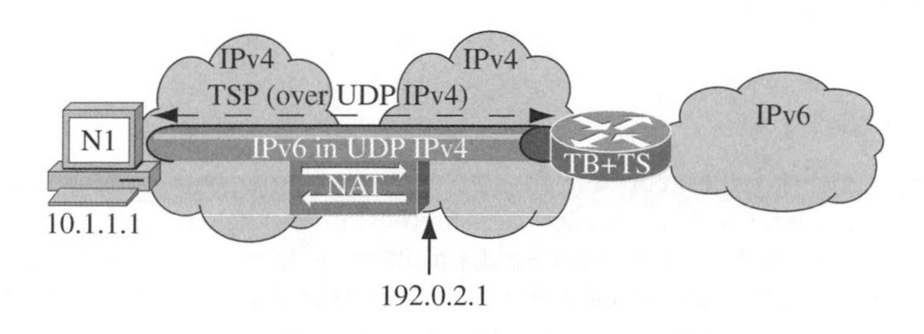

Figure 16.31 TSP session over UDP-IPv4

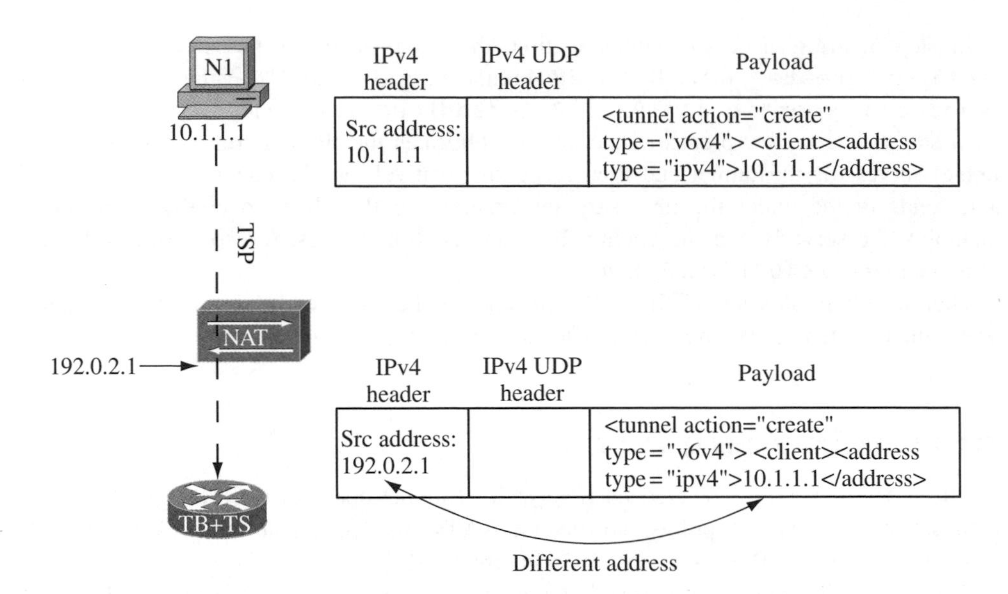

Figure 16.32 TSP NAT traversal detection

The client requests a tunnel (`<tunnel>`) to be created (`action="create"`) using IPv6 in IPv4 encapsulation (`type="v6v4"`), advertising its IPv4 address (`<address type="ipv4">10.1.1.1</address>`). The broker detects the client is behind a NAT, and offers instead the IPv6 in UDP IPv4 encapsulation.

```
<tunnel action="info" type="v6udpv4" lifetime="1440">
 <server>
  <address type="ipv4">192.0.1.1</address>
  <address type="ipv6">3ffe:b00:0:1::</address>
 </server>
 <client>
  <address type="ipv4">10.1.1.1</address>
  <address type="ipv6">3ffe:b00:0:1::1</address>
 </client>
</tunnel>
```

This IPv6 in UDP IPv4 tunnel will then take place over the same UDP channel that was established for the TSP connection.

16.2.9.9 Keepalive

To detect any network failure between the two endpoints, a keepalive mechanism is implemented where the client and the broker exchange a null packet periodically. This enables the TSP client to detect the failure, remove the route in its kernel and re-establish the tunnel by contacting again the broker or another broker. The keepalive is also used to maintain the NAT port mapping so that the UDP connection can be used as the encapsulation method for IPv6 traffic through the NAT.

16.2.9.10 TSP Client Tunnel Endpoint DNS Registration

The TSP client may have its tunnel endpoint address registered in the DNS by the broker. The broker confirms the registration using the <address type= "dn"> statement in the info reply, as shown below.

```
<tunnel action="info" type="v6udpv4" lifetime="1440">
 <server>
  <address type="ipv4">192.0.1.1</address>
  <address type="ipv6">3ffe:b00:0:1::</address>
 </server>
 <client>
  <address type="ipv4">10.1.1.1</address>
  <address type="ipv6">3ffe:b00:0:1::1</address>
  <address type="dn">client1.freenet6.net</address>
 </client>
</tunnel>
```

In this example, the client tunnel endpoint 3ffe:b00:0:1::1 is known in the DNS by client1.freenet6.net using a AAAA record.

16.2.9.11 TSP Client is a Router and Requests a Prefix

When the TSP client is a router with a network behind, it needs an IPv6 prefix to be used in its IPv6 network behind it. Figure 16.33 shows such a situation.

In this case, the TSP client includes in its tunnel request the <router> statement with the length of the requested prefix (<prefix>). The following example requests a tunnel with a /48 IPv6 prefix.

```
<tunnel action="create" type="v6v4">
 <client>
  <address type="ipv4">192.0.2.1</address>
  <router>
   <prefix length="48"/>
  </router>
 </client>
</tunnel>
```

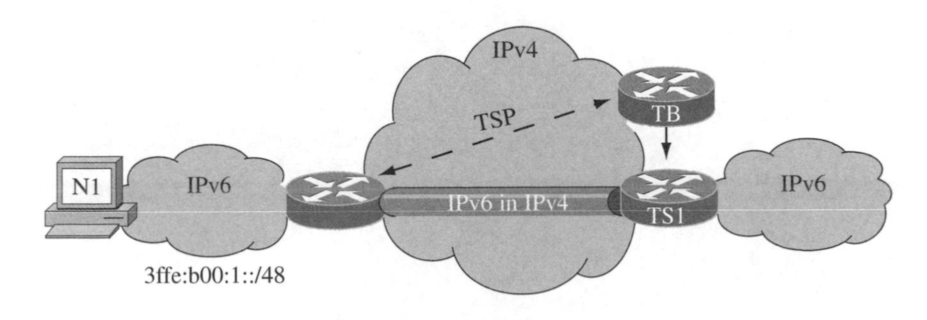

Figure 16.33 TSP client with a network behind

The tunnel broker then allocates a prefix from its prefix pool and responds with the information of the tunnel and prefix. The following example shows the tunnel broker answer to the above client request, where the 3ffe:b00:1::/48 prefix is allocated to this user tunnel.

```
<tunnel action="info" type="v6v4" lifetime="1440">
 <server>
  <address type="ipv4">192.0.1.1</address>
  <address type="ipv6"> 3ffe:b00:0:1::</address>
 </server>
 <client>
  <address type="ipv4">192.0.2.1</address>
  <address type="ipv6"> 3ffe:b00:0:1::1</address>
  <router>
   <prefix length="48">3ffe:b00:1::</prefix>
  </router>
 </client>
</tunnel>
```

The TSP client router then uses the /48 prefix to number the network behind. Figure 16.34 shows an example where two links are attached to the TSP client router.

The router receives the 3ffe:b00:1::/48 from the broker and then assigns and advertises in router advertisements one /64 for each of his connected interface (3ffe:b00:1:1::/64, 3ffe:b00:1:2::/64), enabling the nodes behind to be autoconfigured.

16.2.9.12 TSP Client is a Router and Requests a DNS Delegation

When a TSP client is a router, it requests and gets an IPv6 prefix from the broker, as discussed above. When an IPv6 prefix is owned by an entity, this entity may also want to own the DNS reverse tree under ip6.arpa corresponding to the IPv6 prefix, as discussed in Section 8.2. For example, if the IPv6 prefix is 3ffe:b00:1::/48, then the DNS reverse tree is 0.0.0.1.0.0.b.0.e.f.f.3.ip6.arpa.

Together with the request for delegation of an IPv6 prefix, a TSP client controlling a DNS server may request for the delegation of the DNS reverse tree corresponding to the IPv6 prefix.

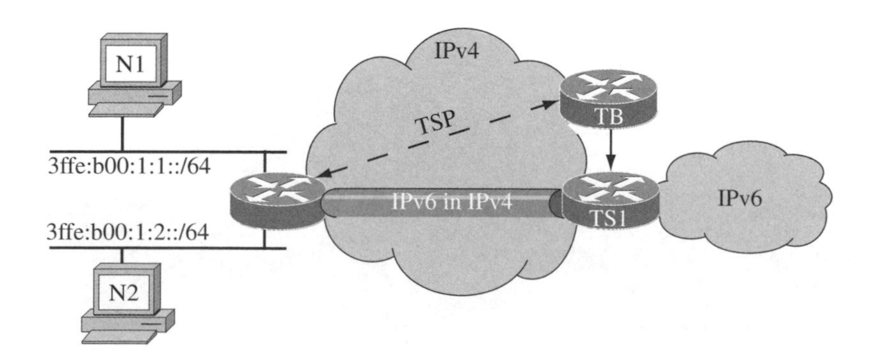

Figure 16.34 TSP client router advertising /64 prefixes on links

This TSP client sends within the XML router statement the list of DNS servers, using the `dns_server` statement, that will be authoritative for the DNS reverse tree. Using the same basic TSP request in the previous section, the TSP request now includes a `dns_server` section, as follows.

```
<tunnel action="create" type="v6v4">
 <client>
  <address type="ipv4">192.0.2.1</address>
  <router>
   <prefix length="48"/>
   <dns_server>
    <address type="ipv4">192.0.2.3</address>
    <address type="ipv4">192.0.2.4</address>
    <address type="ipv6">3ffe:c00::1</address>
   </dns_server>
  </router>
 </client>
</tunnel>
```

In this example, the TSP client requests an IPv6 in IPv4 tunnel and an /48 IPv6 prefix. It is also requesting the delegation of the DNS reverse tree corresponding to the allocated prefix, to the DNS servers listed in the request.

After allocating the prefix, the broker configures the DNS server to delegate the allocated prefix DNS reverse tree to the listed DNS servers in the request message of the client. Then it answers the query with the information, to confirm the delegation is done.

```
<tunnel action="info" type="v6v4" lifetime="1440">
 <server>
  <address type="ipv4">192.0.1.1</address>
  <address type="ipv6"> 3ffe:b00:0:1::</address>
 </server>
 <client>
  <address type="ipv4">192.0.2.1</address>
  <address type="ipv6"> 3ffe:b00:0:1::1</address>
  <router>
   <prefix length="48">3ffe:b00:1::</prefix>
   <dns_server>
    <address type="ipv4">192.0.2.3</address>
    <address type="ipv4">192.0.2.4</address>
    <address type="ipv6">3ffe:c00::1</address>
   </dns_server>
  </router>
 </client>
</tunnel>
```

16.2.9.13 Node Moving in the IPv4 Network

Since the user is authenticated, the broker saves the information of the tunnel attached to the user. When the TSP client re-contacts the server, it sends a create tunnel request. When the node has changed its IPv4 address because it moved, the tunnel request contains the new IPv4 address. The broker then changes the IPv4 address of the client endpoint and replies to the client with the tunnel info, confirming the change in the IPv4 client address, but no change in the IPv6 addresses and prefixes. For the client node, the IPv6 addresses and connectivity are then permanent and remain stable even when the IPv4 address changes. 6to4, ISATAP and Teredo change the IPv6 address when the node changes its IPv4 address. Keeping the IPv6 address stable enables the established connections to continue, the node to advertise its IPv6 address and prefix in directories such as DNS, LDAP or others, and the IPv6 applications, which exchange IP addresses such as multimedia applications and VOIP, to continue to work while the IPv4 point of attachment changes.

In Figure 16.35, Node N1 moves from A to B, and changes its IPv4 address from 192.0.2.1 to 192.0.3.1.

When N1 is at point A, it sends to the Tunnel Broker (TB) the TSP tunnel request using 192.0.2.1 as the source address of the tunnel endpoint, as shown below.

```
<tunnel action="create" type="v6v4">
 <client>
  <address type="ipv4">192.0.2.1</address>
 </client>
</tunnel>
```

The broker responds with the tunnel information.

```
<tunnel action="info" type="v6v4" lifetime="1440">
 <server>
  <address type="ipv4">192.0.1.1</address>
  <address type="ipv6"> 3ffe:b00:0:1::</address>
 </server>
 <client>
  <address type="ipv4">192.0.2.1</address>
  <address type="ipv6"> 3ffe:b00:0:1::1</address>
 </client>
</tunnel>
```

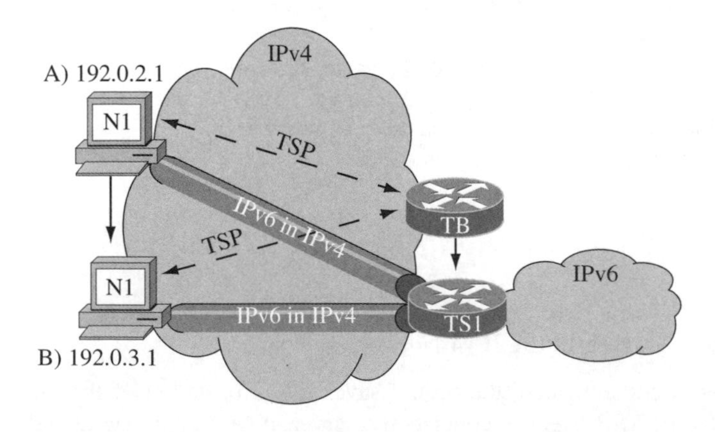

Figure 16.35 Tunnel re-establishment of a moving node

The tunnel is established between N1 and R1 in Figure 16.35.

Then Node N1 moves to B where its IPv4 address changes to 192.0.3.1. The change of address triggers the TSP client on the node to re-contact the broker for a new TSP session and re-establish the tunnel. N1 sends the following modified request, with its new IPv4 address.

```
<tunnel action="create" type="v6v4">
 <client>
   <address type="ipv4">192.0.3.1</address>
 </client>
</tunnel>
```

The broker responds with the tunnel information.

```
<tunnel action="info" type="v6v4" lifetime="1440">
 <server>
  <address type="ipv4">192.0.1.1</address>
     <address type="ipv6"> 3ffe:b00:0:1::</address>
 </server>
 <client>
   <address type="ipv4">192.0.3.1</address>
   <address type="ipv6"> 3ffe:b00:0:1::1</address>
 </client>
</tunnel>
```

The broker sends the same IPv6 addresses as before because it authenticates the client each time a TSP session is established with the broker.

16.2.9.14 IPv4 in IPv6 Tunnels

TSP is generic enough to also handle IPv4 in IPv6 tunnels, discussed in Section 17.1.4.

16.2.9.15 Ubiquitous IP Connectivity

The TSP handling of IPv6 in IPv4, IPv6 in IPv4 with NAT traversal and IPv4 in IPv6 enables the client to be connected to both IP networks in all cases, so the user will not experience delays or lack of functionality because the applications use one or the other IP protocol. Over time, there will be IPv4-only applications that cannot be upgraded, dual IP applications and IPv6-only applications, with their respective servers. TSP enables a node to use all applications whatever its default connectivity. Mobile nodes changing their connectivity such as laptops, PDAs, cell phones and WIFI devices are connected to both IP networks whichever IP protocol is supported by the network point of attachment.

Figure 16.36 shows an example of such a situation. The node moves from A where IPv4 connectivity is available, to B where IPv4 connectivity is available behind a NAT, to C where IPv6 connectivity is available.

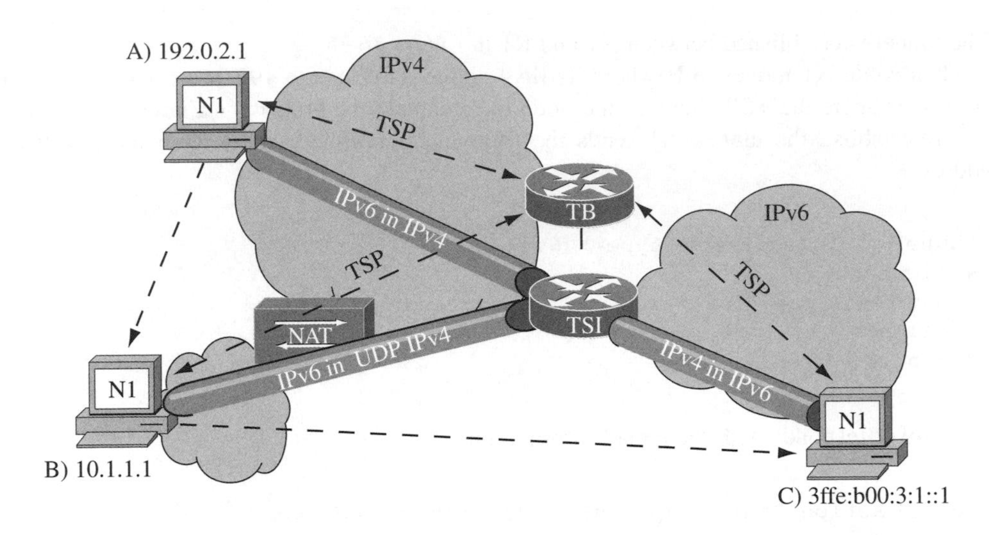

Figure 16.36 TSP providing ubiquitous IP connectivity

With TSP tunnel brokers, the node negotiates IPv6 connectivity for the first two cases and IPv4 for the last one. In all cases, after the tunnel is established, the node is connected to both IP networks seamlessly. IPv4 in IPv6 encapsulation is discussed in Section 17.1.4.

16.2.9.16 Broker Redirect

In some situations, a broker is not able to fulfill the tunnel request of a client. Such situations are listed below:

- The maximum number of tunnels is reached on the broker.
- The client is requesting a /48 prefix but the server AAA policies do not allow the provision of such prefix lengths.
- The client TSP protocol version number is higher than the one supported by the broker.

If configured with other brokers information, the broker suggests to the client to connect to the other brokers. An example of such a transaction, where the maximum number of tunnels is achieved, follows.

```
<tunnel action="create" type="v6v4">
 <client>
   <address type="ipv4">192.0.2.1</address>
 </client>
</tunnel>
```

The broker cannot fulfill the request due to the maximum number of tunnels reached. It responds with a list of possible brokers. In the example below, it lists two brokers, one by its IPv4 address, the other by its hostname.

```
1301 No tunnels available
<tunnel action="list" type="broker">
 <broker>
  <address type="ipv4">1.2.3.4</address>
 </broker>
 <broker>
  <address type="dn">tspbroker.isp1.com</address>
 </broker>
</tunnel>
```

The client then closes the current TSP session with the broker and starts a new session with one of the listed brokers.

This capability could also be used to offer redundancy and backup service. As with an NAS when all modems are busy, the call is forwarded to another POP, and the broker could redirect the client to other brokers in the same administrative domain or elsewhere.

16.2.9.17 Requirements

The requirements to deploy the TSP tunnel broker solution are:

- The node is dual-stack.
- The node implements the TSP protocol as a TSP client and supports configured tunnels.
- The TSP tunnel broker implements the TSP protocol.
- The TSP tunnel broker has access to a tunnel server.
- The tunnel server is dual-stack, supports configured tunnels and is connected to the IPv6 network. A tunnel server may be implemented within the same software as the tunnel broker.
- Hosts behind a TSP router client do not need to support or know about TSP and may only be IPv6 enabled.

16.2.9.18 Limitations

If the network distance between the TSP client and the tunnel server is large, then the round trip time might be high. However, this limitation also applies to the 6to4 relay. Nearer tunnel servers can be deployed using an anycast IPv4 address as with the 6to4 relay.

The NAT traversal technique used in the TSP tunnel broker solution ensures that for all kinds of NATs the tunnel can be established. Teredo does not work for symmetric NATs, while TSP tunnel broker works with symmetric NATs. However, for NAT traversal, the tunnel broker and the tunnel server must be co-located, since the TSP channel opened over UDP to the broker is used for both TSP and IPv6 traffic. This co-location makes it more difficult to split the load between multiple servers.

16.2.9.19 Applicability

The TSP tunnel broker is applicable as an IPv6 network access server over the IPv4 network, as a IPv6 VPN server, or as a Mobility server, similar to the MobileIP Home Agent.

TSP tunnel broker as a network access server

The TSP tunnel broker provides IPv6 connectivity over IPv4 networks, for nodes and networks, for mobile devices, through NAT, and with AAA. A TSP tunnel broker can be compared to an IPv4 network access server (NAS), as shown in Figure 16.37.

A typical NAS provides IPv4 connectivity to remote PCs using modems, where the point-to-point protocol (PPP) is used as a control channel to establish the connection between the PC and the NAS and as the data channel to carry IPv4 packets through the telephone network onto the IPv4 network.

Mapped into the TSP tunnel broker model, the PPP control channel is now the TSP protocol and the PPP data channel is the encapsulation of IPv6 in IPv4 packets, carrying IPv6 packets through the IPv4 network onto the IPv6 network. Figure 16.38 shows the TSP tunnel broker.

Table 16.11 compares the functionalities of the two models: PSTN-IPv4 NAS and IPv6-IPv4 TSP tunnel broker.

As Table 16.11 shows, a TSP tunnel broker can be considered as an IPv6 access server.

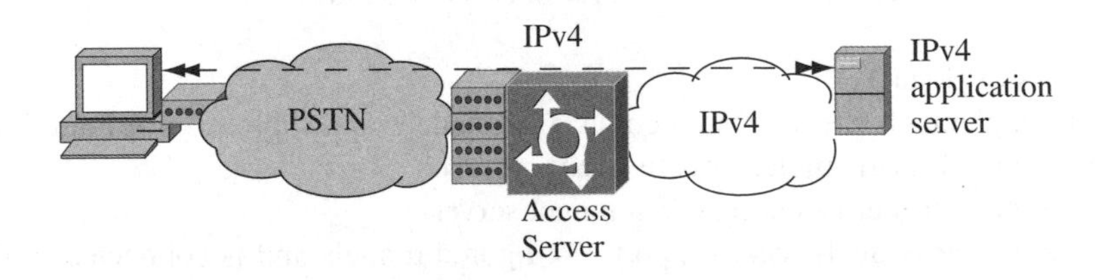

Figure 16.37 PSTN-IPv4 Network Access Server

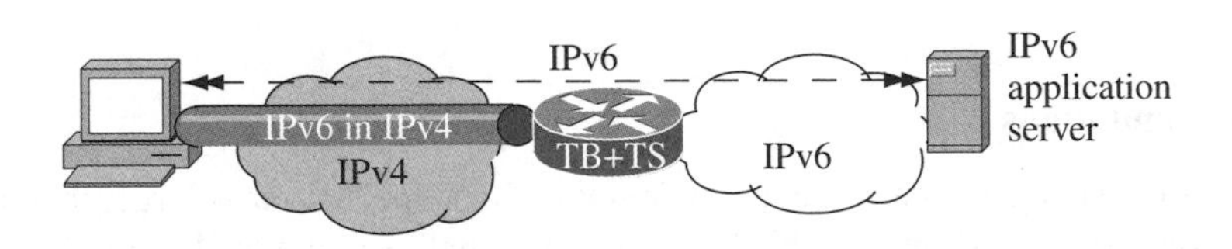

Figure 16.38 Tunnel Broker as an IPv6-IPv4 NAS

Table 16.11 Comparison between NAS and TSP tunnel broker

Functionality	PSTN-IPv4 NAS	TSP tunnel broker
Client	PC	PC or router
Access network (layer 2)	Telephone network	IPv4 network
Interface to the access network	Modem	Tunnel virtual interface
Control channel	PPP	TSP
Data channel	PPP	IPv6 in IPv4 encapsulation
Access device	Network Access Server	TSP Tunnel Broker

TSP tunnel broker as a VPN server

The overlay of IPv6 on a layer 3 network (IPv4) makes the TSP tunnel broker behave as an IPv6 VPN server, providing IPv6 virtual private networking over IPv4 networks, such as the public IPv4 Internet.

Figure 16.39 shows an IPv4 VPN server where the IPsec layer, including the IKE protocol, provides both the control channel and the encapsulation for IPv4 VPNs.

Figure 16.40 shows the TSP tunnel broker acting as an IPv6 VPN server.

TSP tunnel broker as a mobility server

Providing IPv6 in IPv4 tunnels, IPv6 in UDP-IPv4 tunnels traversing NATs and IPv4 in IPv6 tunnels, the TSP tunnel broker enables a mobile node or mobile network to stay connected in all circumstances, as described in Section 16.2.9.15 on ubiquitous IP. The TSP tunnel broker could be considered in MobileIP terminology as a Home Agent serving mobile nodes or networks with tunnels over different IP networks. Figure 16.41 shows a TSP tunnel broker acting as a mobility server. Mobile IP is discussed in Chapter 11.

16.2.10 Teredo

'Teredo navalis' is the Latin name for a little saltwater critter that digs wormholes in wooden boats. Teredo [Huitema, 2005] enables nodes to tunnel IPv6 over IPv4 through NATs. The analogy of the author is related to digging a hole in the NAT boat. The critter only lives in clear water, which also has the analogy of IPv6 being a clear and transparent solution for the Internet[2].

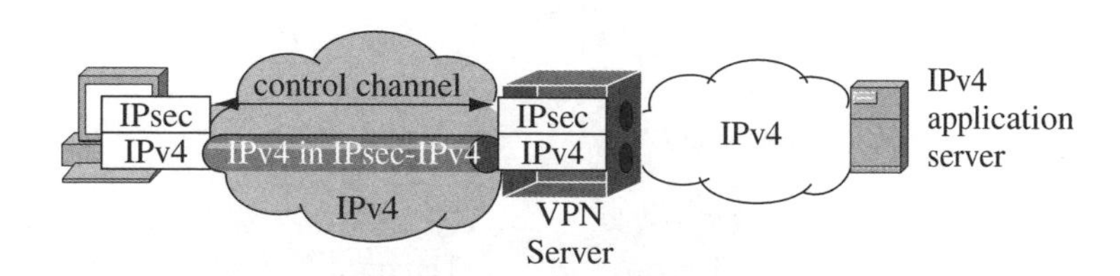

Figure 16.39 IPv4 VPN server

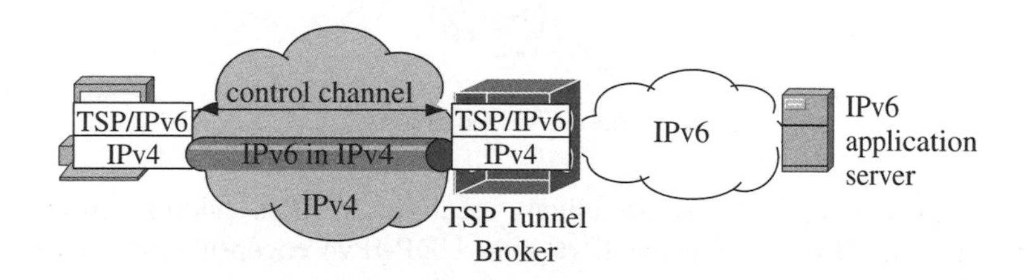

Figure 16.40 TSP Tunnel Broker as IPv6 VPN server

[2] The author started by naming the transition mechanism 'shipworm', but people commented that it had a negative image. 'Teredo' and 'shipworm' are two names for the same mechanism.

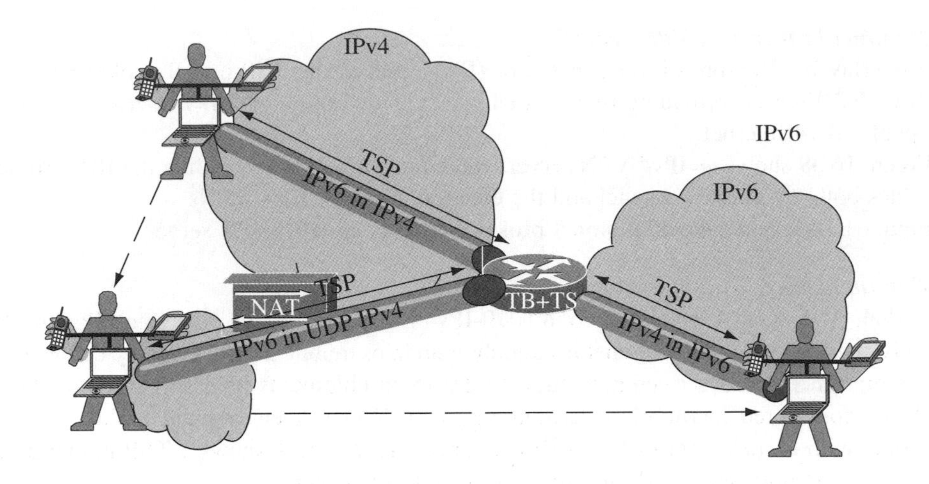

Figure 16.41 TSP tunnel broker as mobility server

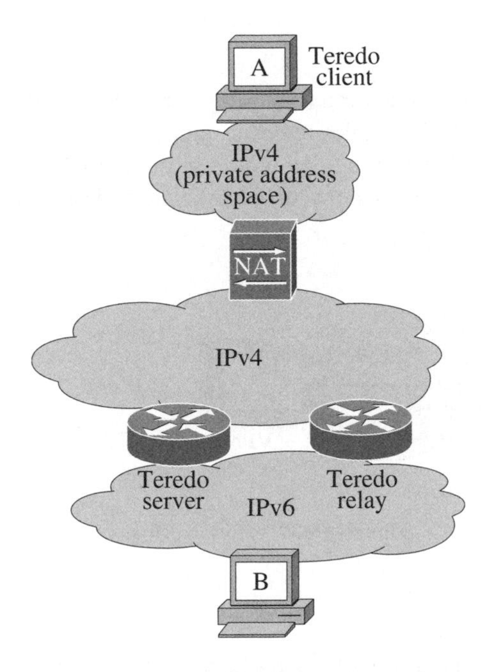

Figure 16.42 Teredo components

A node implementing the Teredo client, receives an IPv6 address from a Teredo server and traverses IPv4 NAT using IPv6 over UDP-IPv4 encapsulation, as discussed in Section 16.2.8.

Teredo is based on the following components, as shown in Figure 16.42:

- Teredo client: node A in the figure;
- Teredo server;
- Teredo relay.

16.2.10.1 Addressing

As with 6to4, Teredo uses a special prefix to provide an IPv6 address to nodes. At the time of writing, this prefix is not yet assigned by IANA, and only will be when Teredo is a standard-track RFC.

A Teredo address, as shown in Figure 16.43, has the following components:

- Teredo IANA assigned prefix: 32 bits;
- Teredo server IPv4 address: 32 bits;
- flags: 16 bits (described in Table 16.12);
- port number of the external NAT mapping of the client: 16 bits;
- IPv4 address of the external NAT mapping of the client: 32 bits.

Each component of a Teredo address is used to establish the tunneling.

The flags field is used to identify the NAT type[3] Two values are possible, thus the use of 1 bit, named the Cone bit, as listed in Table 16.12.

This Cone bit in the address enables Teredo to optimize the path when possible.

It is reported by the Teredo author that some NAT implementations look inside IPv4 packets to find IPv4 addresses and change them to the mapped external address. To avoid this processing, the IPv4 address and port number put in the interface identifier are obfuscated by applying an exclusive OR with 0xFFFFFFFFFFFF.

16.2.10.2 Tunneling Mechanism

The Teredo rationale is to perform path optimizations when certain types of NAT are in the path. The first step for a client is to find the type of NAT and to get an IPv6 address.

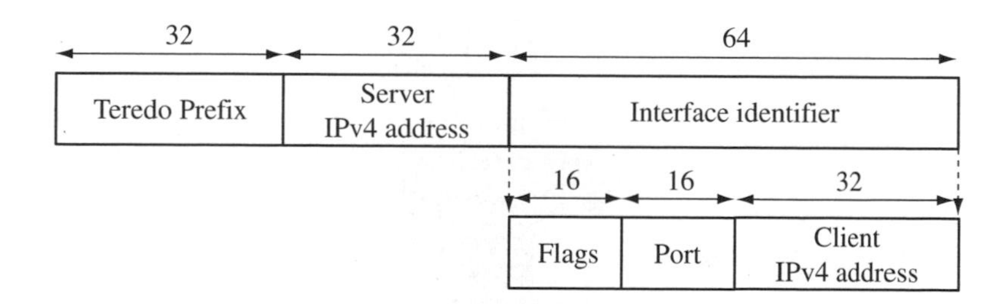

Figure 16.43 Teredo address format

Table 16.12 Cone bit in the flags field of the Teredo address

Value	Description
0x0000	NAT type is cone
0x8000	NAT type is not cone

[3] NAT types are described in the book Web site: http://www.ipv6book.ca.

Finding the NAT type and getting an address

To distinguish between NAT types, a mapping must be triggered by an outgoing packet to a target address and a reply using another address should be sent back. If the reply reaches the internal node, then the NAT is a cone. If the reply from the other address is discarded by the NAT, then it is either a restricted, port restricted or symmetric NAT. When a cone NAT is present, the Teredo client can make some path optimizations.

To discover the NAT type, the Teredo client first goes through a series of tests with the Teredo server. The Teredo client must be preconfigured with the two IPv4 addresses of the Teredo server. The Teredo protocol uses modified versions of IPv6 router solicitation (RS) and router advertisement (RA) messages, named here Teredo RS (TRS) and Teredo RA (TRA), over UDP-IPv4.

Through multiple packet exchanges, this process accomplishes two goals for the Teredo client: identifying the NAT type and getting an IPv6 address. The RA sent by the Teredo server contains the IPv6 prefix for the Teredo client. The IPv6 /64 prefix is made with the Teredo prefix (32 bits) and the Teredo server IPv4 address (32 bits), as shown in Figure 16.43. The Teredo client autoconfigures itself by generating the interface identifier part using the following components: flags (16 bits), the port number of the external NAT mapping (16 bits) and the IPv4 address of the external NAT mapping (32 bits), the last two being received inside the TRA as a special field called origin, within the TRA packet.

Sending IPv6 packets using Teredo

Figure 16.44 shows an example of Teredo components:

- Teredo client (C1);
- NAT (N1);
- Teredo server (S1);

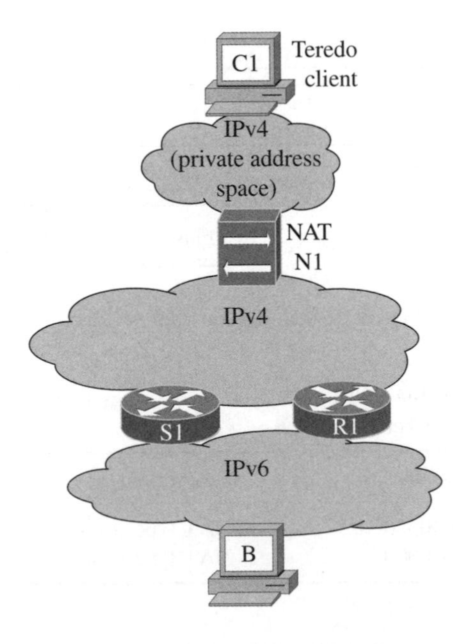

Figure 16.44 Network example of Teredo components

- Teredo relay (R1);
- Non-Teredo IPv6 node B.

This figure is used for the next flow descriptions.

Sending an IPv6 packet from a Teredo client to a non-Teredo IPv6 node
When a Teredo client sends an IPv6 packet to a non-Teredo IPv6 destination, many packet exchanges are completed before sending the packet.

Table 16.13 describes the steps of Figure 16.45. Step 0 is holding the IPv6 packet to be sent to the IPv6 destination: the first packets to new destinations are buffered until the next steps are completed. Steps 1 to 18 are done to open a new hole in the NAT of the Teredo client to reach directly the Teredo relay used by the IPv6 destination. However, the Teredo relay is unknown to the Teredo client, and will only be known when the IPv6 destination sends an IPv6 packet to the Teredo client. The assumption here is that the total path from the Teredo client to NAT to IPv4 Internet to Teredo relay to IPv6 Internet to the IPv6 destination is much shorter than the path from the Teredo client to NAT to IPv4 Internet to Teredo server to IPv6 Internet to the IPv6 destination.

Steps 1 to 5 are a ping sent from Teredo client C1 to the non-Teredo IPv6 destination node B. By replying to the ping (step 6), the IPv6 destination reaches its nearest Teredo relay. Steps 7 to 15 punch the new hole in the NAT for the relay. Steps 16 to 18 are the reply ping packet, which tells the Teredo client that the hole is punched. Steps 19 to 23 are the IPv6 packet sent from the Teredo client to the IPv6 destination, using the direct mapping to the Teredo relay.

Since most of the IPv6 Internet nodes will probably not be Teredo nodes, each new connection to a new destination node initiated by the Teredo client starts this whole process again.

Sending an IPv6 packet from a non-Teredo IPv6 node to a Teredo client
When a non-Teredo IPv6 node sends a packet to a Teredo node, the IPv6 source node does not do any special processing. The IPv6 packet will reach the closest Teredo relay announcing the Teredo prefix on the IPv6 Internet. Given that the relay has no prior opened mapping on the NAT in front of the Teredo node, a bubble round trip is used to open a new mapping from the Teredo client to the relay. Then the IPv6 original packet is sent to the Teredo client, through the NAT. Since most of the IPv6 Internet nodes are not Teredo nodes, each new connection to a Teredo node initiated by an IPv6 internet node starts this whole process again.

Sending an IPv6 packet from a Teredo client to another Teredo client
When a Teredo client sends an IPv6 packet, the IPv6 destination address is checked against the Teredo prefix. If the destination is another Teredo client, then both clients initiate bubbles to open their respective NATs for direct communication between the NATs without going through Teredo relays or servers.

If the two Teredo clients are on the same link, Teredo has a discovery mechanism [Huitema, 2005] using an IPv4 multicast address to send bubbles between the two Teredo client.

If the two Teredo clients are not on the same link but behind the same NAT, the direct communications between the two, after the initial discovery process, involve the routing of packets on the external loopback virtual interface of the NAT. Not all NAT implementations

Table 16.13 Teredo client packet to non-Teredo node

	Description
0	The Teredo client wants to send an IPv6 packet to an IPv6 destination, which does not have a Teredo address. The Teredo client holds the IPv6 packet in a buffer before sending.
1	The Teredo client sends a IPv6 ping to the destination, encapsulated in UDP-IPv4, with the IPv4 destination address of its Teredo server (S1).
2	The source IPv4 address and port number are mapped by the NAT (N1).
3	The IPv6 ping packet encapsulated in UDP-IPv4 is received by the Teredo server S1.
4	The Teredo server decapsulates the IPv6 ping packet.
5	The Teredo server S1 sends the IPv6 ping packet directly to B over the IPv6 network.
6	B sends the IPv6 ping reply to the C1 address.
7	Based on IPv6 routing, the nearest Teredo relay advertising the Teredo prefix receives the IPv6 ping reply with destination C1. The relay holds the IPv6 reply ping packet in a buffer.
8	Teredo relay R1 sends an IPv6 in UDP-IPv4 bubble to C1, with destination IPv4 address of S1.
9	Teredo server S1 receives the bubble and forwards it to C1.
10	NAT N1 receives the bubble and forwards it to C1 remapping the destination IPv4 address and port.
11	Teredo client C1 receives the bubble.
12	Teredo client C1 replies with an IPv6 in UDP-IPv4 bubble to the Teredo relay R1, without going through Teredo server S1.
13	NAT N1 creates a new mapping for connection going to the Teredo relay R1.
14	Bubble is forwarded to Teredo relay R1. Relay knows now that a specific mapping on the Teredo client NAT (N1) is made for sending packets from the relay directly to the client without going through the Teredo server (S1).
15	IPv6 reply ping packet queued from step 7 is now encapsulated over UDP-IPv4.
16	Encapsulated packet is sent by relay to C1 through NAT N1.
17	NAT N1 forwards the IPv6 ping reply packet in UDP-IPv4 to C1, using the new mapping enabled in step 13.
18	Ping reply is received by Teredo client C1. Now the client knows that a new path is established between the client and the relay through a specific mapping done in the NAT.
19	IPv6 packet queued in step 0 is now sent over UDP-IPv4, with the destination IPv4 address of Teredo relay R1.
20	The source IPv4 address and port number are mapped by the NAT (N1).
21	Encapsulated packet is received by Teredo relay R1.
22	Teredo relay R1 decapsulates the IPv6 packet.
23	Teredo relay R1 sends the native IPv6 packet on the IPv6 network to the IPv6 destination B.

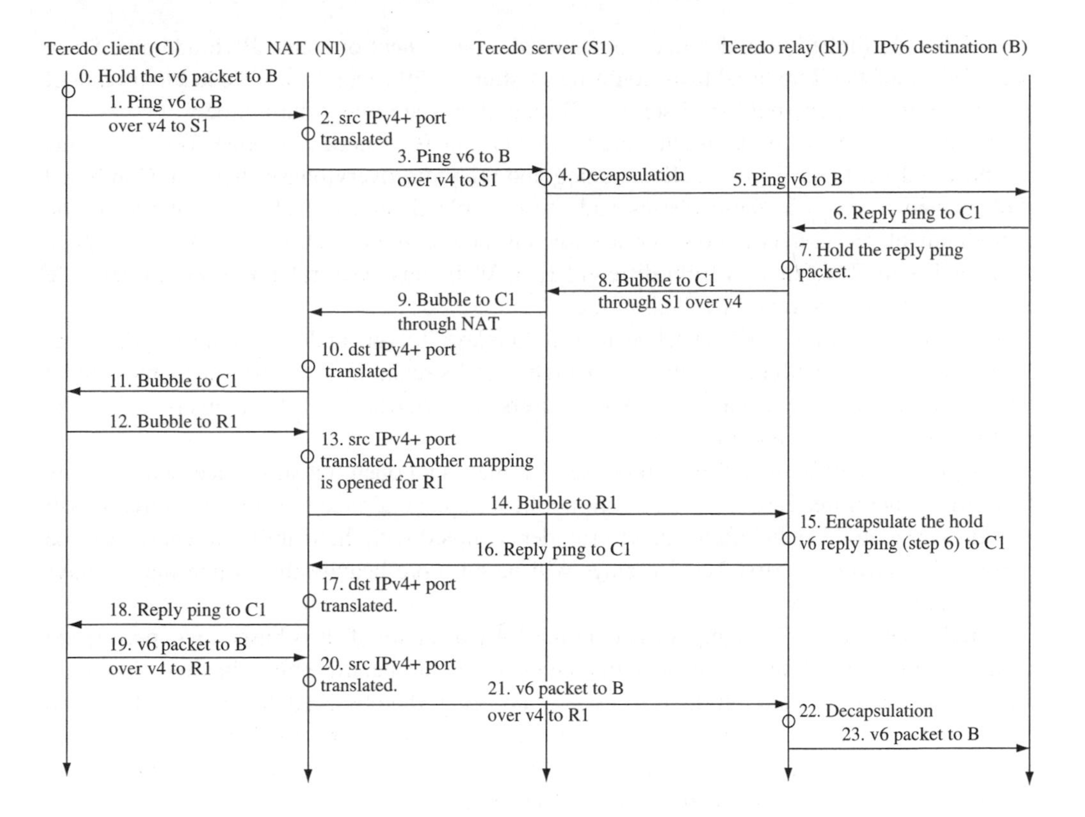

Figure 16.45 Flow of Teredo client packet to non-Teredo node

do this process, and if this is the case Teredo nodes cannot communicate at all between them. If a provider for the home is offering IPv4 connectivity with private address space behind a NAT, then all homes using Teredo cannot communicate IPv6 with each other.

16.2.10.3 Requirements

The requirements to deploy Teredo are:

- IPv6 nodes must be dual-stack.
- IPv6 nodes implement Teredo.
- A Teredo server connected to the IPv4 Internet and IPv6 Internet is available for the Teredo clients.
- Many Teredo relays are available and well dispersed in the IPv4 Internet and IPv6 Internet.
- No symmetric NAT exists between the Teredo node and the Teredo server.
- The Teredo server IPv4 addresses must be statically configured in all Teredo clients.

16.2.10.4 Limitations

Teredo addressing embeds the Teredo server IPv4 address, the Teredo client mapped IPv4 address and the Teredo client mapped port. This dependency on the IPv4 address makes the use of Teredo limited to nodes that do not need a permanent or stable IPv6 address. Since the lifetime of the Teredo address could be as short as 30 seconds, a Teredo client cannot use its address for any published service. Teredo clients are really dumb nodes.

Initial communications between Teredo clients and IPv6 nodes or other Teredo clients are buffered while the Teredo hole punching and relay discovery phases happen. Combined with possible delays in these phases and the possible timeouts at all steps, it means the perception of the user is a slow service for any new communication. When a Web page is loaded with 50 references from different IPv6 Web sites, each reference would generate many hole punching, round-trip bubbles, etc.

The buffering of initial IPv6 packets in Teredo relays, servers and clients means additional important memory buffers for relays. For high speed streaming or peer-to-peer multimedia file exchange between a large number of users, this buffering will adversely affect the performance of the Teredo platform.

For providers, offering a Teredo relay service means an open transit service, with no way to charge. Therefore, Teredo relays might not be a popular service to offer given that it will require substantial bandwidth for these providers coupled with the inability to charge for the service. Moreover, the first Teredo relays will be hit more heavily than when many relays are available.

Teredo makes a lot of assumptions about the NAT behavior. This is based on experimental results from trying different vendors implementations at specific points in time. Not only were all vendors not tried, but there is no guarantee that the observed behaviors will remain in the subsequent releases of those NAT products. Teredo might not function because of some new or non-observed NAT behavior. These behaviors have never been standardized, but an effort was made to document some [RFC3022].

Teredo does not work if a symmetric NAT is in the path between the Teredo client and the Teredo server. A Teredo client in this situation gets no IPv6 connectivity. Symmetric NATs are known to be used in enterprises [RFC3489] because they implement the most secure way to translate IP addresses and ports.

The deployment of Teredo will not be very efficient if only a few Teredo relays are available. Poor response time and long paths will be the result. Teredo deployment depends on the relative high availability and connectivity of a significant number of Teredo relays that are well dispersed in both the IPv4 Internet and the IPv6 Internet.

Teredo is a pretty complex protocol, which means there are possible security issues in the protocol or implementation, interoperability problems between implementations, difficulty with making the implementation efficient, etc.

Also, Teredo assumes that the Teredo relay will always be the best path between two endpoints. This is not always the case and traffic engineering cannot be done with Teredo.

Another problem is that Teredo puts states everywhere in the Teredo network: clients, servers and relays keep a lot of states of current connections, bubbles in progress, etc. Adding states in the network results in a network with much less reliability in the case of the failure of one of its components. No state in the network is the basic design assumption of IP.

Finally, Teredo clients behind the same NAT but not on the same link might not be able to communicate. In this case, the Teredo clients have no way to communicate using IPv6.

16.2.10.5 Applicability

Given the above limitations, Teredo applicability is restricted. It works for isolated IPv6 nodes behind some types of NAT where there is no need for a stable IPv6 address. It does not work for mobile hosts that are connecting behind different kinds of NAT as they travel. It does not provide a prefix to a mobile network.

The Teredo deployment needs IANA to assign a Teredo prefix. This prefix will be put in all Teredo implementations: clients, servers and relays. Until then, it will not be a standardized service deployable on any large scale.

16.3 Tunneling IPv6 in GRE-IPv4

Generic Route Encapsulation (GRE) [RFC2784] is a specification for encapsulating one protocol packet into another protocol packet. The inner packet is put in the payload of the outer packet with a GRE header, as shown in Figure 16.46. IPv6 with GRE in IPv4 is shown in Figure 16.47.

The IPv4 header protocol field has the value of 47, identifying that the payload of the IPv4 packet is a GRE packet. Inside the GRE header, the Protocol Type field contains the layer 2 code points of the inner protocol. As shown in Figure 6.3, the IPv6 layer 2 protocol code point is 0x86dd, so the GRE protocol type field has a value of 0x86dd to identify the inner packet as IPv6.

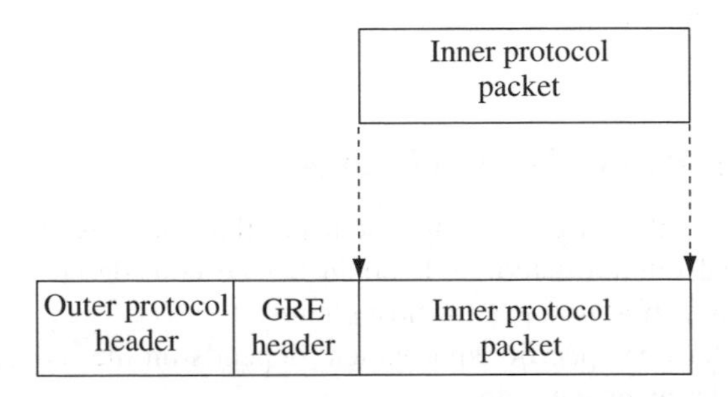

Figure 16.46 GRE encapsulation

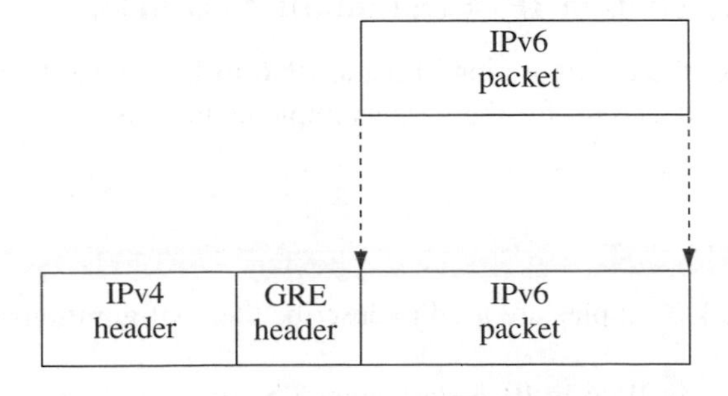

Figure 16.47 GRE encapsulation of IPv6 packets in IPv4

16.3.1 Requirements

The two tunnel endpoints are IPv4, IPv6 and GRE enabled, and are reachable by the other party.

16.3.2 Limitations

GRE tunnels have the same limitations as static tunnels, described in Section 16.2.4. Manual configuration of the tunnel endpoints is tedious if:

- many tunnels have to be done;
- the IP addresses have to be changed for many tunnels.

A GRE tunnel over IPv4 uses the IPv4 addresses of both endpoints, so it cannot traverse a NAT.

16.3.3 Applicability

GRE is useful when multiple protocols have to be tunneled within the same tunnel. In these environments, adding IPv6 as another protocol over the same established tunnels is useful. For example, if a GRE tunnel is established to carry IS-IS traffic, then the same tunnel could be used for the IPv6 traffic. Otherwise, direct IPv6 in IPv4 tunneling is more efficient, given the GRE header tax added to each packet.

16.4 Comparing IPv6 in IPv4 Solutions

Table 16.14 summarizes the differentiating features of the solutions discussed in this chapter. This table is an initial reference tool and should not be considered as comprehensive. All solutions enable IPv6 in IPv4 tunnels for host clients.

The best solution for one specific environment depends on the requirements. Chapter 23 describes typical deployment scenarios.

16.5 Configuring IPv6 in IPv4 Dominant Networks

This section describes the configuration of static IPv6 in IPv4 tunnels, 6to4, ISATAP, TSP tunnel broker, GRE and Teredo for the various implementations.

16.5.1 Examples

The following network examples are used to describe the configurations of hosts and routers implementations.

Figure 16.48 shows an IPv6 in IPv4 static tunnel between node N1 and router R1, where N1 has 192.0.2.1 and 3ffe:b00:1:1::1 addresses and R1 has 192.0.3.1 and 3ffe:b00:1:1::2.

Table 16.14 Comparison of the different IPv6 in IPv6 in IPv4 solutions

	Static tunnels	6to4	ISATAP	Tunnel broker with TSP	Teredo
NAT traversal	NO	NO	NO	YES	YES
Network clients	NO	YES	NO	YES	NO
IPv6 prefix delegation	NO	YES*	NO	YES	NO
IPv6 address not dependent on IPv4 address	YES	NO	NO	YES	NO
Without authentication	YES	YES	YES	YES	YES
With authentication	NO	NO	NO	YES	NO
Accounting	NO	NO	NO	YES	NO
Change of IPv4 address does not change IPv6 address or prefix	YES	NO	NO	YES	NO
The tunnel is kept up even when the IPv4 address changes	NO	NO	NO	YES	NO
Applicability	Very few tunnels	Small networks	Inside an enterprise network	Provider, enterprise and NAT traversal	Isolated hosts behind NAT
One enabled gateway enables the nodes behind to get connectivity without needing the transition mechanisms implemented in the nodes	NO	YES	NO	YES	NO
The border router does not need to support the transition mechanism	YES	NO	YES	YES	YES
IPv4 in IPv6	NO	NO	NO	YES	NO
Redundancy	NO	NO	NO	YES	SOME
Load balancing	NO	NO	NO	YES	SOME

* The IPv6 prefix, however, is bound to the IPv4 address. Issues with IPv4 address bindings are discussed in Section 16.2.7.6.

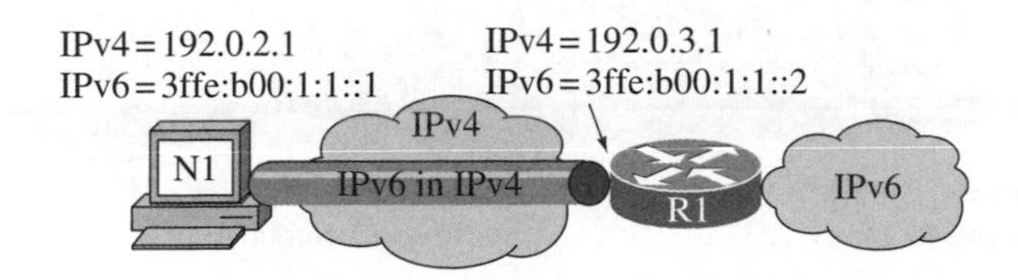

Figure 16.48 Static tunnel example

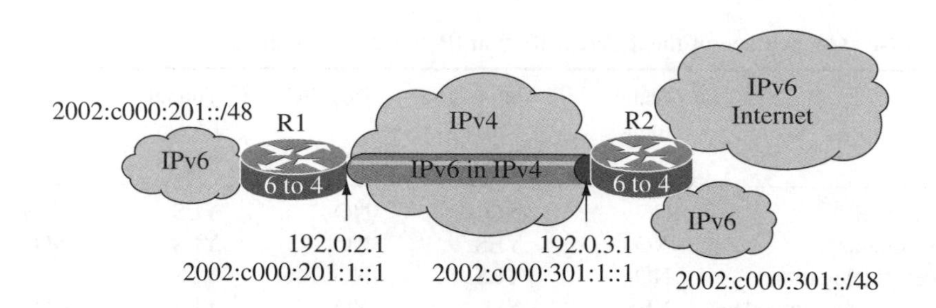

Figure 16.49 6to4 example

Figure 16.49 shows the 6to4-enabled router R1 with 192.0.2.1 address. Based on the R1 IPv4 address, the 6to4 prefix is then 2002:c000:201::/48. The 6to4 relay is R2 with 192.0.3.1 address.

Figure 16.50 shows an ISATAP tunnel configured on N1 with 192.0.2.1. The ISATAP router is R1 with 192.0.3.1 address. The ISATAP prefix advertised is 3ffe:b00:1:1::/64.

Figure 16.51 shows an IPv6 in IPv4 tunnel between node N1 and the TSP tunnel broker TB1, where the tunnel broker and server are co–located. N1 has 192.0.2.1 and TB1 has 192.0.3.1.

After the TSP transaction, N1 has 3ffe:b00:1:1::1 as its tunnel endpoint IPv6 address and the broker has 3ffe:b00:1:1::2 as its IPv6 address for this tunnel.

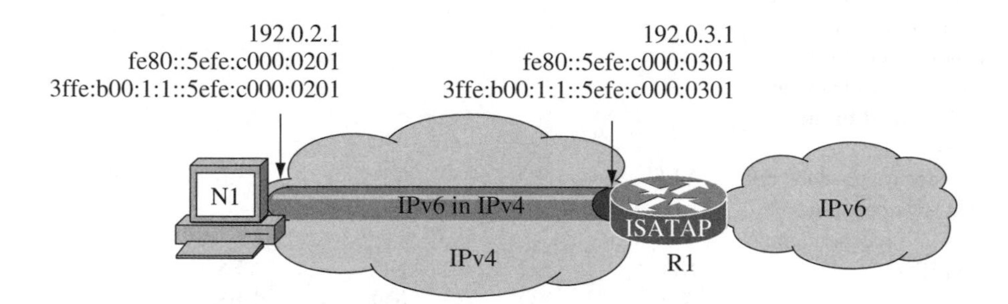

Figure 16.50 ISATAP example

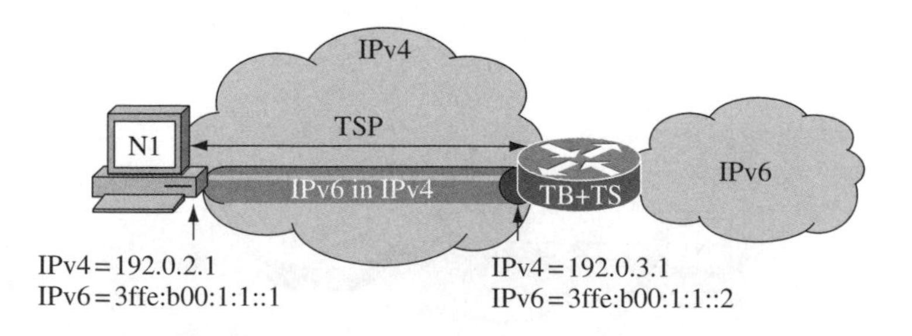

Figure 16.51 Tunnel broker example

16.5.2 FreeBSD

FreeBSD supports static tunnels through the '`gif`' interface, 6to4 through the '`stf`' interface and TSP tunnel broker through the tsp/freenet6 client.

16.5.2.1 Static IPv6 in IPv4 Tunnels

IPv6 in IPv4 tunnels in FreeBSD is implemented as `gif` interfaces. These interfaces can also be used for IPv4 in IPv6, IPv4 in IPv4 and IPv6 in IPv6 encapsulations. They are clonable devices using the `create` argument of the `ifconfig` command. The `tunnel` argument specifies the IPv4 outer source and destination addresses. The IPv6 inner addresses are defined by the `inet6` argument.

The following commands configure an IPv6 in IPv4 static tunnel, as in Figure 16.48.

```
# ifconfig gif0 create
# ifconfig gif0 tunnel 192.0.2.1 192.0.3.1 up inet6
3ffe:b00:1:1::1 3ffe:b00:1:1::2 prefixlen 64 alias
```

To store this configuration, use the `ipv6_ifconfig_gif0` variable in the `/etc/rc.conf` file.

To forward all IPv6 traffic through this static tunnel, use the `route add default` command, as shown below.

```
# route add -inet6 default 3ffe:b00:1:1::2
```

16.5.2.2 6to4

6to4 is implemented as a virtual interface named '`stf`' in FreeBSD. The whole 6to4 network is seen as a virtual link by setting the prefix length to 16. Setting the prefix length to /48 would enable communications to within the site only. The following command configures the 6to4 interface, as in Figure 16.49.

```
# ifconfig stf0 inet6 2002:c000:201:1::1 prefixlen 16 alias
```

To save this configuration permanently, set the `stf_interface_ipv4addr` variable in the `/etc/rc.conf` file to the IPv4 address used for the 6to4 prefix.

```
# cat /etc/rc.conf
stf_interface_ipv4addr="192.0.2.1"
```

To forward all IPv6 traffic through the 6to4 relay, use the `route add default` command pointing to the IPv6 address of the 6to4 relay, as shown below.

```
# route add -inet6 default 2002:c000:301:1::1
```

16.5.2.3 TSP Tunnel Broker Client

The TSP tunnel broker client is available on FreeBSD in the ports/net/freenet6, or from the freenet6 Web site (http://www.freenet6.net) or from Hexago directly.

The TSP tunnel broker client on FreeBSD uses the 'gif' interface for the IPv6 in IPv4 tunnels and the 'tun' interface for the IPv6 in UDP IPv4 tunnels. The 'tspc.conf' file, used to configure the client, is discussed in Section 16.5.7.2. In the 'tspc.conf' file, the variable-value pairs shown in Table 16.15 (from Table 16.20) are specific to FreeBSD.

The TSP client is run in the command line as 'tspc'. Its configuration file is usually placed in '/usr/local/etc/tspc.conf'.

For additional information on the TSP tunnel broker client, see Section 16.5.7.2.

16.5.3 Linux

Linux supports static tunnels through the 'sit' interface, 6to4 through a generic interface, ISATAP through the 'is' interface and TSP tunnel broker through the tsp/freenet6 client.

16.5.3.1 Static IPv6 in IPv4 Tunnels

Linux has the 'ifconfig', 'route' and 'ip' commands (part of the iproute package) to create tunnels.

The following commands configure an IPv6 in IPv4 static tunnel using interface sit0, as in Figure 16.48.

```
# ip tunnel add sit0 mode sit ttl 64 remote 192.0.3.1
# ifconfig sit0 up
# ifconfig sit0 add 3ffe:b00:1:1::1/128
```

To forward all IPv6 traffic through this static tunnel, use the route add command, as shown below.

```
# route -A inet6 add 2000::/3 dev sit0
```

Note the use of 2000::/3 as the default route, since Linux kernels do not like IPv6 default routes.

Table 16.15 TSP Tunnel Broker Client Configuration for FreeBSD

Variable	Value	Description
template	freebsd	script template file
if_tunnel_v6v4	gif0	interface name for IPv6 in IPv4 tunnels
if_tunnel_v6udpv4	tun	interface name for IPv6 in UDP IPv4 tunnels

16.5.3.2 6to4

6to4 is implemented as a virtual interface in Linux, below named 'tun6to4'. The whole 6to4 network is seen as a virtual link by setting the prefix length to 16. The following commands configure the 6to4 interface, as in Figure 16.49.

```
# /sbin/ip tunnel add tun6to4 mode sit ttl 64 remote any local
192.0.2.1
# /sbin/ip link set dev tun6to4 up
# /sbin/ip -6 addr add 2002:c000:201:1::1/16 dev tun6to4
# /sbin/ip -6 route add 2000::/3 via ::192.88.99.1 dev tun6to4
metric 1
```

Note the use of the 192.88.99.1 6to4 relay anycast address as the default gateway.

16.5.3.3 ISATAP

ISATAP is implemented in Linux using the 'is' interface and the tunnel mode 'isatap'. The following commands configure the ISATAP interface, as in Figure 16.50.

```
# ip tunnel add is0 mode isatap local 192.0.2.1 ttl 64
# ip link set is0 up
# ip addr add 3ffe:b00:1:1::5efe:c000:0201/64 dev is0
```

16.5.3.4 TSP Tunnel Broker Client

The TSP tunnel broker client is included in many Linux distributions, or from the freenet6 Web site (http://www.freenet6.net) or from Hexago directly.

The TSP tunnel broker client on Linux uses the 'sit' interface for the IPv6 in IPv4 tunnels and the 'tun' interface for the IPv6 in UDP IPv4 tunnels. The 'tspc.conf' file, used to configure the client, is discussed in Section 16.5.7.2. In the 'tspc.conf' file, the variable-value pairs shown in Table 16.16 (from Table 16.20), are specific to Linux.

The TSP client is run in the command line as 'tspc'.

For additional information on the TSP tunnel broker client, see Section 16.5.7.2.

16.5.4 Solaris

Solaris supports static tunnels through the 'ip.tun' interface, 6to4 through the 'ip.6to4tun' interface and TSP tunnel broker through the tsp/freenet6 client.

Table 16.16 TSP Tunnel Broker Client Configuration for Linux

Variable	Value	Description
`template`	linux	script template file
`if_tunnel_v6v4`	sit0	interface name for IPv6 in IPv4 tunnels
`if_tunnel_v6udpv4`	tun	interface name for IPv6 in UDP IPv4 tunnels

16.5.4.1 Static IPv6 in IPv4 Tunnels

Static IPv6 in IPv4 tunnels are configured using the 'ifconfig' command with the 'tsrc', 'tdst' and 'addif' parameters. The parameters are put in the '/etc/hostname6.ip.tun0' file to save the tunnel configuration permanently. The files '/etc/hostname6.ip.tun1', tun2, etc. are used for additional tunnels. The 'tsrc' and 'tdst' are the tunnel source and destination address of the outer packet (i.e. IPv4 for an IPv6 in IPv4 tunnel). The 'up' parameter makes the interface live. This also configures the link-local addresses by embedding the IPv4 tunnel addresses in the last 32 bits of the link-local prefix. Static IPv6 addresses are assigned to the tunnel endpoints with the 'addif' parameter. The following file configures an IPv6 in IPv4 static tunnel, as in Figure 16.48.

```
# cat /etc/hostname6.ip.tun0
 tsrc 192.0.2.1 tdst 192.0.3.1
 addif 3ffe:b00:1:1::1 3ffe:b00:1:1::2 up
```

The 'ifconfig ip.tun0' is used to look at the tunnel interface configuration.

```
# ifconfig ip.tun0
```

16.5.4.2 6to4

6to4 is implemented as a virtual interface named 'ip.6to4tun' in Solaris. The following file configures the 6to4 interface, as in Figure 16.49.

```
# cat /etc/hostname6.ip.6to4tun0
 tsrc 192.0.2.1 2002:c000:201:1::1/64 up
```

To forward all IPv6 traffic through the 6to4 relay, use the route add default command pointing to the IPv6 address of the 6to4 relay, as shown below.

```
# route add -inet6 default 2002:c000:301:1::1
```

16.5.4.3 TSP Tunnel Broker Client

The TSP tunnel broker client is available on Solaris from the freenet6 Web site (http://www.freenet6.net) or from Hexago directly.

The TSP tunnel broker client on Solaris uses the 'ip.tun' interface for the IPv6 in IPv4 tunnels. The 'tspc.conf' file, used to configure the client, is discussed in Section 16.5.7.2. In the 'tspc.conf' file, the variable-value pairs shown in Table 16.17 (from Table 16.20), are specific to Solaris.

The TSP client is run in the command line as 'tspc'.

For additional information on the TSP tunnel broker client, see Section 16.5.7.2.

Table 16.17 TSP Tunnel Broker Client Configuration for Solaris

Variable	Value	Description
`template`	solaris	script template file
`if_tunnel_v6v4`	ip.tun0	interface name for IPv6 in IPv4 tunnels

16.5.5 Windows

Windows supports static IPv6 in IPv4 tunnels, 6to4, ISATAP, Teredo and TSP tunnel broker client. It identifies interfaces by index numbers, where the index is incremented each time a new physical or logical interface is configured. This makes it difficult to predict the index number for a specific interface. The examples used in this section should therefore be adapted to your environment.

16.5.5.1 Static IPv6 in IPv4 Tunnels

A tunnel interface is always created by default, and is usually interface #2. To configure a static IPv6 in IPv4 tunnel, the `add route` and `add address` statements in the `netsh interface ipv6` command are used.

The following file configures an IPv6 in IPv4 static tunnel, as in Figure 16.48.

```
c> netsh interface ipv6 add route prefix=::/0 interface=2
nexthop=::192.0.3.1 publish=yes
c> netsh interface ipv6 add
address interface=2
address=3ffe:b00:1:1::1
```

The '`netsh interface ipv6 add v6v4tunnel`' command can also be used for the same purpose.

16.5.5.2 6to4

By default, Windows creates 6to4 addresses for all public IPv4 addresses assigned to all interfaces. It also creates a 2002::/16 route to forward to the 6to4 tunneling pseudo-interface (index=3). It also does an automatic DNS query to find a 6to4 relay router. The '`6to4 set relay`' statement is used to set a 6to4 relay manually.

```
c> netsh interface ipv6 6to4 set relay 192.0.3.1
```

16.5.5.3 ISATAP

Similar to 6to4, by default, Windows creates ISATAP link-local addresses for all public IPv4 addresses assigned to all interfaces. It also does an automatic DNS query to find an ISATAP router by querying '`_ISATAP.example.org`' for Windows XP or '`ISATAP.example.org`' for Windows2003, where example.org is the domain name of the site. The '`isatap set router`' statement is used to set an ISATAP router manually.

```
c> netsh interface ipv6 isatap set router 192.0.3.1
```

For Windows to be an ISATAP router, configure the forwarding on the ISATAP interface using the '`set interface forwarding`' statement and define the ISATAP prefix using the '`set route`' statement, as shown below.

```
c> netsh interface ipv6 set interface 2 forwarding=enabled
advertise=enabled
c> netsh interface ipv6 set route 3ffe:b00:0:1::/64 2
publish=yes
```

16.5.5.4 Teredo

By default, Windows enables Teredo if the node is not part of a Microsoft domain. To set the Teredo server, use the `set teredo client` statement. The following command sets the Teredo server to 192.0.3.1.

```
c> netsh interface ipv6 set teredo client 192.0.3.1
```

When the node is part of a Microsoft domain, Teredo is disabled by default. It could be enabled by using the `set teredo enterpriseclient` statement, as shown below.

```
c> netsh interface ipv6 set teredo enterpriseclient 192.0.3.1
```

Client ports and refresh interval can also be set.

16.5.5.5 TSP Tunnel Broker Client

The TSP tunnel broker client is available on Windows from the freenet6 Web site (http://www.freenet6.net) or from Hexago directly.

The TSP tunnel broker client on Windows uses the interface numbered 2 for the IPv6 in IPv4 tunnels and a new interface for IPv6 in UDP IPv4 tunnels. The '`tspc.conf`' file, used to configure the client, is discussed in Section 16.5.7.2. In the '`tspc.conf`' file, the variable-value pairs shown in Table 16.18 (from Table 16.20) are specific to Windows.

The TSP client is run in the command line as '`tspc`' or installed as a Windows service. For additional information on the TSP tunnel broker client, see Section 16.5.7.2.

Table 16.18 TSP Tunnel Broker Client Configuration for Windows

Variable	Value	Description
`template`	windows	script template file
`if_tunnel_v6v4`	2	interface name for IPv6 in IPv4 tunnels
`if_tunnel_v6udpv4`	6	interface name for IPv6 in UDP IPv4 tunnels

16.5.6 Cisco

All Cisco IOS tunnel interfaces are defined using the '`interface tunnel`' statement. Each type of tunnel is defined with the '`tunnel mode`' statement. The static IPv6 in IPv4 tunnel has the '`ipv6ip`' mode, 6to4 has the '`ipv6ip 6to4`' mode , ISATAP has the '`ipv6ip isatap`' mode and GRE has the '`gre ip`' mode.

16.5.6.1 Static IPv6 in IPv4 Tunnels

Cisco IOS implements IPv6 in IPv4 tunnels using the '`interface tunnel`' statement under which the '`tunnel mode`' is set to '`ipv6ip`'.

The inner source IPv6 address is defined by the '`ipv6 address`' statement and the outer IPv4 source and destination addresses are defined by the '`tunnel source`' and '`tunnel destination`' statements respectively.

The following statements configure an IPv6 in IPv4 static tunnel, as in Figure 16.48.

```
configure terminal
interface tunnel 0
 ipv6 address 3ffe:b00:1:1::1/128
 tunnel source 192.0.2.1
 tunnel destination 192.0.3.1
 tunnel mode ipv6ip
```

16.5.6.2 6to4

6to4 is implemented as a new '`tunnel mode`', named '`ipv6ip 6to4`', under a tunnel interface.

The following statements configure the 6to4 interface, as in Figure 16.49.

```
configure terminal
interface tunnel 0
 ipv6 address 2002:c000:201:1::1/64
 tunnel source 192.0.2.1
 tunnel mode ipv6ip 6to4
```

Since 6to4 tunnels are created on-demand based on the IPv6 destination address, there is no '`tunnel destination`' command used for 6to4 tunnels.

To forward all incoming 6to4 packets through the 6to4 interface, a specific 6to4 route must be installed towards the 6to4 interface, as shown below.

```
configure terminal
route 2002::/16 tunnel 0
```

16.5.6.3 ISATAP

ISATAP is implemented as a new '`tunnel mode`', named '`ipv6ip isatap`', under a tunnel interface.

The following statements configure the ISATAP interface, as in Figure 16.50.

```
configure terminal
interface tunnel 0
 ipv6 address 3ffe:b00:1:1::/64 eui-64
 tunnel source 192.0.2.1
 tunnel mode ipv6ip isatap
```

The ISATAP specific address is a specific EUI-64 form. By identifying this tunnel interface as ISATAP and identifying the /64 prefix of the address, IOS automatically computes the ISATAP full address (3ffe:b00:1:1::5efe:c000:0201).

16.5.6.4 TSP Tunnel Broker Client

A template for Cisco IOS configuration is available in the TSP tunnel broker client. The client can be run on any platform such as Windows, FreeBSD and Linux and configured to generate the Cisco IOS configuration, by specifying template=cisco.

The TSP tunnel broker client for Cisco IOS uses the '`tunnel`' interface for the IPv6 in IPv4 tunnels. The '`tspc.conf`' file, used to configure the client, is discussed in Section 16.5.7.2. In the '`tspc.conf`' file, the variable-value pairs shown in Table 16.19 (from Table 16.20) are specific to Cisco IOS.

For additional information on the TSP tunnel broker client, see Section 16.5.7.2.

16.5.6.5 Static IPv6 in GRE IPv4 Tunnels

A static IPv6 in GRE IPv4 tunnel is very similarly configured to a static IPv6 in IPv4 tunnel. The only difference is the tunnel encapsulation mode specified as '`tunnel mode gre ip`'. The following statements configure a GRE tunnel as in Figure 16.48.

```
configure terminal
interface tunnel 0
 ipv6 address 3ffe:b00:1:1::1/64
 tunnel source 192.0.2.1
 tunnel destination 192.0.3.1
 tunnel mode gre ip
```

16.5.7 Hexago

The Migration Broker is a TSP tunnel broker implementation. Its operating system, HexOS, is configured by a CLI similar to other vendors.

Table 16.19 TSP Tunnel Broker Client Configuration for Cisco

Variable	Value	Description
`template`	cisco	script template file
`if_tunnel_v6v4`	tunnel 0	interface name for IPv6 in IPv4 tunnels
`proxy_client`	yes	client is not the tunnel end-point

16.5.7.1 Static Tunnels

Static tunnels are configured using the '`tunnel`' interface, where IPv6 addresses are defined by the '`ipv6 address`' statement and IPv4 addresses are defined by the '`tunnel source`' and '`tunnel destination`' statements.

The following statements configure an IPv6 in IPv4 static tunnel, as in Figure 16.48, where the tunnel broker is node N1 and the other tunnel endpoint is R1.

```
configure terminal
interface tunnel 0
 ipv6 address 3ffe:b00:1:1::1/64
 tunnel source 192.0.2.1
 tunnel destination 192.0.3.1
```

Since the tunnel broker automates the establishment of static tunnels through the use of the TSP protocol, static tunnels on the broker are rarely used.

Figure 16.52 shows an example where a static tunnel on the tunnel broker is used to reach an IPv6 connected router when the tunnel broker is not directly connected to the IPv6 network.

The tunnel broker connects TSP clients such as N1 through the leftmost IPv4 network. The broker has a static IPv6 in IPv4 tunnel with R1 through the center IPv4 network to reach the IPv6 network on the right. A better scenario would be to co-locate the tunnel broker with R1 to access the IPv6 network directly.

16.5.7.2 TSP Tunnel Broker Client

TSP is a control protocol to establish and maintain static tunnels. The TSP client is used on the tunnel broker client to get the information of the tunnel. When the information is received, it creates the static tunnel on its operating system. The TSP client code is mostly identical for all client platforms. However, the creation of the static tunnel, which is operating system dependent, is done by a shell script called by the TSP client. The shell script is targetted to specific operating systems and takes care of all specificities. This separation enables fast and easy additions to new operating systems, which is reflected by the community contributions for many operating systems.

The TSP client is either available as part of the operating system distributions, such as Linux or FreeBSD, as downloadable software from the Web site of the tunnel broker service such as Freenet6 (http://www.freenet6.net), or directly from Hexago (http://www.hexago.com).

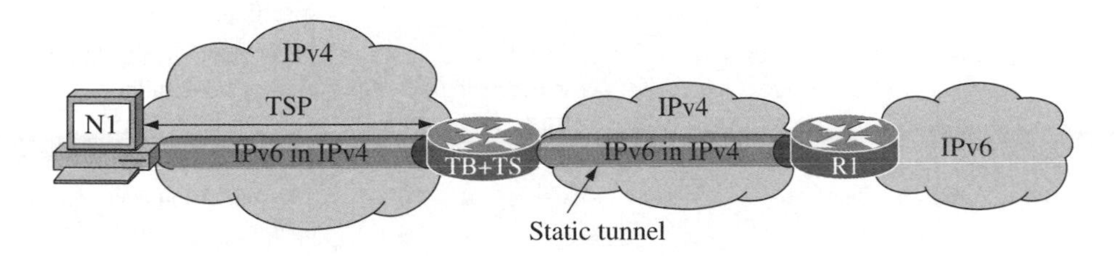

Figure 16.52 Static tunnel on Tunnel Broker

The TSP client is a single application 'tspc' with one configuration file 'tspc.conf' and shell scripts, available in the template directory, for each operating system. The configuration file is a text file with comments using the '#' character. Each statement has the format variable = value, as in rc.conf files in FreeBSD. Most statements are optional and are listed in Table 16.20.

Table 16.20 TSP client tspc.conf key statements

Variable	Default value	Possible values	Description
auth_method	any	any digest-md5 plain anonymous	The authentication used for the TSP session. The digest-md5 method is the most secure, where the password is not sent. The plain method sends userid and password. The anonymous method has no userid and no password. With any, the client uses the most secure method based on its capabilities and broker authentication capabilities.
userid	anonymous	anonymous *string*	The user identification string.
passwd		*string*	The password for the userid.
template		checktunnel cisco darwin freebsd linux netbsd openbsd solaris windows	The script file used to create the tunnel. The value is the name of the script file in the template directory which will be called by the TSP client at the end of the TSP session, to create the tunnel.
server	broker. freenet6.net	*ip_address* *hostname* *ip_address:port* *hostname:port*	The IP address or hostname (full domain name when appropriate) of the broker. A TSP port number can be specified.
tunnel_mode	v6anyv4	v6v4 v6udpv4 v6anyv4 v4v6	The tunnel encapsulation mode (see details in Table 16.21).
if_tunnel_v6v4		*string*	The tunnel interface on the operating system of the client used for IPv6 in IPv4 encapsulation.

if_tunnel_v4udpv4		*string*	The tunnel interface on the operating system of the client used for IPv6 inUDP IPv4 encapsulation.
host_type	host	host router	Shows if the client is a host or a router. In router mode, the TSP client receives a prefix from the broker.
prefixlen	0	0 48 64	The length of the prefix desired by the client.
if_prefix		*string*	The interface on the operating system of the client used to send router advertisements with the received prefix from the broker.
dns_server		*string*	The fully qualified domain name of the DNS server for the reverse DNS delegation of the prefix.

Table 16.21 Tunnel encapsulation mode keywords in tspc.conf

Tunnel mode keyword	Description
v6v4	IPv6 in IPv4 encapsulation, using IPv4 protocol 41: does not work through NAT.
v6udpv4	IPv6 in UDP IPv4 encapsulation: works through NAT.
v6anyv4	IPv6 in any IPv4 encapsulation. The tunnel broker will suggest the right encapsulation method to the client based on its findings if a NAT is in the path. If a NAT is found, then `v6udpv4` is proposed to the client, otherwise, `v6v4` is proposed.
v4v6	IPv4 in IPv6 encapsulation.

Tunnel encapsulation modes are listed in Table 16.21. When '`v6anyv4`' is sent by the client, the broker tests if the client is behind a NAT and responds with the right encapsulation mode.

A minimal tspc.conf file for anonymous use on a FreeBSD client to the freenet6 service is shown below.

```
# cat tspc.conf
server=broker.freenet6.net
```

In this case, all defaults are applied and the requested tunnel has anonymous access. Without a configuration file, one can use the command line to specify the broker, as shown below.

```
# tspc broker.freenet6.net
```

The '**tspc**' program has some arguments to the command line, as described in Table 16.22.

Table 16.22 Arguments to the '**tspc**' program

Argument	Description
−v	Sets the verbose level and debugging information sent on the screen. −vvv gives the most debugging information, such as the TSP XML content.
−vv	
−vvv	
−h	Shows the list of options

When the TSP client finishes its transaction with the broker, it calls the shell script, named by the template variable in the tspc.conf, in tsp_dir/template directory. All the information needed to configure the tunnel is given as environment variables from the tspc program to the shell script.

The TSP client command can be put in the DCHP client scripts so that every time the DHCP client gets a new IPv4 address, the TSP tunnel request is sent and the tunnel is re-established.

16.5.7.3 TSP Tunnel Broker

The TSP tunnel broker is the main function of the Hexago Migration Broker. It is implemented with the following logical modules: tunnel broker, tunnel server, TSP listener and Authentication, and Authorization and Accounting (AAA), as shown in Figure 16.53.

The high-level interactions between the modules are described in Table 16.23.

These four modules are configured each with a specific configuration object in the CLI. One can define multiple tunnel brokers within the same Migration Broker, each could instantiate multiple tunnel servers and multiple TSP listeners. TSP listeners instantiate an AAA model.

The following examples describe a basic but complete configuration of the Migration Broker for offering the tunnel broker service. Since the objects are referring to each other,

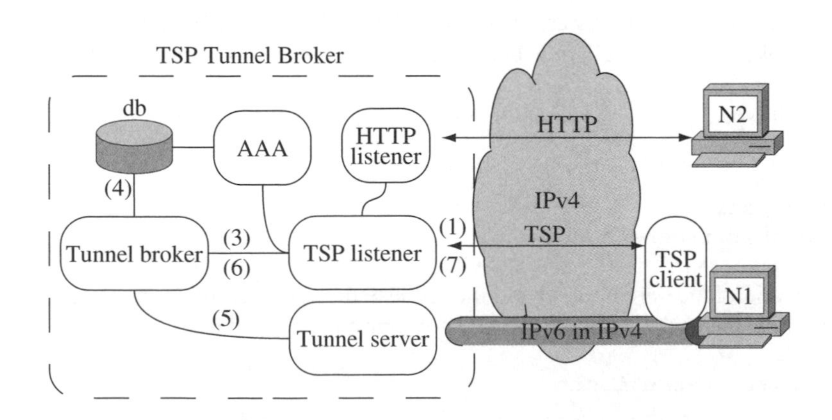

Figure 16.53 TSP tunnel broker modules

Table 16.23 Interactions between TSP tunnel broker modules

Module	Description
1 TSP listener	TSP client connects to the TSP listener on the broker.
2 AAA	TSP listener passes the received user credentials to the AAA module for authentication verification. The AAA module compares the user credentials using the database.
3 Tunnel broker	If authentication is successful, the TSP listener forwards the user tunnel request to the broker. The broker assigns resources for that tunnel: tunnel interface on the tunnel server and IPv6 addresses.
4 Tunnel broker	The tunnel broker saves the tunnel information in the database.
5 Tunnel server	The tunnel broker then configures the tunnel server for that tunnel.
6 TSP listener	The tunnel broker sends the tunnel info to the TSP listener which forwards it to the TSP client through the TSP session.
7 TSP listener	TSP session is closed.

creating a hierarchy, the easiest way to configure the objects is to define them in order: pool of addresses, tunnel server, aaa model, tsp listener and finally tunnel broker.

The '`ipv6 local pool`' statement defines the range of IPv6 addresses that will be allocated by the tunnel broker. A pool is defined for tunnel endpoints and one is defined for the prefix allocation. In the configuration example below, P1 is the tunnel endpoint pool consisting of /128 addresses taken out of a /64. P2 is defined as a pool of /48 prefixes taken out of a /40 and will be used for prefix allocation.

```
ipv6 local pool P1 3ffe:b00:fe00::/64 128
ipv6 local pool P2 3ffe:b00:ff00::/40 48
```

The '`tunnel server`' statement defines a tunnel server which will be the endpoint of the client tunnels. The '`interface range`' statement defines the number of tunnel interfaces this tunnel server can create. The tunnel endpoint is an IPv4 address defined by the '`ip address`' statement. The IPv6 address pools are instantiated by the `ipv6 pool` statement. The pool for endpoint addresses is referenced by '`ipv6 pool endpoints`' and the pool for prefixes is referenced by the '`ipv6 pool prefix`' statement.

```
tunnel server TS1
 interface range 0 2000
 ip address 192.0.1.1
 ipv6 pool endpoints P1
 ipv6 pool prefix P2
 tunnel mode v6v4
 tunnel mode v6udpv4
```

The tunnel server declares its encapsulation mode capabilities using the 'tunnel mode' statement. In the above example, the tunnel server has IPv6 in IPv4 (v6v4) and IPv6 in UDP IPv4 (v6udpv4) encapsulation modes.

AAA is defined by the 'aaa model' statement. Within that object, the AAA authentication and authorization are defined. The A1 AAA model below defines that all users should connect through the digest-md5 authentication mechanism. Other possible authentication mechanisms are 'anonymous' and 'plain'.

```
aaa model A1
 aaa authentication digest-md5
```

The TSP listener is defined by the 'tsp' statement and is given an arbitrary name. The process listens on an IPv4 address defined by the 'ip address' statement. This address must be already configured on the Migration Broker. The listener also instantiates the previously defined AAA model A1 by the 'set aaa model A1' statement to establish which authentication method and policy is applied to connecting TSP users.

```
tsp TSP1
 set aaa model A1
 ip address 192.0.1.1
 tunnel mode v6v4
 tunnel mode v6udpv4
```

The TSP listener must declare which encapsulation modes are offered to the TSP client, using the 'tunnel mode' statement.. In the example above, the TSP listener is offering the IPv6 in IPv4 (v6v4) and IPv6 in UDP IPv4 (v6udpv4) encapsulation modes to TSP clients. In most cases the TSP listener encapsulation modes should be the sum of the tunnel server encapsulation modes attached to the tunnel broker. However, for policy reasons, one might configure a TSP listener to offer only one encapsulation mode and another TSP listener to offer another encapsulation mode.

The tunnel broker object is the last object defined that glues together all the TSP listeners and the tunnel servers. The dtunnel range command defines the maximum number of active tunnels offered on this broker.

```
tunnel broker TB1
 dtunnel range 0 1000
 set tsp TSP1
 set tunnel server TS1
```

16.5.8 Juniper

JunOS supports static IPv6 in IPv4 tunnels and IPv6 in GRE IPv4.

16.5.8.1 Static IPv6 in IPv4 Tunnels

On Juniper platforms, tunnels are supported with the Tunnel PIC. The IP-in-IP tunnels are configured with the 'ip' interface and the GRE tunnels with the 'gr' interface.

The following statements configure an IPv6 in IPv4 static tunnel, as in Figure 16.48. Note that the IPv6 address of the remote tunnel endpoint is not specified.

```
interfaces {
 ip-0/0/1 {
  unit 0 {
   tunnel {
    source 192.0.2.1;
    destination 192.0.3.1;
   }
   family inet6 {
    address 3ffe:b00:1:1::1/126;
   }
  }
 }
}
```

A GRE tunnel for IPv6 is similarly configured using the 'gr' interface name.

```
interfaces {
 gr-0/0/1 {
  unit 0 {
   tunnel {
    source 192.0.2.1;
    destination 192.0.3.1;
   }
   family inet6 {
    address 3ffe:b00:1:1::1/126;
   }
  }
 }
}
```

16.6 Summary

This chapter describes the various techniques and tools for migrating to IPv6 over current IPv4 networks, such as dual-stacks, static IPv6 in IPv4 tunnels, 6to4, ISATAP, TSP tunnel broker and Teredo. Considerations on tunneling are discussed. No single technique works in all cases. The reader should refer to Section 16.4 to help identify the technique best suited for his or her network.

16.7 References

Blanchet, M. and F. Parent, "IPv6 Tunnel Broker with the Tunnel Setup Protocol(TSP)", draft-blanchet-v6ops-tunnelbroker-tsp-03 (work in progress), August 2005.

Huitema, C., "Teredo: Tunneling IPv6 over UDP through NATs", draft-huitema-v6ops-teredo-05 (work in progress), April 2005.

[RFC2222] Myers, J., 'Simple Authentication and Security Layer (SASL)', RFC 2222, October 1997.

[RFC2245] Newman, C., 'Anonymous SASL Mechanism', RFC 2245, November 1997.

[RFC2473] Conta, A. and Deering, S., 'Generic Packet Tunneling in IPv6 Specification', RFC 2473, December 1998.

[RFC2663] Srisuresh, P. and Holdrege, M., 'IP Network Address Translator (NAT) Terminology and Considerations', RFC 2663, August 1999.

[RFC2784] Farinacci, D., Li, T., Hanks, S., Meyer, D. and Traina, P., 'Generic Routing Encapsulation (GRE)', RFC 2784, March 2000.

[RFC2831] Leach, P. and Newman, C., 'Using Digest Authentication as a SASL Mechanism', RFC 2831, May 2000.

[RFC2893] Gilligan, R. and Nordmark, E., 'Transition Mechanisms for IPv6 Hosts and Routers', RFC 2893, August 2000.

[RFC2960] Stewart, R., Xie, Q., Morneault, K., Sharp, C., Schwarzbauer, H., Taylor, T., Rytina, I., Kalla, M., Zhang, L. and V. Paxson, 'Stream Control Transmission Protocol', RFC 2960, October 2000.

[RFC3022] Srisuresh, P. and Egevang, K., 'Traditional IP Network Address Translator (Traditional NAT)', RFC 3022, January 2001.

[RFC3053] Durand, A., Fasano, P., Guardini, I. and Lento, D., 'IPv6 Tunnel Broker', RFC 3053, January 2001.

[RFC3056] Carpenter, B. and Moore, K., 'Connection of IPv6 Domains via IPv4 Clouds', RFC 3056, February 2001.

[RFC3068] Huitema, C., 'An Anycast Prefix for 6to4 Relay Routers', RFC 3068, June 2001.

[RFC3489] Rosenberg, J., Weinberger, J., Huitema, C. and Mahy, R., 'STUN – Simple Traversal of User Datagram Protocol (UDP) Through Network Address Translators (NATs)', RFC 3489, March 2003.

[RFC3964] Savola, P. and C. Patel, 'Security Considerations for 6to4', RFC 3964, December 2004.

Templin, F., Gleeson, T., Talwar, M. and Thaler, D., 'Intra-Site Automatic Tunnel Addressing Protocol (ISATAP)', draft-ietf-ngtrans-isatap-16 (work in progress), October 2003.

17

Deploying IPv6 Dominant Networks with IPv4 Support

Chapter 16 discussed deploying IPv6 in IPv4 dominant networks, which is the common case for the initial deployment of IPv6. However, the inverse case happens in new networks: here, IPv6 is deployed as the dominant IP protocol and IPv4 legacy support is needed, but there is no intention or need to deploy an IPv4 fully routed network. In this context, the IPv4 in IPv6 tunneling methods, such as static tunnels, DSTM and TSP tunnel broker are used and are discussed in this chapter.

17.1 Tunneling IPv4 in IPv6

As in Chapter 16, tunneling is used to overlay one IP protocol over another. In this section, the IPv4 in IPv6 tunnels are described.

17.1.1 IPv4 in IPv6 Encapsulation

IPv4 packets are encapsulated in the payload of the IPv6 packet, as shown in Figure 17.1.

The Next Header field value in the IPv6 header, identifying an IPv4 packet inside the IPv6 packet payload, is 4, as for IPv4 in IPv4 tunnels.

17.1.2 IPv4 in IPv6 Static Tunnels

IPv4 in IPv6 static tunnels are configured at each tunnel endpoint by the IPv4 and IPv6 address of the other endpoint, similar to IPv6 in IPv4 tunnels in Section 16.2.4. Figure 17.2 shows an example of an IPv4 in IPv6 static tunnel between two routers.

Migrating to IPv6: A Practical Guide to Implementing IPv6 in Mobile and Fixed Networks Marc Blanchet

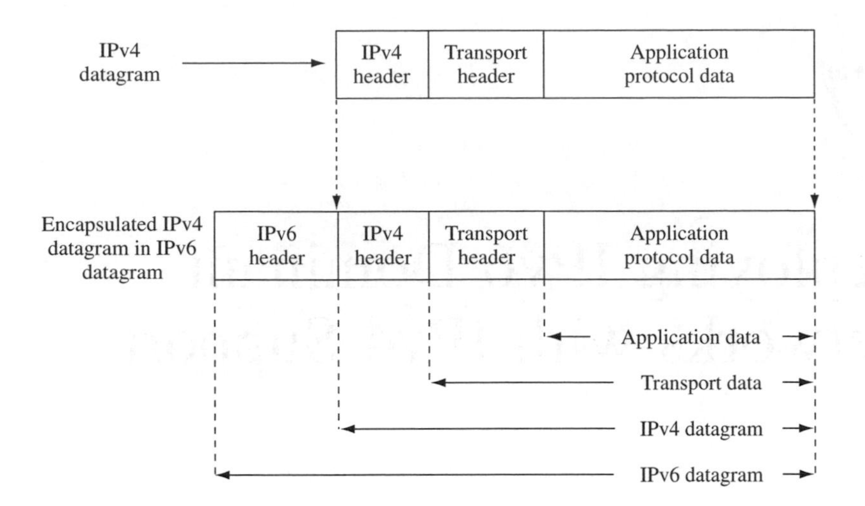

Figure 17.1 IPv4 in IPv6 encapsulation

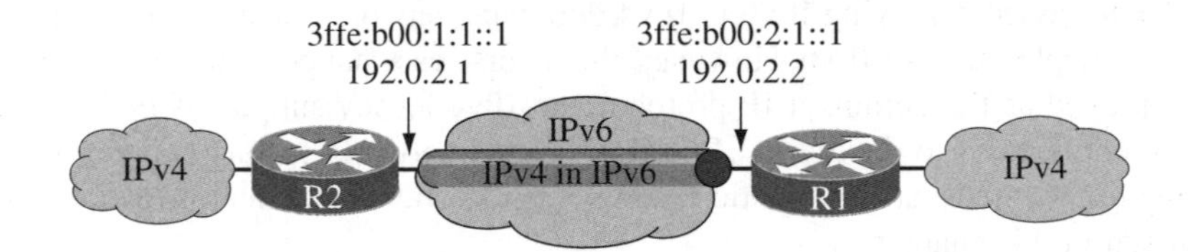

Figure 17.2 IPv4 over IPv6 tunnel between two routers

17.1.2.1 Requirements

The following requirements have to be met in this router to router configuration:

- Routers R1 and R2, the tunnel endpoints, are dual stack.
- IPv6 is the network between the tunnel endpoints.
- Tunnel endpoints must have a reachable address from each other.

17.1.2.2 Limitations

IPv4 in IPv6 static tunnels suffer the same limitations of manual configuration as the IPv6 in IPv4 tunnels described in Section 16.2.4.

17.1.2.3 Applicability

IPv4 in IPv6 static tunnels are applicable when a very small number of tunnels are needed. The Dual Stack Transition Mechanism (DSTM) and TSP Tunnel Broker are designed to automate the setup of IPv4 in IPv6 tunnels.

17.1.3 DSTM with DHCPv6

DSTM [Bound, 2003] automates the configuration of IPv4 in IPv6 tunnels and defines three components: the DSTM client, the DSTM server and the DSTM border router, shown in Figure 17.3.

The DSTM client is a dual-stack node connected to an IPv6-only network. The DSTM server is either a DHCPv6 server or a TSP server (discussed in Section 17.1.4). Using DHCPv6, the DSTM client makes a DHCPv6 request over the IPv6 network to the DSTM server acting as a DHCPv6 server. The client requests an IPv4 address and the IPv6 address of the tunnel endpoint on the DSTM border router. The DSTM border router is the gateway to the IPv4 network.

The DSTM specification [Bound, 2003] do not specify the protocol between the DSTM server and the DSTM border router. When the DSTM server allocates IPv4 addresses and tunnel endpoints, it must make sure the DSTM border router is configured to accept the tunnels. The DSTM server functionality can also be co-located with the DSTM border router, as shown in Figure 17.4.

DSTM provides non-NAT connections for IPv4 applications on the DSTM clients. If the network and applications are mostly migrated to IPv6, DSTM provides a good alternative to support legacy IPv4 applications that need IPv4 reachability.

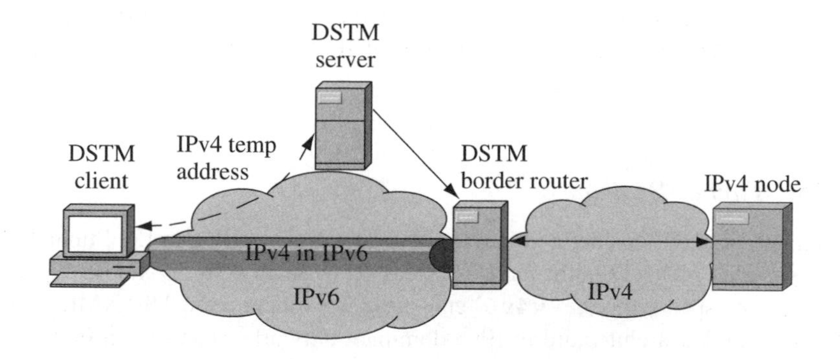

Figure 17.3 DSTM components

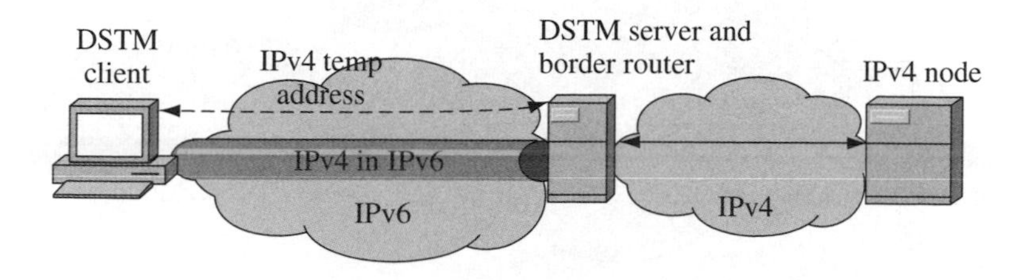

Figure 17.4 DSTM combined server and border router

17.1.3.1 Requirements

The requirements for DSTM with DHCPv6 are:

- Nodes are dual stack and implement the DSTM and DHCPv6 client.
- A DSTM server acting as a DHCPv6 server is available on the IPv6 network.
- The DHCPv6 server implements the added functionality needed for DSTM.
- DHCPv6 relay forwarding is enabled on the IPv6 network.
- The border router to the IPv4 network is DSTM enabled.
- The DSTM server can configure the DSTM border router remotely.
- A sufficiently large IPv4 address pool is available on the DSTM server to serve all concurrent IPv4 clients.

17.1.3.2 Limitations

DSTM only provides IPv4 in an IPv6 network. If the node is mobile and moving to an IPv4-only network, DSTM does not enable IPv6 connectivity.

17.1.3.3 Applicability

The other way to provide IPv4 and IPv6 connectivity is to deploy a full dual-stack network, where all routers are forwarding and routing IPv4 and IPv6. DSTM is useful when such an environment is not possible, either because of the lack of IPv4 address space and the need not to use NAT, or by the very dominance of IPv6 deployment over the legacy IPv4 applications.

17.1.4 TSP Tunnel Broker

The TSP tunnel broker [Blanchet and Parent, 2005] establishes IPv6 in IPv4 tunnels, with or without the presence of NATs in the path, as discussed in Section 16.2.9. It also supports IPv4 in IPv6 tunnels, by specifying the '`v4v6`' encapsulation mode in the TSP XML negotiation. The TSP tunnel broker architecture in IPv6 dominant networks is shown in Figure 17.5.

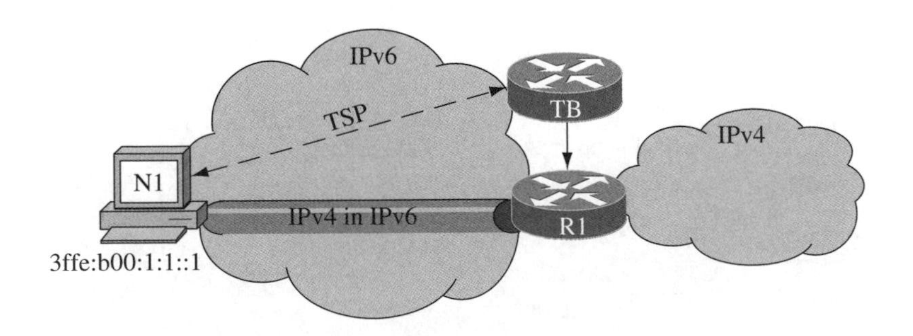

Figure 17.5 IPv4 in IPv6 tunnels with TSP

If the TSP client is on an IPv6-only network, it contacts the tunnel broker at its IPv6 address and requests an IPv4 in IPv6 tunnel. In the tunnel request, the client sends its IPv6 address and the 'v4v6' tunnel encapsulation mode, as shown below.

```
<tunnel action="create" type="v4v6">
  <client>
    <address type="ipv6">3ffe:b00:1:1::1</address>
  </client>
</tunnel>
```

The broker responds with the IPv4 in IPv6 tunnel information.

```
<tunnel action="info" type="v4v6" lifetime="1440">
  <server>
    <address type="ipv4">192.0.2.1</address>
    <address type="ipv6"> 3ffe:b00:1:1::2</address>
  </server>
  <client>
    <address type="ipv4">192.0.2.2</address>
    <address type="ipv6"> 3ffe:b00:1:1::1</address>
  </client>
</tunnel>
```

The client and the tunnel server configures their IPv4 in IPv6 tunnel endpoints and then IPv4 traffic is encapsulated in IPv6 from the client to the tunnel server.

DSTM and TSP tunnel broker can be used together [Bound, 2003] where the DSTM server is the TSP tunnel broker.

17.1.4.1 Requirements

The requirements to deploy the TSP tunnel broker solution are:

- The node is dual-stack.
- The node implements the TSP protocol and supports configured tunnels.
- The TSP tunnel broker implements the TSP protocol.
- The TSP tunnel broker has access to a tunnel server.
- The tunnel server is dual-stack, supports configured tunnels and is connected to the IPv4 network. A tunnel server may be implemented within the tunnel broker.
- Hosts behind a TSP router client do not need to support or know about TSP and may only be IPv4 enabled.

17.1.4.2 Limitations

In IPv4 in IPv6 tunnel mode, the TSP tunnel broker shares the same limitations as the IPv6 in IPv4 tunnels, described in Section 16.2.9.

17.1.4.3 Applicability

The TSP tunnel broker can be used as an IPv4 in IPv6 VPN server or mobility server, where it provides ubiquitous IP to dual stack nodes in any context: IPv4, IPv4 behind NAT and IPv6 networks.

17.2 IP Packet and Transport Translation

Some deployment scenarios involve IPv6-only nodes. For example, new networks with new applications might require IPv6-only functionalities, prohibiting the use of IPv4. The simplicity of running only one IP protocol might drive some networks to be IPv6-only. Legacy IPv4 nodes would never be upgraded to IPv6, not necessarily for technical reasons. For example, a good printer will continue to print while the vendor might decide not to provide a new version of the IP driver to support IPv6. Together, we have IPv6-only nodes and IPv4-only nodes. If these two kinds of nodes need to interact with each other, some translation at the IP packet layer is necessary. This generic situation is shown in Figure 17.6.

When one node is querying the IP address corresponding to the name of the other node, the querier will never get a response. For example, the IPv6 node is querying the IPv6 address of the IPv4 node, which it does not have. The IPv4 node is querying the IPv4 address of the IPv6 node, which it does not have. The translation process should intercept the DNS query packets, convert them to the other IP address format and then send the modified query packet. When the DNS reply arrives, it should be reconverted. The translation process must allocate temporary addresses to the nodes for the query to work. The IPv6 node is querying the AAAA record for the other node. The translator changes that query to an A record and sends it. The reply is received by the translator and reconverts it to an AAAA record answer. By reconverting, it must allocate a temporary IPv6 address for the IPv4 node. An obvious limitation here is that the translator must be placed between the two nodes in all cases, since it needs to intercept all DNS queries. However, if one node is trying to reach the other node using its IP address, then no DNS request is sent, which means no connection.

In summary, translation involves translating IP headers of packets, and intercepting and translating the DNS queries.

The translation can be done at the IP layer, at the transport layer or at the application layer. Some proposals such as the Network Address Translation-Protocol Translation (NAT-PT) and Transport Relay Translator (TRT) have been defined but are deprecated or have limited use. They are described in the book Web site (http://www.ipv6book.ca).

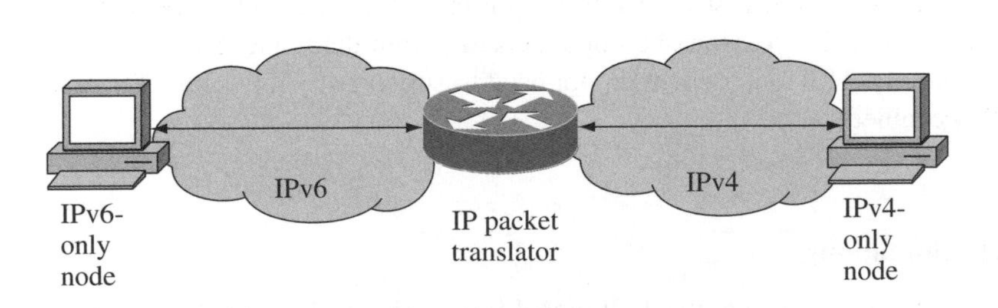

Figure 17.6 IPv6-only and IPv4-only nodes

17.3 Configuring IPv4 in IPv6 Dominant Networks

This section describes the configuration of static IPv4 in IPv6 tunnels and the TSP tunnel broker for the various implementations.

17.3.1 FreeBSD

FreeBSD supports static tunnels through the 'gif' interface and the TSP tunnel broker client through the tsp/freenet6 client.

17.3.1.1 Static IPv4 in IPv6 Tunnels

IPv4 in IPv6 tunnels in FreeBSD are implemented as gif interfaces. These interfaces can also be used for IPv6 in IPv4 (as described in Section 16.5.2), IPv4 in IPv4 and IPv6 in IPv6 encapsulations. They are clonable devices using the create argument of the ifconfig command. The tunnel argument specifies the IPv6 outer source and destination addresses. The IPv4 inner addresses are defined by the inet argument.

The following commands configure an IPv4 in IPv6 static tunnel.

```
# ifconfig gif0 create
# ifconfig gif0 tunnel 3ffe:b00:1:1::1 3ffe:b00:1:1::2 inet
192.0.2.1 192.0.3.1 alias up
```

To store this configuration, use the ipv6_ifconfig_gif0 variable in the /etc/rc.conf file.

To forward all IPv4 traffic through this static tunnel, use the route add default command, as shown below.

```
# route add default 192.0.3.1
```

17.3.1.2 TSP Tunnel Broker Client

The TSP tunnel broker client is available on FreeBSD in the ports/net/freenet6, or from the freenet6 Web site (http://www.freenet6.net) or from Hexago directly.

The TSP tunnel broker client on FreeBSD uses the 'gif' interface for the IPv4 in IPv6 tunnels. The 'tspc.conf' file, used to configure the client, is discussed in Section 16.5.7.2 and specifics on IPv4 in IPv6 tunnels are discussed in Section 17.3.4.2. In the 'tspc.conf' file, the variable-value pairs in Table 17.1 (taken from Table 16.20) are specific to FreeBSD.

Table 17.1 TSP Tunnel Broker Client Configuration for FreeBSD

Variable	Value	Description
template	freebsd	Script template file
if_tunnel_v4v6	gif0	Interface name for IPv4 in IPv6 tunnels

The TSP client is run in the command line as 'tspc'. Its configuration file is usually placed in '/usr/local/etc/tspc.conf'.

For additional information on the TSP tunnel broker client, see Sections 16.5.7.2 and 17.3.4.2.

17.3.2 Solaris

Solaris supports static tunnels through the 'ip6.tun' interface and TSP tunnel broker through the tsp/freenet6 client.

17.3.2.1 Static IPv4 in IPv6 tunnels

Static IPv4 in IPv6 tunnels are configured using the 'ifconfig' command with the 'tsrc', 'tdst'. The parameters are put in the '/etc/hostname6.ip6.tun0' file to save the tunnel configuration permanently. The files '/etc/hostname6.ip6.tun1', tun2, etc., are used for additional tunnels. The 'tsrc' and 'tdst' are the tunnel source and destination address of the outer packet (i.e. IPv6 for an IPv4 in IPv6 tunnel). The 'up' parameter makes the interface live. Static IPv4 addresses are assigned to the tunnel endpoints.

The following file configures an IPv4 in IPv6 static tunnel.

```
# cat /etc/hostname6.ip6.tun0
 tsrc 3ffe:b00:1:1::1 tdst 3ffe:b00:1:1::2
 192.0.2.1 192.0.3.1 up
```

The 'ifconfig ip6.tun0' is used to look at the tunnel interface configuration.

```
# ifconfig ip6.tun0
```

17.3.2.2 TSP Tunnel Broker Client

The TSP tunnel broker client is available on Solaris from the freenet6 Web site (http://www.freenet6.net) or from Hexago directly.

The TSP tunnel broker client on Solaris uses the 'ip6.tun' interface for the IPv4 in IPv6 tunnels. The 'tspc.conf' file, used to configure the client, is discussed in Section 16.5.7.2 and specifics on IPv4 in IPv6 tunnels are discussed in Section 17.3.4.2. In the 'tspc.conf' file, the variable-value pairs, in Table 17.2 (taken from Table 16.20) are specific to Solaris.

The TSP client is run in the command line as 'tspc'.

Table 17.2 TSP Tunnel Broker Client Configuration for FreeBSD

Variable	Value	Description
template	solaris	Script template file
tunnel_mode	v4v6	IPv4 in IPv6 tunnels
if_tunnel_v4v6	ip6.tun0	Interface name for IPv4 in IPv6 tunnels

For additional information on the TSP tunnel broker client, see Sections 16.5.7.2 and 17.3.4.2.

17.3.3 Cisco

Cisco IOS tunnel interfaces are defined using the 'interface tunnel' statement. Each type of tunnel is defined with the 'tunnel mode' statement. The static IPv4 in IPv6 tunnel has the 'ipipv6' mode.

17.3.3.1 Static IPv4 in IPv6 tunnels

Cisco IOS implements IPv4 in IPv6 tunnels using the 'interface tunnel' statement under which the 'tunnel mode' is set to 'ipipv6'.

The inner source IPv4 address is defined by the 'ip address' statement and the outer IPv6 source and destination addresses are defined by the 'tunnel source' and 'tunnel destination' statements respectively.

The following statements configure an IPv4 in IPv6 static tunnel.

```
configure terminal
interface tunnel 0
  ip address 192.0.2.1 255.255.255.255
  tunnel source 3ffe:b00:1:1::1
  tunnel destination 3ffe:b00:1:1::2
  tunnel mode ipipv6
```

17.3.4 Hexago

The Migration Broker is a TSP tunnel broker implementation. Configuring HexOS is discussed in Section 16.5.7. This section describes only the changes for IPv4 in IPv6 tunnels.

17.3.4.1 Static IPv4 in IPv6 tunnels

Static tunnels are configured using the 'tunnel' interface, where IPv4 addresses are defined by the 'ip address' statement and IPv6 addresses are defined by the 'tunnel source' and 'tunnel destination' statements.

The following statements configure an IPv4 in IPv6 static tunnel.

```
configure terminal
interface tunnel 0
  ip address 192.0.2.1
  tunnel source 3ffe:b00:1:1::1
  tunnel destination 3ffe:b00:1:1::2
```

Since the tunnel broker automates the establishment of static tunnels through the use of the TSP protocol, static tunnels on the broker are mostly used between distant brokers and routers.

17.3.4.2 TSP Tunnel Broker Client

The TSP tunnel broker client is described in Section 16.5.7.2. The only change from the IPv6 in IPv4 tunnels is to set the '`tunnel_mode`' variable to '`v4v6`' in the `tspc.conf` file, as shown in Table 17.3.

17.3.4.3 TSP Tunnel Broker

The TSP tunnel broker is described in Section 16.5.7.3. The changes in the configuration are listed in Table 17.4.

17.3.5 Juniper

JunOS supports static IPv4 in IPv6 tunnels through the '`ip`' interface.

17.3.5.1 Static IPv4 in IPv6 tunnels

On Juniper platforms, tunnels are supported with the Tunnel PIC. The IP-in-IP tunnels are configured with the '`ip`' interface.

The following statements configure an IPv4 in IPv6 static tunnel. Note that the IPv4 address of the remote tunnel endpoint is not specified.

Table 17.3 TSP Tunnel Broker Client Configuration for IPv4 in IPv6 tunnels

Variable	Value	Description
`tunnel_mode`	v4v6	IPv4 in IPv6 tunnels

Table 17.4 TSP Tunnel Broker Changes for IPv4 in IPv6 tunnels

IPv6 in IPv4	IPv4 in IPv6	Description
ipv6 local pool	ip local pool	Pools of addresses used for assignements of tunnel endpoints.
tunnel server		Under tunnel server mode.
ip address	ipv6 address	The tunnel server endpoint is an IPv6 address specified by ipv6 address.
ipv6 pool endpoints	ip pool endpoints	The tunnel server pool is an IPv4 address pool specified by ip pool endpoints.
tsp		Under the TSP mode.
ip address	ipv6 address	The TSP listener address is an IPv6 address specified by ipv6 address.
tunnel mode v6v4	tunnel mode v4v6	The TSP listener offers the IPv4 in IPv6 tunnels.

```
interfaces {
  ip-0/0/1 {
    unit 0 {
      tunnel {
        source 3ffe:b00:1:1::1;
        destination 3ffe:b00:1:1::2;
      }
      family inet {
        address 192.0.2.1/31;
      }
    }
  }
}
```

17.4 Summary

This chapter describes techniques used when an IPv6 dominant network is deployed and IPv4 support is required. Techniques such as DSTM and TSP Tunnel Broker involve the use of IPv4 in IPv6 tunneling.

17.5 References

Blanchet, M. amd F. Parent, 'IPv6 Tunnel Broker with the Tunnel Setup Protocol (TSP)', draft-blanchet-v6ops-tunnelbroker-tsp-03 (work in progress), August 2005.

Bound, J., 'Dual Stack Transition Mechanism', draft-bound-dstm-exp-00 (work in progress), August 2003.

18

Migrating with Application Level Gateways

Chapters 16 and 17 describe transition mechanisms applied at the IP level, which makes them independent of the applications. When a limited number of applications need to be migrated, application level gateways (ALG) that receive and decode the application protocol connections over one IP protocol and then restart another connection over the other IP protocol, can be considered and are described in this chapter.

18.1 Application Level Gateway

Store and forward applications are good candidates for application level gateways. For example, an SMTP server can listen on one side on a TCP-IPv4 socket for incoming mail. When mail arrives, it is stored locally and then submitted in the outgoing queue. The outgoing socket can be on TCP-IPv6. Figure 18.1 shows an example in which host H1 uses IPv6 to its SMTP server.

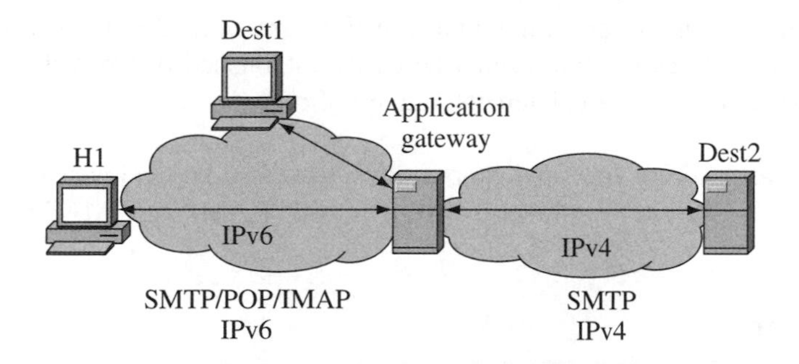

Figure 18.1 SMTP application level gateway

Migrating to IPv6: A Practical Guide to Implementing IPv6 in Mobile and Fixed Networks Marc Blanchet

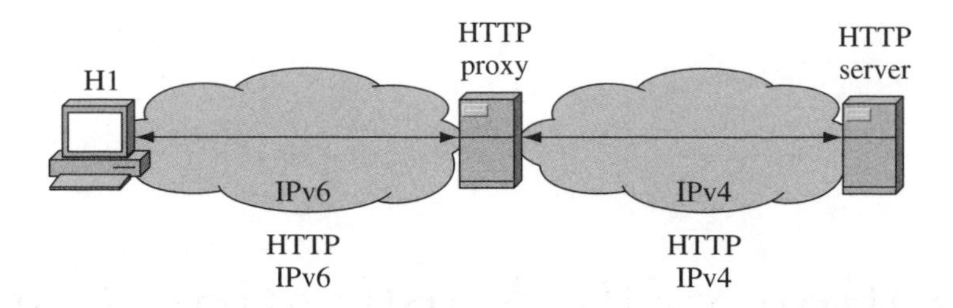

Figure 18.2 HTTP application specific proxy

Dest1 and Dest2 hosts are on IPv6 and IPv4 respectively. The SMTP server stores the incoming and outgoing mail temporarily. Since this architecture is already used in mail deployments, it is easy to upgrade it for IPv6.

18.2 Application Specific Proxy

Applications that can be easily proxied such as HTTP are good candidates for application specific proxies. For example, an HTTP proxy listening on the IPv6 side for HTTP requests can proxy the requests on the IPv4 side to the HTTP server. Figure 18.2 shows an example of this architecture.

If a site is already using an HTTP proxy for IPv4, this architecture for IPv6 transition is easy to deploy.

18.3 Considerations of Application Level Gateways

The ALGs described in the previous sections share the same limitations:

- Scalability: depending on the application protocol, having all enterprise traffic going to a single server requires processing power and bandwidth.
- Reliability: ALGs are a single point of failures.
- Limited number of applications: not all applications are defined to work with a proxy.
- Security: A security association cannot be easily established between the end node and the server because the ALG is intercepting the trafic.

For all these reasons, very few applications are targeted to be deployed with ALGs. The application that fits well in this model is mail with SMTP, POP and IMAP protocols.

18.4 Summary

Application-level gateways are used for migration applications that are designed to be proxied. However, in all cases, important limitations of these architectures decrease their usefulness in deployments, with the exception of mail.

19

Transport Protocols

IPv6 is a new design of layer 3 of the Internet protocol suite. A requirement of the design is to reuse the current transport protocols (layer 4) as they stand, including the Transmission Control Protocol (TCP) and the User Datagram Protocol (UDP). Recent work on other transport protocols such as the Stream Control Transmission Protocol (SCTP) [RFC2960] is not discussed.

19.1 Checksum

As discussed in Section 3.7.1, the IP packet checksum is removed from the IPv6 header under the assumption that the transport protocol above IP must have a checksum. This transport protocol checksum is calculated over the following fields [RFC2460]:

- Source address;
- Destination address;
- Next Header field;
- Transport-layer payload length.

Figure 19.1 illustrates the pseudo-header used to calculate the checksum.

The Next Header field contains the identifier of the transport protocol. If extension headers are present in the IP header as described in Section 3.5, then the Next Header field is taken from the last extension header in the chain. The address fields contain the source and destination addresses. If a routing header is used when the destination address is not the final destination, as described in Section 9.2, the checksum is still computed over the final destination address.

All transport protocols, such as TCP, UDP and ICMP, compute the checksum for IPv6 packets. For IPv4, UDP checksum is optional and TCP checksum is mandatory.

Migrating to IPv6: A Practical Guide to Implementing IPv6 in Mobile and Fixed Networks Marc Blanchet

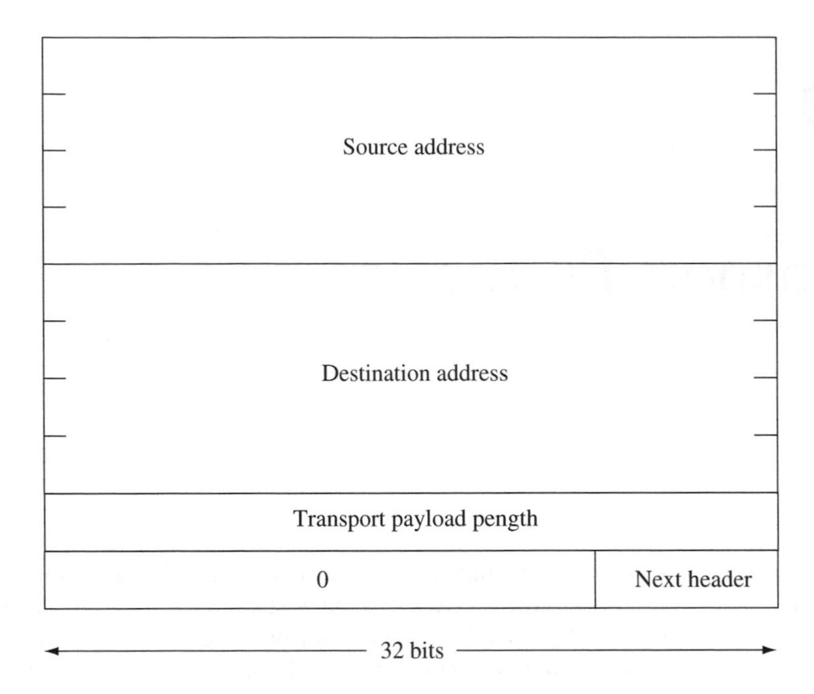

Figure 19.1 Transport checksum

19.2 Transmission Control Protocol (TCP)

TCP has not changed for IPv6. However, parallel to the engineering of IPv6, TCP has evolved with optimizations based on years of research, such as Selective acknowledgment (SACK)[RFC2018] and Explicit Congestion Notification (ECN) [RFC3168]. Some of these optimizations are useful for wired and wireless networks. The latter creates additional constraints and requirements which are discussed in Chapter 12.

TCP design is based on the following assumptions:

- packet loss is due to network congestion;
- all hosts are participating to equally share the bandwidth;
- the network is a black-box: transport congestion control is done by the end systems.

This section discusses ECN since it modifies the definition of the IPv6 header.

19.2.1 Explicit Congestion Notification (ECN)

Active queue management mechanisms such as Random Early Detection (RED) [Floyd and Jacobson, 1993] enable the router to detect congestion before its queue overflows, causing all subsequent packets to be dropped. ECN is a way for routers to signal congestion, based on their early detection, to the end nodes before the router starts dropping packets. End nodes supporting ECN decrease their transport window size to participate in the reduction of traffic, thus avoiding the congestion, and consequently avoiding the packet drops.

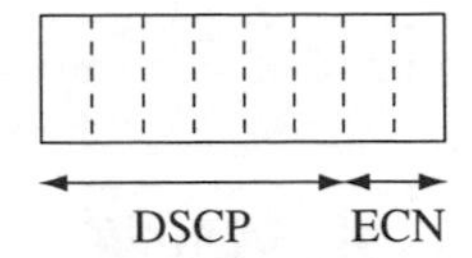

Figure 19.2 ECN bits in the Traffic class field of the IPv6 header

The signaling is done by bits in both the IP header and the TCP header. The two ECN bits in the IP header are in the traffic class field, as shown in Figure 19.2.

The ECN bit values are defined in Table 19.1. An ECN-capable node puts '10' in the ECN bits of the traffic class field of the IPv6 header when it sends packets. A non-ECN-capable node puts '00'. A router detecting early congestion within its active queue modifies the IPv6 header by setting the ECN bits to '11' if the ECN bits were set to '10' by the sending node.

ECN bits are also defined in the TCP header: the ECN-echo bit (ECE) and the Congestion Window Reduced (CWR) bit, as shown in Figure 19.3.

The ECE bit is used by a node to publish its support of the ECN mechanism to the other node. The CWR bit is used by a node to acknowledge the reception of the congestion signal.

Figure 19.4 shows an example of an IPv6 network with ECN-capable nodes N1 and N2 and ECN-capable router R1 in the IPv6 network. N1 to N2 traffic goes through R1.

Figure 19.5 shows the ECN relevent exchanges between ECN-capable nodes. When N1 and N2 establish their TCP connection, N1 publishes its support of ECN to the other party (step 1) by setting the ECE bit in the TCP header. N2 acknowledges to N1 its own support of ECN (step 2) by setting the ECE bit. Then traffic flows freely between the two nodes, all packets being marked with the ECN capability bits ('10') in the IP header. When the early congestion is detected by the ECN-capable router R1, R1 modifies the queued packet for N2 by changing the ECN IP bits to '11', indicating early congestion, and forwards it to N2. N2 then signals back to N1 with the ECE TCP bit in the next ACK TCP packet. N1 receiving this ECE TCP bit acts as if congestion happened by decreasing its TCP window size.

Table 19.1 ECN values in the Traffic class field

Value	Description
00	No ECN capability
10	ECN capable node
01	ECN capable node[47]
11	Router signaling congestion to ECN capable nodes

Note: 10 and 01 are equivalent; typically, '10' is used. See RFC3168 for discussion on the use of these bits.

Figure 19.3 TCP header ECN bits

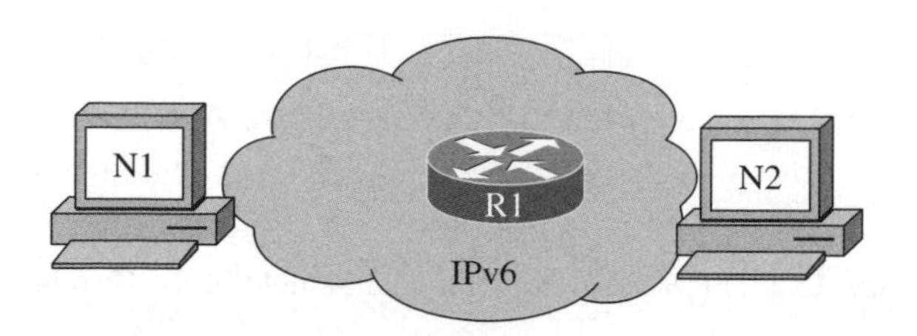

Figure 19.4 ECN network example

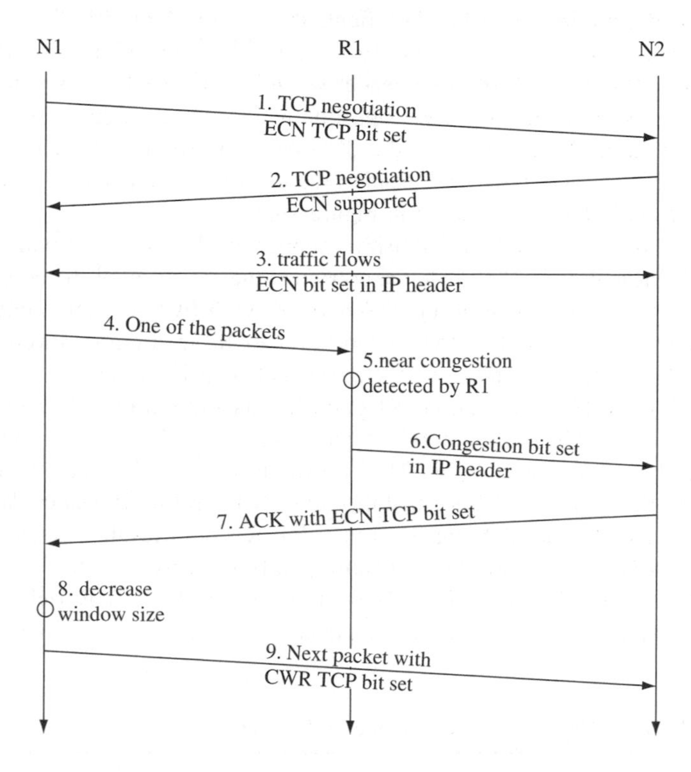

Figure 19.5 ECN process

N1 acknowledges to N2 the reception of the congestion signal by setting the TCP CWR bit. Thereafter, the normal TCP congestion mechanism applies.

This ECN mechanism is in fact a way for routers that anticipate congestion to signal traffic slow down to ECN-capable nodes in order to avoid the congestion. However, the real effect of ECN will only be seen when enough nodes and routers are ECN-capable. ECN-capable nodes mean modifications to the IP and TCP stacks and ECN-capable routers mean modifications to the forwarding engine. Note that the ECN mechanism is defined for both IPv4 and IPv6.

19.3 User Datagram Protocol (UDP)

No modification is made to UDP for IPv6, except the mandatory checksum computed over the pseudo-header, as shown in Figure 19.1.

19.4 Internet Control Message Protocol (ICMP)

ICMP is not a transport protocol like UDP or TCP: it does not transport application data and protocols. But from the perspective of the IPv6 header, ICMP has a checksum computed over the pseudo-header just as TCP and UDP compute checksums. ICMP, as a layer above IPv6, isolates the implementation from being bound to the layer 2 as in IPv4 ICMP. ICMP is detailed in Chapter 7.

19.5 Summary

The transport protocols are not changed from IPv4 to IPv6. However, transport checksums are now mandatory. Over the years, TCP has been enhanced by many optimizations, such as ECN. ECN is a mechanism for routers to signal the early detection of congestion to nodes by modifying the ECN bits in the traffic class field of the IPv6 header.

19.6 References

Floyd, S., and Jacobson, V., "Random Early Detection gateways for Congestion Avoidance", IEEE/ACM Transactions on Networking, V.1 N.4, August 1993.

[RFC2018] Mathis, M., Mahdavi, J., Floyd, S. and Romanow, A., 'TCP Selective Acknowledgment Options, RFC 2018, October 1996.

[RFC2460] Deering, S. and Hinden, R., 'Internet Protocol, Version 6 (IPv6) Specification', RFC 2460, December 1998.

[RFC2960] Stewart, R., Xie, Q., Morneault, K., Sharp, C., Schwarzbauer, H., Taylor, T., Rytina, I., Kalla, M., Zhang, L. and Paxson, V., 'Stream Control Transmission Protocol', RFC 2960, October 2000.

[RFC3168] Ramakrishnan, K., Floyd, S. and Black, D., 'The Addition of Explicit Congestion Notification (ECN) to IP', RFC 3168, September 2001.

20

Network Management

Network management is carried out using various tools and protocols. The most used is the Simple Network Management Protocol (SNMP). SNMP is a protocol between agents located inside managed nodes and management stations, to query and set network management variables, as shown in Figure 20.1.

Considerations on changes to support IPv6 for SNMP transport and the MIBs are described.

20.1 SNMP Transport

SNMP uses UDP over IPv4 for transport [RFC3417]. Transport mappings have been redefined[RFC3419] for IPv4 and IPv6 as well as additional transport protocols to UDP, such as TCP and SCTP. The transport is now specified with a set of variables, as shown in Table 20.1.

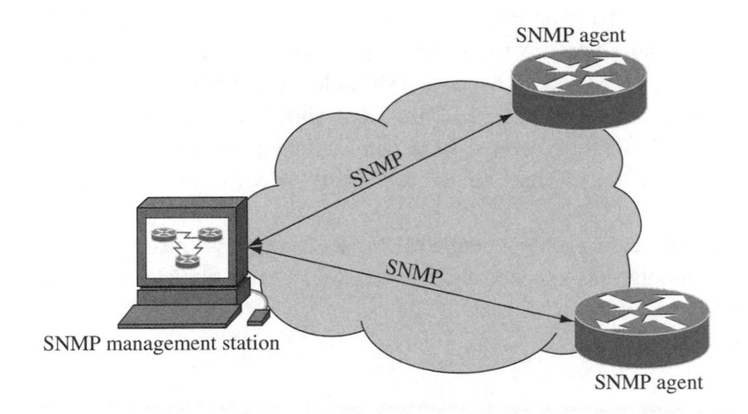

Figure 20.1 SNMP basic architecture

Migrating to IPv6: A Practical Guide to Implementing IPv6 in Mobile and Fixed Networks Marc Blanchet

Table 20.1 IP version independent SNMP transport mapping variables

Object name	Some possible values	Description
`TransportDomain`	`transportDomainUdpIpv4`, `transportDomainUdpIpv6`, `transportDomainTcpIpv6`	Transport domain is the combination of the transport protocol (UDP, TCP, SCTP) and the IP protocol used for SNMP transport.
`TransportAddressType`	`udpIpv4`, `udpIpv6`, `tcpIpv6`	The type of transport address used in the `TransportAddress` variable.
`TransportAddress`	*IPv6 address with transport port*	The transport address itself, containing both the IP address and the transport port number.

As with DNS, the transport protocol is not related to the information queried. For example, a management station can use SNMP over IPv4 to query IPv6 variables.

20.2 Management Information Base (MIB)

SNMP uses the Management Information Base (MIB) to structure the data that can be managed in network devices. MIB schemas [RFC2578] were based on IPv4 addresses, specifying an IPv4 address field with type '`IpAddress`'. Table 20.2 shows an IP forwarding variable '`ipCidrRouteDest`' used to identify a destination route and defined as an IPv4 address.

All address-related MIB variables are now IP version independent [RFC4001]. For any address field, two variables are used: the first is used to define the type of the address ('`InetAddressType`') used in the second variable, defined as '`InetAddress`', a generic type holding both kind of addresses. Table 20.3 shows the same IP forwarding variable shown in Table 20.2 converted to the IP version independent method. New variable names were defined in order not to break the current schema, with the '`ip`' prefix being changed to '`inet`'.

This work on IP version independent MIBs [RFC4001] is fairly recent and at the time of writing implementations are not easily available. Many current implementations either support the IPv6 specific variables [RFC2465] or have enterprise specific MIBs, and use UDP IPv4 as transport.

Table 20.2 Example of IPv4 MIB variable

Object name	Object type
`ipCidrRouteDest`	`IpAddress`

Table 20.3 Example of IP version independent MIB variable

Object name	Object type
`inetCidrRouteDestType`	`InetAddressType`
`inetCidrRouteDest`	`InetAddress`

20.3 Other Management Tools

Discussions in IETF and products on the market are starting to replace SNMP with an XML-based framework [Enns, 2004].

20.4 Authentication, Authorization and Accounting using RADIUS

Authentication, authorization and accounting (AAA) are security services used in enterprise and provider networks. Figure 20.2 shows an example of a network access server NAS1 with modems receiving a call from node N1 to reach node N2 on the IPv6 network. N1 and NAS1 use the Point-to-Point Protocol (PPP) to negotiate the parameters of the connection and then establish a data channel to carry IPv6 traffic over the PPP connection (see Section 6.7 for details of IPv6 and PPP).

During the PPP negotiation phase, N1 sends its credentials, such as username and password, to NAS1, which in turn is configured to use an AAA server S1. S1 receives the authentication data and compares them with the database. The database may contain additional information for these credentials, such as the IP address of N1. The AAA protocol used between NAS1 and S1 is the Remote Authentication Dial In User Service (RADIUS) protocol [RFC2865]. Other protocols are used for the same purpose, such as Lightweight Directory Access Protocol (LDAP) [RFC2251], TACACS [RFC1492] and DIAMETER[Calhoun, 2003], but are not covered here.

Radius is enhanced [RFC3162] to support IPv6 as a transport for Radius packets, and as a supported IP protocol in its dictionary. This enables an NAS to request or receive an IPv6 address from the Radius server for the connecting node.

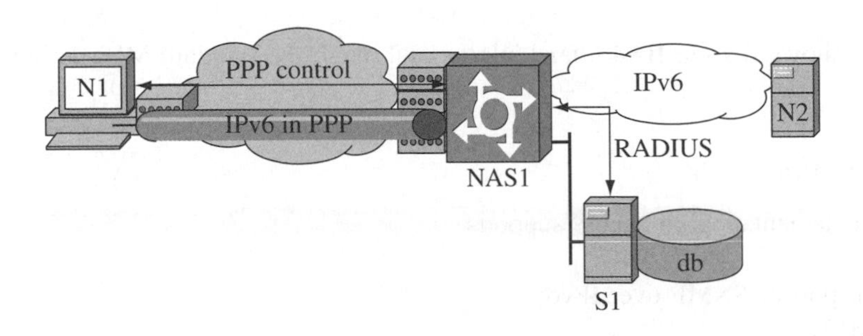

Figure 20.2 Network access server with Radius

20.5 Configuring SNMP on Hosts and Routers

SNMP configurations are shown for Cisco, Hexago and Juniper.

20.5.1 Cisco

SNMP implementation on Cisco IOS supports:

- IPv6 MIB;
- agents for IPv6.

The example below shows a basic SNMP configuration of R1 where a SNMP management station (3ffe:b00:0:4::b) can make queries to the router and will receive traps.

```
configure terminal
snmp-server community public
snmp-server enable traps
snmp-server host 3ffe:b00:0:4::b
```

Cisco has defined an enterprise MIB for IPv6.

20.5.2 Hexago

SNMP implementation on HexOS supports:

- IPv6 MIB for tunnels;
- agents for IPv6.

The example below shows a basic SNMP configuration of R1 where a SNMP management station (192.0.2.1) can make queries to the router and will receive traps.

```
configure terminal
snmp-server community public
snmp-server enable traps
nmp-server host 192.0.2.1
```

HexOS shows IPv6 in IPv4 tunnel interfaces through the standard MIB definitions.

20.5.3 Juniper

SNMP implementation on JunOS supports:

- the transport of SNMP over IPv6;
- IPv6 MIB;
- agents for IPv6.

The example below shows a basic SNMP configuration of R1 where a SNMP management station (3ffe:b00:0:4::b) can make queries to the router and will receive traps.

```
snmp {
 community PUBLIC {
  clients {
    3ffe:b00:0:4::b/128;
  }
 }
 trap-group GR1 {
   targets {
    3ffe:b00:0:4::b;
  }
 }
}
```

Juniper has also defined an enterprise MIB [JUNIPERMIB] for IPv6 to extend the `ifTable` for IPv6 support.

20.6 Summary

SNMP is the protocol used to manage network devices through a management station. While it uses UDP over IPv4 as transport, it can now be transported over IPv6. The MIBs have evolved from IPv4-only to IP version independent.

20.7 References

Calhoun, P., 'Diameter Base Protocol', draft-ietf-aaa-diameter-17 (work in progress), January 2003.

[JUNIPERMIB] Juniper Enterprise Specific MIB: ipv6 MIB Extension, http://www.juniper.net/techpubs/software/junos/junos55/swconfig55-net-mgmt/html/mib-jnx-ipv6.txt

Enns, R. 'NETCONF Configuration Protocol', draft-ietf-netconf-beep-02 (work in progress), February 2004.

[RFC1492] Finseth, C., 'An Access Control Protocol, Sometimes Called TACACS', RFC 1492, July 1993.

[RFC2251] Wahl, M., Howes, T. and Kille, S., 'Lightweight Directory Access Protocol (v3)', RFC 2251, December 1997.

[RFC2465] Haskin, D. and Onishi, S., 'Management Information Base for IP Version 6: Textual Conventions and General Group', RFC 2465, December 1998.

[RFC2578] McCloghrie, K., Perkins, D., Schoenwaelder, J., Case, J., McCloghrie, K., Rose, M. and Waldbusser, S., 'Structure of Management Information Version 2 (SMIv2)', STD 58, RFC 2578, April 1999.

[RFC2865] Rigney, C., Willens, S., Rubens, A. and Simpson, W., 'Remote Authentication Dial In User Service (RADIUS)', RFC 2865, June 2000.

[RFC3162] Aboba, B., Zorn, G. and Mitton, D., 'RADIUS and IPv6', RFC 3162, August 2001.

[RFC3417] Presuhn, R., 'Transport Mappings for the Simple Network Management Protocol (SNMP)', STD 62, RFC 3417, December 2002.

[RFC3419] Daniele, M. and Schoenwaelder, J., 'Textual Conventions for Transport Addresses', RFC 3419, December 2002.

[RFC4001] Daniele, M., Haberman, B., Routhier, S. and Schoenwaelder, J., 'Textual Conventions for Internet Network Addresses', RFC 4001, February 2005.

21

Porting Applications

This chapter presents the changes to the C Language Application Programming Interface (API) for applications to support IPv6. The intended audience is programmers who need to know the required changes to make an application IPv6 ready. This chapter does not discuss the Unix sockets, the advanced API used for controlling raw IP packets and multicast. For the interested reader, excellent references [Hagino, 2004; Stevens *et al.*, 2004] are provided at the end of the chapter.

This chapter describes the key considerations when porting an application to IPv6, then details the new data structures and functions. Best practices are listed and an example of porting a small IPv4 client and server to IPv6 is described. This chapter uses FreeBSD as the platform, but the examples should be highly portable, as indeed is the API.

21.1 Introduction

Applications using IPv4 network sockets need to be converted to IPv6. The conversion can be very simple when the network programming code is in a single location and does not do complex networking, or can be very challenging for kernel and complex networking applications. The well-known socket API, defined initially in BSD Unix, is a defacto standard. With IPv6, the new socket API, derived from the BSD one, is an IETF and IEEE standard, making it much more portable and easy for programmers.

The new socket API adds new structures and new functions and does not change IPv4 structures and functions. An operating system providing the old and the new API will not break any application linked with the old API.

21.2 Considerations

From a programmer's point of view, the important structural changes are described in this section.

Migrating to IPv6: A Practical Guide to Implementing IPv6 in Mobile and Fixed Networks Marc Blanchet

21.2.1 IP Protocol Version Independence

Before starting to make the porting effort, the programmer has to make a decision on whether the program will be IPv4-only, separated IPv4 and IPv6, or IP protocol version independent. The latter is recommended for most cases. IP protocol version independence is where the code is agnostic to which IP protocol the connections will use. In some cases, programmers might want to do specific tasks for one IP protocol as against the other. The IP protocol version independent code, discussed in this chapter, makes the code agnostic to the IP protocol version, which in the end makes the design simpler.

21.2.2 Multiple Addresses

An IPv6-enabled stack handles multiple IP addresses on a single interface. A typical example might be one IPv4 address, one global IPv6 address and one link-local IPv6 address. For both clients and servers, the code must take into account the multiple addresses. For example, when querying the DNS for the server addresses, the client code should loop through all the received IP addresses until one is answering.

21.2.3 Scoped Addresses

IPv6 addresses are scoped as described in Section 4.3.2. A fully specified address must include its scope and its usage in this scope. For example, a link-local address is limited to the link; however, a link-local address is ambiguous given that it could apply to any interface on the host. A link-local address must be tagged as link-local and be bound to a specific interface when it is used for a connection. A program must carry a data structure for addresses where the IP address itself, the scope and the interface are identified. Moreover, the text representations of IPv6 addresses with scope and interface have a special syntax. Conversion routines between binary and text representations are included in the API.

21.2.4 Address Memory Space

IPv4 addresses are 32 bits long, making them easy to store in `int` variables. IPv6 addresses are 128 bits long and require additional information, such as their scope (link, site, etc). IPv6 addresses can only be handled properly using a struct. The new structs are defined to hold either IPv4 and IPv6 addresses in a combined and transparent way.

21.2.5 URL and Text Representation of IP Addresses

As discussed in Section 4.2.3, the usual URL includes a ':' to separate the IPv4 address and the port number. A new URL syntax is defined [RFC3986] for IPv6 addresses where brackets '[]' are used to enclose them. A program parsing IP addresses with port numbers should be modified to handle the new URL syntax.

21.3 Structures

This section describes the most used structures defined in the new API.

21.3.1 Struct addrinfo

The `addrinfo` structure is a replacement of the `hostent` structure. It holds connection information used in handling name to IP address resolution, where the DNS query for IP address returns multiple IP addresses. It includes a linked list of addresses and the type of socket. It is defined in `/usr/include/netdb.h`. `Struct addrinfo` is used by the `getaddrinfo()` function.

```
struct addrinfo {
  int ai_flags;
  int ai_family;
  int ai_socktype;
  int ai_protocol;
  size_t ai_addrlen;
  char *ai_canonname;
  struct sockaddr *ai_addr;
  struct addrinfo *ai_next;
};
```

Table 21.1 lists the elements with the possible values and a description.

Table 21.1 `Struct addrinfo` values

Element	Possible values	Comment
	AI_PASSIVE	Get address to use bind().
	AI_CANONNAME	Fill `ai_canonname`.
	AI_NUMERICHOST	Prevent name resolution on that address.
	AI_ALL	IPv6 and IPv4-mapped.
`ai_flags`	AI_V4MAPPED_CFG	Accept IPv4-mapped if kernel supports.
	AI_ADDRCONFIG	Only if any address is assigned.
	AI_V4MAPPED	Accept IPv4-mapped IPv6 address.
	AI_DEFAULT	AI_V4MAPPED_CFG | AI_ADDRCONFIG (special recommended flags for getipnodebyname).
`ai_family`	AF_INET	IPv4 address family.
	AF_INET6	IPv6 address family.
`ai_socktype`	SOCK_STREAM	TCP transport.
	SOCK_DGRAM	UDP transport.
	SOCK_SEQPACKET	SCTP transport.
`ai_protocol`	IPPROTO_IPV4	IPv4 protocol.
	IPPROTO_IPV6	IPv6 protocol.
`ai_addrlen`		Length of struct pointed by `ai_addr`.
`ai_canonname`		Hostname string.
`ai_addr`		IP Address itself.
`ai_next`		Pointer to next structure in the linked list of addresses.

An IPv4-mapped address is a way to insert an IPv4 address in an IPv6 address format. This was defined so programmers would only care about IPv6 addresses, even when they were using IPv4. However, this feature raised some security concerns [Metz and Hagino, 2003; Hagino, 2004].

21.3.2 Struct sockaddr_in6

The sockaddr_in6 structure is the IPv6 version of sockaddr_in. As shown below, it holds the IP address, port number of a connection, IPv6 flow label and scope of the address.

```
struct sockaddr_in6 {
  uint8_t sin6_len;
  sa_family_t sin6_family;
  in_port_t sin6_port;
  uint32_t sin6_flowinfo;
  struct in6_addr sin6_addr;
  uint32_t sin6_scope_id;
};
```

Table 21.2 lists the elements with the possible values and a description.

The sockaddr_in6 structure is used in socket calls as place holder for IPv6 addresses. However, sockaddr_in6 is specific to IPv6 and, as such, is not IP protocol version independent and should be avoided.

21.3.3 Struct sockaddr_storage

The sockaddr_storage is a struct defined for casting either sockaddr_in or sockaddr_in6. This struct should be used when making a program IP protocol version independent.

21.3.4 Definitions

Some definitions were converted from IPv4 to IPv6. Table 21.3 lists those used most often.

These definitions are used in bind function calls.

Table 21.2 Struct sockaddr_in6 values

Element	Possible values	Comment
sin6_len		Length of the struct itself. This field is not present on all platforms.
sin6_family	AF_INET6	IP address family.
sin6_port		Transport port number.
sin6_flowinfo		IPv6 header flow information (see Section 3.4.3).
sin6_addr		IPv6 address itself.
sin6_scope_id		The scope of the IPv6 address (e.g.: link-local, site, etc.).

Table 21.3 Definitions

IPv4	IPv6	Comment
INADDR_ANY	`in6addr_any`	The wildcard address used by servers to bind on all addresses.
INADDR_LOOPBACK	`in6addr_loopback`	The loopback address used to specify the local host using internal connections not going on the wire.

21.4 Functions

This section describes the most used functions defined in the new API.

21.4.1 Getaddrinfo

The `getaddrinfo` function is the replacement of `gethostbyname`. It queries the DNS for the IP addresses of a hostname. The result is a linked list of addresses provided by the `res` pointer. This list should be traversed by the calling program.

```
int getaddrinfo(
  const char *nodename,
  const char *servname,
  const struct addrinfo *hints,
  struct addrinfo **res
);
```

Table 21.4 lists the arguments of `getaddrinfo`.

21.4.2 Getnameinfo

The `getnameinfo` function is a replacement of the `gethostbyaddr, inet_addr` and `inet_ntoa` functions. It queries the DNS for the hostname of an IP address. The result is the hostname string in the node element.

```
int getnameinfo(
  const struct sockaddr *sa,
  socklen_t salen,
```

Table 21.4 `Getaddrinfo` arguments

Argument	Description
`nodename`	The string of the hostname.
`servname`	The string of the service name.
`hints`	An `addrinfo` structure used to give requirements for the DNS resolution.
`res`	The result of the DNS query as a linked list of IP addresses.

Table 21.5 `Getnameinfo` arguments

Argument	Description
`sa`	The IP address of the socket.
`salen`	The size of the `sa` struct.
`node`	Pointer to a buffer. This will hold the hostname string on the return of the function.
`nodelen`	The length of the node buffer.
`service`	Pointer to a buffer. This will hold the service name string on the return of the function.
`servicelen`	The length of the service buffer.
`flags`	Flags that influence the DNS resolution actions.

```
    char *node,
    socklen_t nodelen,
    char *service,
    socklen_t servicelen,
    int flags
);
```

Table 21.5 lists the arguments of `getnameinfo`.

`Getnameinfo` is also used to convert IP addresses from binary to text and vice versa by using specific flags in the call.

21.4.3 Macros

The following definitions are macros used to test IPv6 addresses. MC stands for multicast.

```
int IN6_IS_ADDR_UNSPECIFIED  (const struct in6_addr *);
int IN6_IS_ADDR_LOOPBACK     (const struct in6_addr *);
int IN6_IS_ADDR_MULTICAST    (const struct in6_addr *);
int IN6_IS_ADDR_LINKLOCAL    (const struct in6_addr *);
int IN6_IS_ADDR_SITELOCAL    (const struct in6_addr *);
int IN6_IS_ADDR_V4MAPPED     (const struct in6_addr *);
int IN6_IS_ADDR_V4COMPAT     (const struct in6_addr *);

int IN6_IS_ADDR_MC_NODELOCAL (const struct in6_addr *);
int IN6_IS_ADDR_MC_LINKLOCAL (const struct in6_addr *);
int IN6_IS_ADDR_MC_SITELOCAL (const struct in6_addr *);
int IN6_IS_ADDR_MC_ORGLOCAL  (const struct in6_addr *);
int IN6_IS_ADDR_MC_GLOBAL    (const struct in6_addr *);
```

21.5 Change Table

Table 21.6 lists the most used structures, functions and definitions for IPv4 and the new versions for IPv6.

Table 21.6 API changes from IPv4 to IPv6

IPv4	IPv6	Section	IP protocol version independence
hostent	addrinfo	21.3.1	yes
sockaddr_in	sockaddr_in6	21.3.2	no
sockaddr_in	sockaddr_storage	21.3.3	yes
INADDR_ANY	in6addr_any	21.3.4	no
INADDR_LOOPBACK	in6addr_loopback	21.3.4	no
gethostbyname	getaddrinfo	21.4.1	yes
gethostbyaddr	getnameinfo	21.4.2	yes
inet_addr	getnameinfo	21.4.2	yes
inet_ntoa	getnameinfo	21.4.2	yes

21.6 Best Practice

The best practices in programming network applications are:

- make your code IP protocol version independent;
- use data structures to hold addresses;
- loop through all IP addresses.

21.7 Basic Example

This section describes the conversion of simple TCP client and server programs. For easy reading, error handling is not provided. These programs were compiled and verified on FreeBSD 5.2, however, they should work fine on most platforms.

The following code implements a TCP client over IPv4 to a server on port 2525. The client prints the line of data received from the server and quits. It uses '`localhost`' as the destination host for simplicity. This code represents the networking code of a minimal IPv4 client application.

```
#include <sys/socket.h>
#include <netinet/in.h>
#include <netdb.h>

int main(void)
{
  int sockfd;
  struct hostent *he;
  char buf[1000];
  int numbytes;
  struct sockaddr_in their_addr;

  he = gethostbyname("localhost");
  memset(&their_addr, 0, sizeof(their_addr));
  their_addr.sin_family = AF_INET;
```

```
  their_addr.sin_port = htons(2525);
  their_addr.sin_addr = *((struct in_addr *)he->h_addr);
  sockfd = socket(AF_INET, SOCK_STREAM, IPPROTO_TCP);
  connect(sockfd, (struct sockaddr *) &their_addr,
    sizeof(their_addr));
  numbytes = recv(sockfd,buf,sizeof(buf),0);
  buf[numbytes] = '\0';
  printf("%s",buf);
  close(sockfd);
}
```

The same client code is then converted to work as IP protocol independent. It implements a loop going through the linked list of the IP addresses returned from `getaddrinfo` and tries to connect until one is found answering on the required port.

```
#include <sys/socket.h>
#include <netinet/in.h>
#include <netdb.h>

int main(void)
{
  int sockfd;
  struct addrinfo hints,*res, *res0;
  char buf[1000];
  int numbytes;

  memset(&hints,0,sizeof(struct addrinfo));
  hints.ai_socktype = SOCK_STREAM;
  getaddrinfo("localhost", "2525", &hints, &res0);
  for (res = res0; res ; res = res->ai_next) {
    sockfd = socket(res->ai_family, res->ai_socktype,
      res->ai_protocol);
    if (connect(sockfd, res->ai_addr, res->ai_addrlen)==0)
      break;
  }
  numbytes = recv(sockfd,buf,sizeof(buf),0);
  buf[numbytes] = '\0';
  printf("%s",buf);
  close(sockfd);
}
```

The following code implements the TCP server over IPv4 on port 2525, serving the above client. When receiving a new connection, the server prints 'Hello, world!' and quits. It binds on all IPv4 addresses on the host.

```
#include <sys/types.h>
#include <sys/socket.h>
#include <netinet/in.h>

#define HW_STR "Hello World!\n"
```

```
int main(void)
{
  int sockfd;
  struct sockaddr_in my_addr;
  struct sockaddr_in their_addr;
  int sin_size;
  int new_fd;

  sockfd = socket(AF_INET, SOCK_STREAM, IPPROTO_TCP);
  my_addr.sin_family = AF_INET;
  my_addr.sin_port = htons(2525);
  my_addr.sin_addr.s_addr = INADDR_ANY;
  memset(&(my_addr.sin_zero), '\0',8);
  bind(sockfd, (struct sockaddr *) &my_addr,
    sizeof(my_addr));
  listen(sockfd,1);
  sin_size = sizeof(their_addr);
  new_fd = accept(sockfd, (struct sockaddr *) &their_addr,
    &sin_size);
  send(new_fd, HW_STR,sizeof(HW_STR),0);
  close(new_fd);
}
```

When executing this server program, `netstat` shows the 2525 port opened on TCP over IPv4 (`tcp4`).

```
# netstat -an
tcp4 0 0 *.2525 *.* LISTEN
```

This server code is converted to IPv6, as shown below. The socket calls, such as `bind`, `listen`, `accept` and `send` are not changed.

```
#include <sys/types.h>
#include <sys/socket.h>
#include <netinet/in.h>

int main(void)
{
  int sockfd;
  struct sockaddr_in6 my_addr;
  struct sockaddr_in6 their_addr;
  int sin_size;
  int new_fd;

  sockfd = socket(AF_INET6, SOCK_STREAM, IPPROTO_TCP);
  memset(&my_addr, 0, sizeof(my_addr));
  my_addr.sin6_family = AF_INET6;
  my_addr.sin6_port = htons(2525);
  my_addr.sin6_addr = in6addr_any;
  bind(sockfd, (struct sockaddr *) &my_addr, sizeof(my_addr));
  listen(sockfd,1);
```

```
  sin_size = sizeof(their_addr);
  new_fd = accept(sockfd, (struct sockaddr *) &their_addr,
    &sin_size);
  send(new_fd, "Hello, world!\n",
    sizeof("Hello, world!\n") 0);
  close(new_fd);
}
```

When executing this program, `netstat` shows the 2525 port opened on TCP over IPv6 (`tcp6`).

```
# netstat -an
tcp6 0 0 *.2525 *.* LISTEN
```

As shown, this server only listens on IPv6. To have one single program listening on both IPv4 and IPv6, the code should be IP protocol independent, by looping through all addresses of the host and binding each address.

21.8 Summary

Porting an application to IPv6 is in most cases relatively straightforward. New data structures and functions are provided to make the application IP protocol version independent, which is the recommended method.

21.9 References

Hagino, J-I., *IPv6 Network Programming*, Elsevier, 2004.

Metz, C. and Hagino, J., 'IPv4-Mapped Addresses on the Wire Considered Harmful', draft-itojun-v6ops-v4mapped-harmful-02 (work in progress), October 2003.

Stevens, W.R., Fenner, B., Rudoff, A.M., *Unix Network Programming vol. 1,* Addison-Wesley, 2004.

[RFC3986] Berners-Lee, T., Fielding, R. and L. Masinter, 'Uniform Resource Identifier (URI): Generic Syntax', STD 66, RFC 3986, January 2005.

21.10 Further Reading

[RFC3493] Gilligan, R., Thomson, S., Bound, J., McCann, J. and Stevens, W., 'Basic Socket Interface Extensions for IPv6', RFC 3493, February 2003.

[RFC3542] Stevens, W., Thomas, M., Nordmark, E. and Jinmei, T., 'Advanced Sockets Application Program Interface (API) for IPv6', RFC 3542, May 2003.

22

Configuration and Usage of IPv6-enabled Open Source Software

This chapter is targeted to the system and network administrators enabling IPv6 on their systems. It discusses IPv6 support and configuration for Apache, Sendmail, Postfix, SSH, XFree86, MRTG and Dovecot.

22.1 Apache Web Server

Setting up the Apache [Apache] Web server for IPv6 requires two basic steps. First, the server has an IPv6 address and that address is registered in the DNS using AAAA records, as described in Chapter 8. Second, the Web server is configured to listen on that IPv6 address.

In the httpd.conf configuration file of the Apache Web server, the `Listen` directive is used to configure the address and port on which the Web server listens. The IPv6 address is enclosed by brackets as with URL described in Section 4.2.3. For example, to listen on the `3ffe:b00:1:1::1` address, the `Listen` directive is:

```
# cat httpd.conf
Listen [3ffe:b00:1:1::1]
```

Similarly, virtual hosting is specified by the `NameVirtualHost` and `VirtualHost` directives, which also use the bracket enclosing for IPv6 addresses, as shown below. `NameVirtualHost` can only take one IP address per line. The following example is for a Web server on both `192.0.2.1` and `3ffe:b00:1:1::1` addresses.

Migrating to IPv6: A Practical Guide to Implementing IPv6 in Mobile and Fixed Networks Marc Blanchet

```
# cat httpd.conf
NameVirtualHost 192.0.2.1
NameVirtualHost [3ffe:b00:1:1::1]
```

However, the corresponding `VirtualHost` directive must list all the named addresses within the directive, as shown below for `192.0.2.1` and `3ffe:b00:1:1::1`.

```
# cat httpd.conf
<VirtualHost 192.0.2.1 [3ffe:b00:1:1::1]>
</VirtualHost>
```

22.2 Sendmail

Sendmail [Sendmail] is enabled to listen to both IPv4 and IPv6 on all configured addresses by defining two DAEMON_OPTIONS statements: one for IPv4 and one for IPv6 by setting the `Family` variable to "`inet6`" in the `DAEMON_OPTIONS` m4 statement, as shown below.

```
DAEMON_OPTIONS('Name=IPv4, Family=inet')dnl
DAEMON_OPTIONS('Name=IPv6, Family=inet6')dnl
```

To enable sendmail to listen to a specific IPv6 address, specify the address with the `Addr` variable, as shown below for `3ffe:b00:1:1::1`.

```
DAEMON_OPTIONS('Name=IPv6, Family=inet6,
Addr=3ffe:b00:1:1::1')dnl
```

In sendmail configuration files such as `mailertable`, `access`, and `relay-domains`, IPv6 addresses are specified by prefixing them with the keyword IPv6 followed by a colon, as shown below for `3ffe:b00:1:1::1`. Note that no bracket is used.

```
IPV6:3ffe:b00:1:1::1
```

A technical white paper [HEXAGOSENDMAIL] describes the setup of IPv6 mail with sendmail and bind.

22.3 Postfix

Postfix version 2.2 [Postfix] is the first version supporting IPv6. By default, IPv6 is not enabled. To enable, set the `inet_protocols` variable to `all` in the `main.cf` configuration file, as shown below.

```
# cat main.cf
inet_protocols=all
```

To specify the local interface address for outgoing SMTP connections, set the `smtp_bind_address6` to the IPv6 address, as shown below.

```
# cat main.cf
smtp_bind_address6 = 3ffe:b00:1:1::1
```

IPv6 addresses in the `mynetworks` and `debug_peer_list` variables must be enclosed in brackets, as shown below.

```
mynetworks = [3ffe:b00:1:1::]/64
```

22.4 SSH

By default [SSHDCONFIG], the SSH daemon listens to all configured IPv6 addresses. The `ListenAddress` statement in the sshd configuration file (`sshd_config`) is used to restrict to a specific IPv6 address, as shown below.

```
ListenAddress 3ffe:b00:1:1::1
ListenAddress [3ffe:b00:1:1::1]:2222
```

22.5 XFree86

XFree86 version 4.4 [XFREE86_4_4] is the first release to support IPv6. If IPv6 is configured on the host, the X server will listen on IPv6 by default.

The `DISPLAY` environment variable is used to redirect the display of the applications to a remote X server. The family name `inet6` with a slash is used to restrict the transport to IPv6, as shown below.

```
% setenv DISPLAY inet6/host1.example.org:0
```

The `xhost` command is used to control the remote access to the X server. The argument of the `xhost` command is a hostname. If the hostname has IPv4 and IPv6 addresses in the DNS, then all the returned addresses are added to the access control list. If only the IPv6 addresses are to be added, prefix the hostname with the IPv6 family name `inet6` and a colon, as shown below.

```
% xhost inet6:host1.example.org
```

22.6 MRTG

By default, MRTG [MRTG] does not use IPv6 as transport for SNMP. To enable IPv6 transport, the MRTG configuration file must contain the following statement.

```
EnableIPv6: Yes
```

The `cfgmaker` command can also be used to generate the MRTG configuration file. To enable IPv6 transport, add the `– enable-ipv6` argument to the `cfgmaker` command, as shown below.

```
# cfgmaker —enable-ipv6
```

22.7 Dovecot

Dovecot is an IMAP/POP server [DOVECOT]. To enable dovecot to listen on an IPv6 address, set the `listen` variable in the `dovecot.conf` configuration file to the IPv6 address enclosed in brackets, as shown below.

```
listen = [3ffe:b00:1:1::1]
```

To listen on all IPv6 interfaces, set the `listen` variable to `[::]` as shown below.

```
listen = [::]
```

22.8 Summary

The support, configuration and use of IPv6 in some open-source software is presented. In most cases, IPv6 addresses are typed with enclosed brackets, as discussed in Section 4.2.3.

22.9 References

Apache HTTP Server version 2.1, http://httpd.apache.org/docs-2.1/en/

Dovecot 1.0 Configuration file, http://www.dovecot.org/doc/dovecot-example.conf

[HEXAGOSENDMAIL] Deploying an IPv4/IPv6 mail exchange server with Sendmail, August 2003, http://eng.hexago.com/papers/data/hexago-technical-note-1001.pdf

MRTG: Multi Router Traffic Grapher, Tobias Oetiker, http://people.ee.ethz.ch/~oetiker/webtools/mrtg/mrtg-ipv6.html

Postfix IPv6 Support, http://www.postfix.org/IPV6_README.html

Sendmail 8.12 documentation, http://www.sendmail.org/m4

[SSHDCONFIG] OpenSSH SSH daemon configuration file, http://www.openbsd.org/cgi-bin/man.cgi?query=sshd_config

[XFREE86_4_4] Documentation for XFree86™ version 4.4.0, The XFree86 Project, Inc, 29 February 2004, http://www.xfree86.org/4.4.0

23

Best Current Practices and Case Studies

There are so many ways to build networks these days, with a large variety of available technologies, that covering the interaction of all possible scenarios with IPv6 could fill a whole book by itself. This chapter discuss the IPv6 Internet and related policies, generic considerations for IPv6 deployment, and some scenarios for IPv6 deployment for single nodes, home, enterprise and provider networks.

23.1 IPv6 Internet Address Space

The IPv6 Internet uses an address space allocated by the Internet Assigned Numbers Authority (IANA) to the regional registries, which then allocates address space to the providers. The IANA started to allocate address space in the 2001::/16 range [IANA, 2005]. IANA seems to allocate incrementally, allocating 2002::/16 for the 6to4 addressing and started 2003::/16 in 2005. For future years, the IPv6 Internet uses the 2000::/8 address space.

23.2 IPv6 Address Policy

Address allocation made by the registries to the providers is based on an address policy. Provider address space is then used to allocate address space to the enterprises, maximizing aggregation in the global routing table.

The Internet Architecture Board (IAB) and the Internet Engineering Steering Group (IESG) published recommendations on the IPv6 address allocation to sites [RFC3177], as listed in Table 23.1.

Following these recommendations, the Regional Internet Registries (RIR) such as Réseaux IP Européens (RIPE), Asia-Pacific Network Information Center (APNIC) and American Registry for Internet Numbers (ARIN), mandated to assign address space to providers,

Migrating to IPv6: A Practical Guide to Implementing IPv6 in Mobile and Fixed Networks Marc Blanchet

Table 23.1 IAB recommendations on IPv6 addresses

Allocated size	Usage
/48	For an end-site. This site can be a home network, or small and large enterprises.
/64	For a single subnet, mostly mobile networks, such as personal area networks. Could be a home network if there is no use of multiple subnets in the future of this home network.
/128	For a single host, with a point-to-point interface, such as PPP.

adopted the policy [RIPE267].[1] The RIRs give a /32 to providers satisfying the following criteria:

- they must be a provider assigning address space to users and organizations: such a provider is called a Local Internet Registry (LIR) in the address policy;
- they must not be an end-site;
- they must plan to provide IPv6 connectivity to organizations, by assigning them a /48;
- they must plan to assign 200 /48 to organizations within 2 years.

The allocation structure is shown in Figure 23.1.

Out of its /32 prefix received from its RIR, each provider in turn assigns /48 to end-sites. The end-site has 16 bits for subnets, each subnet has 64 bits.

A provider may request additional space based on an assignment ratio [RFC3194]. A large organization may request more than /48 if necessary.

The current policy allocates a /48 to home networks to allow 64K links within the home. This is a nice boundary but would consume some significant address space if all homes are connected and the providers are using the assignment ratio to get more address space from registries. To alleviate this issue, a proposal [Narten, 2005] is currently discussed to allocate a /56 prefix to home and small office networks.

The current address policy reinforces the aggregation of routes, which means that the IPv6 Internet BGP routing table should only have a small number of /32 routes, compared to the IPv4 global routes. However, as discussed in Section 9.12, multihoming may change this assumption in the future.

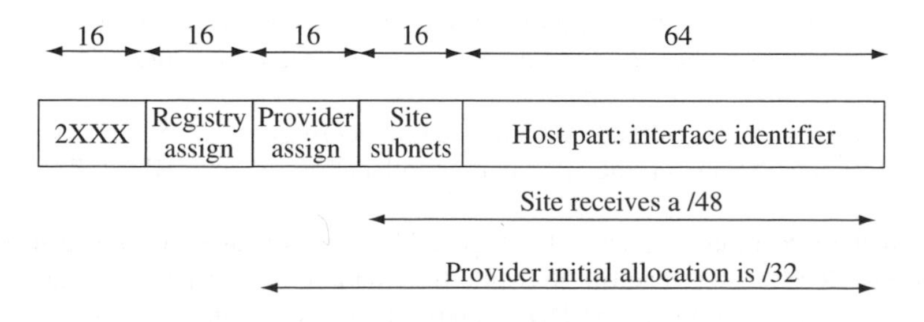

Figure 23.1 Address policy structure

[1] After multiple iterations of both IETF and RIRs.

23.3 IPv6 Address Planning

From the address architecture and policy, a site gets a /48 and a link a /64. Given plenty of address space, one can be lazy and just start assigning numbers incrementally. This section discusses some considerations and best practice for IPv6 address plans.

23.3.1 Optimal Address Plans

A tier-1 provider receives a /32 from the registry. This provider might provide connectivity to enterprises – to very large enterprises and to tier-2 providers – which need /32, /30 and /40 respectively. However, the provider does not know in advance how many enterprises and tier-2 providers will be connected to its network. Moreover, any of these organizations may come back to the provider for additional address space. Therefore, careful address planning enables the provider to manage these changes in the future, without the need to renumber its customers.

An algorithm [RFC3531] was designed to make IPv6 address plans easier. It is an enhancement of the algorithm [RFC1918] used for IPv4 address plans. It consists of using the leftmost bits for the leftmost assignments, the rightmost bits for the rightmost assignments and the center bits for the assignments in between.

For example, the tier-1 provider described above receives a /32 (3ffe:ffff::/32), which leaves room for 16 bits of assignments, because the longest prefix to assign is a /48. The initial forecast of customers for this provider is a maximum of 50 tier-2 providers. Obviously, this number can be larger or smaller, depending on their success over time: 50 providers consume 6 bits.

To make the optimal address plan, the tier-1 provider starts assigning from the left bits of its /32 and defines a virtual boundary (B1) after 6 bits, as shown in Figure 23.2.

Customers of the tier-1 provider start assigning with their center bits (bits C in Figure 23.2). Tier-1 customers are assigning /48 (boundary B2) to their customers. The key benefit of this method is that B1 and B2 boundaries may change later since the bits around the boundaries are assigned last. Changing boundaries later keeps the assignment fully aggregated and does not introduce multiple non-adjacent assignments.

Table 23.2 shows the first assignments that the tier-1 provider gives to its customers (C1, C2, C3, C4, . . .), starting from the left bits. While its defined boundary is /38 (/32 + 6), it assigns /40 to its tier-2 provider customers.

Table 23.3 shows the first /48 assignments C2 gives to its own customers (C2C1, C2C2, C2C3, . . .), using center bits.

The key feature of this method is the ability to change the boundaries later on, while keeping full aggregation and without renumbering. For example, if the bits near the B1 boundary are not yet assigned, C2 can grow to 3ffe:ffff:4000::/39 or decrease to 3ffe:ffff:4000::/41 without any changes to the address plans, and the aggregation is kept intact. Similarly, C2C3 can grow to 3ffe:ffff:4030::/47 or decrease to 3ffe:ffff:4030::/49 without any changes.

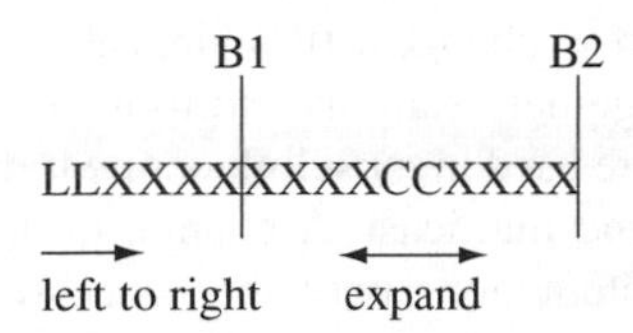

Figure 23.2 Left and center assignment of bits

Table 23.2 Leftmost bits assignments

	Assignments	Bits to be assigned	Resulting assignments
C1	100000	XXXXXXXXXX	3ffe:ffff:8000::/40
C2	010000	XXXXXXXXXX	3ffe:ffff:4000::/40
C3	110000	XXXXXXXXXX	3ffe:ffff:C000::/40
C4	001000	XXXXXXXXXX	3ffe:ffff:2000::/40

Table 23.3 Center bits assignments

	C2 prefix	Assignments	Resulting assignments
C2C1	010000	0000010000	3ffe:ffff:4010::/48
C2C2	010000	0000100000	3ffe:ffff:4020::/48
C2C3	010000	0000110000	3ffe:ffff:4030::/48
C2C4	010000	0000001000	3ffe:ffff:4008::/48

Also, similar assignments can be applied within the /48 of an organization. For example, when an organization has multiple sites and local assignment authorities, the same method is used to ensure the full aggregation even when one site grows faster than forecasted.

This book Web site (http://www.ipv6book.ca) has a tool to create address plans using the leftmost, rightmost or centermost assignments, as described in this section.

23.3.2 Numbering Links

From the addressing architecture (see Chapter 4), an IPv6 /64 prefix is assigned to each link. A /64 is sufficient to accommodate any number of hosts on that link. Since the IPv4 address space is limited, IPv4 subnet numbering implied some guessing on the future number of hosts on each subnet of a network, which makes address plans more difficult and prone to changes in the future.

When making an IPv6 address plan, each subnet can be numbered with any of the prefixes available in the /48 space of the site. The only consideration is the aggregation of the prefixes in the IGP routing table. If the numbering of the links is synchronized with the topology of the network, then the best aggregation is achieved, which makes routing tables smaller and creates more stability in the routing advertisements.

23.3.3 EUI-64 Considerations

EUI-64 in the host part embeds the hardware address of the interface. Whenever the hardware has to change, then the IP address changes. IP addresses are often put in filters, in server configuration, in network management stations and tools, etc.

Routers link-local addresses are used in most link-scope traffic, such as neighbor discovery, routing protocols, mobileIPv6 and multicast. A change in the link-local address of a router interface will break all these interactions until the processes are recovered. To avoid this delay in recovery, link-local address of router interfaces could be statically assigned without the use of the EUI-64. This enables the link-local address to remain stable even if the hardware address changes on the interface.

The role of the infrastructure of a server in a network means that the server's IP addresses are often statically written in the configuration files of other devices, in network management tools, in security filters, etc. If the address uses EUI-64 and the interface is changed, then the IPv6 address changes. A good practice is to assign static addresses to server interfaces that are put statically in other devices configuration files. This is especially true for the DNS servers IPv6 addresses.

23.3.4 Use of Unique Local Address Space

Unique local (Section 4.3.2.4) address space is used inside an organization to assign addresses to local devices that are not reachable from the Internet. Printers, sensors and management networks are examples of its use. These addresses are provider-independent and permanently assigned, so they survive in the event of a change of provider. Moreover, they enable private networks to be connected together as well as sites merging when companies merge.

Care must be taken to avoid the leaking of these addresses on the Internet. Border routers, firewalls and gateways must not route any packets with unique local source or destination address. DNS zone files containing unique local addresses must be of site scoped view.

The address plan for a unique-local /48 may use the optimal assignment method described in Section 23.3.1.

23.4 Incremental Deployment

To have one successful application session between two nodes over IPv6, many blocks between the two application programs must support IPv6. On the source node, the application itself must be converted to IPv6 (see Chapter 21) and the operating system must support IPv6. On the destination node, the application and the operating system must also be IPv6 ready. The whole network between the source and the destination node must also be routing IPv6 packets. Figure 23.3 shows such deployment blocks.

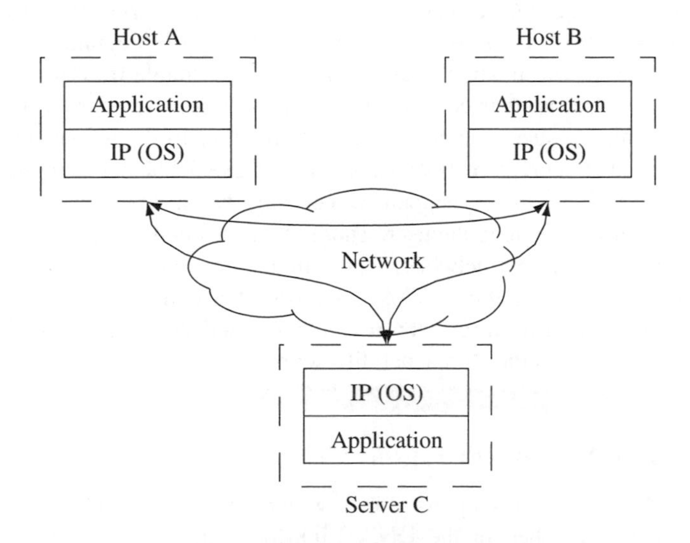

Figure 23.3 IPv6 deployment blocks

Until all these blocks are IPv6-ready, no single IPv6 packet can flow between the two application instances. The upgrade of all applications, all operating systems and the whole network and its services is often not possible, given the overall costs. Moreover, it is likely that some applications, operating systems and network devices will never support IPv6.

The typical strategy to deploy IPv6 is incremental deployment, where blocks are upgraded or converted one at a time. However, to get IPv6 trafic flowing between two IPv6-ready nodes, the network must be upgraded as a whole or an overlay network must be provided to connect the IPv6 nodes over the IPv4 network. As such, a typical deployment in enterprise or provider networks uses a migration technique, as described in Chapter 16.

Additional infrastructure services must also be IPv6-enabled, such as the DNS service, security services such as firewalls and VPN servers, network management systems, etc.

23.5 DNS Considerations

Putting IPv6 addresses in DNS is straightforward, as described in Chapter 8. However, care must be taken on the timing to publish the IPv6 addresses as well as some specific IPv6 addresses.

23.5.1 Publishing the AAAA Record

When an IPv6-enabled node has an A record for its name, the administrator adds an AAAA record for the same name. All the services on that node will be known by a name resolving to an IPv4 or an IPv6 address. However, putting the AAAA record has some underlying implications:

- The IPv6 address is configured and reachable.
- All the services on the node are IPv6 enabled.

If one of the service is not IPv6 enabled on the destination node, then the client node will timeout and fallback to IPv4, which works, but introduces some delay for the user.

As discussed in Section 8.6, a dual-stack node usually first tries IPv6 by querying the DNS for an AAAA record. If the AAAA record exists, an IPv6 connection is tried to the destination. However, the destination node might not be reachable from the source node, for example, when the source node is connected only to some private address space, or when it uses transition techniques such as 6to4 with no reachability to the whole IPv6 Internet (for example without a 6to4 relay). An IPv6 packet will be sent but never replied. After some timeout, a fallback to IPv4 should happen. Depending on the implementation, the timeout might be long and unacceptable to the user. These situations happen more for nomadic users who are connecting to different networks; some might not reach some of the destinations. To avoid this problem, only publish AAAA records of a node when it is reachable from wherever all the nodes that need to reach it can be. Stated differently, the DNS resolution scope should be the same as the IP reachability scope.

23.5.2 Publishing Special IPv6 Addresses

By default, the DNS resolution scope is global, which implies that only the globally scoped addresses should be published in the DNS. Link-local and unique local addresses have smaller scopes. Link-local addresses should not be published in the DNS since they are ambiguous out of the link. Unique local addresses might be published in the DNS if they are only accessible from a private view, only available from the same scope as the unique local address scope.

Temporary addresses (see Section 13.4), generated randomly and usually for a short period of time, may be published in the DNS, since they are global by nature. However, given their relative short lifetime, extra care must be taken to make sure the TTL of the AAAA record is shorter or equal to the lifetime of the address.

23.5.3 TTL Use

When records are cached in intermediary DNS servers, they are deleted from the cache when the TTL is expired. Given that the A and AAAA records are two different records, if they have different TTLs, they will expire at different times in cache servers, which implies incomplete answers to DNS clients from these servers. To avoid this situation, use the same TTL for both A and AAAA records of the same name.

23.6 Routing Considerations

As discussed in Chapter 9, routing has not changed that much in IPv6. However, some hints on topology and policies are discussed in this section.

23.6.1 Topologies

IS-IS only manages one link topology database for all layer 3 protocols. As discussed in Section 9.7.2, if the IPv4 and IPv6 link topologies differs, then IS-IS does not work by default. In such a case, verify the support of separate topologies in the IS-IS implementation of your vendor.

23.6.2 Policies

Wrong routes should be filtered to avoid routing issues. Table 23.4 list a set of guidelines [Blanchet, 2005].

23.7 Security Considerations

An important objective for an organization deploying IPv6 is to have at least the same level of security as in its IPv4 infrastructure. Some considerations on the security model, the filtering policies, the transition mechanisms and the special IPv6 addresses are now discussed.

23.7.1 End-to-End-Model

The security model of IPv6 might be seen as changed from IPv4, since there are no NATs in IPv6. The IPv4 NAT hides nodes not sending traffic to the external networks. Some security

Table 23.4 Guidelines for IPv6 routing policies

Guideline	Context	Description
Filter link-local prefixes (fe80::/16)	IGP, EGP	Link-local addresses are of link-local scope and should not appear in any routing table.
Filter unique-local prefixes (fec0::/16)	IGP, EGP	Unique-local addresses are deprecated and therefore should not be used.
Filter unique-local prefixes (fc00::/8, fd00::/8)	IGP	If a site is using a unique-local prefix, only that prefix should be announced. If the site is connected to other sites using other unique-local prefixes, only these prefixes should be announced. All other unique-local prefixes should not be announced.
Filter unique-local prefixes (fc00::/7)	EGP	A provider should filter all unique-local prefixes.
Filter loopback and unspecified (::/128, ::1/128)	IGP, EGP	Loopback and unspecified addresses are of node or link scope and should not appear in any routing table.
Filter multicast (ff00::/8) in unicast routing	IGP, EGP	Multicast routes should not appear in unicast routing tables.
Filter appropriate multicast scopes in multicast routing	IGP, EGP	Multicast addresses have scope bits. The scope of the routes should be enforced at the borders, in the multicast routing tables.
Filter IPv4-compatible (::/96) and IPv4-mapped (::FFFF:0:0/96) prefixes	IGP, EGP	These addresses are deprecated and should not be advertised.
Filter Default route (::/0)	EGP	A provider should not advertise the default route to its peers, but may advertise it to its customers.
Undefined address space (out of 2000::/3)	IGP, EGP	Undefined address space.
Filter long prefixes (> /48)	EGP	A provider should not advertise any prefix longer than /48, which is a site prefix.

models in IPv4 are based on that assumption. However, peer-to-peer applications and traffic have proven that NATs do not hide nodes if they wish to communicate, and because of the NAT translation, are very difficult to manage and filter securely.

By avoiding translation, IPv6 filtering devices show all traffic with the right source and destination addresses, enabling a full and non-blind approach to filtering the traffic.

As we have seen in recent years, filtering at the node level becomes increasingly important, if not mandatory, due to the nomadic nature of computing and the proliferation of viruses and attacks to nodes. IPv6 does not change that model, instead it enables a better filtering of traffic, given the non-translation.

23.7.2 Policies

As discussed in Section 13.3, IPv6 ICMP filtering should not block some ICMP messages for path MTU discovery to work. Extension headers filtering might pass packets with some headers such as the ones used by MobileIP.

23.7.3 Transition Mechanisms

Many IPv6 transition mechanisms use some form of IP-in-IP or IP over UDP-IP encapsulation. Encapsulation requires the filtering gateways to be smarter in decoding the traffic to filter correctly. If encapsulation techniques are used and traverse the security gateways, the gateways should support inspection of encapsulated packets and should be properly configured.

23.7.4 Special Addresses

Temporary addresses (see Section 13.4) are random in order to provide anonymity. An organization might consider anonymity as a problem in managing the nodes in the network. Disabling the feature might be one way to manage this issue.

In an enterprise network, always prefer to use link-local or unique-local addresses to associate nodes together, since they have a smaller scope than global adddresses.

23.8 Mail Considerations

Backup SMTP servers are configured using the DNS MX record with different levels of preferences. When the primary server is down, e-mails are sent to the secondary server (with the higher preference number in the MX record). When the primary server is backed up, the secondary servers contact the primary to relay the hold e-mails. To ensure this relay mechanism works, an IP protocol must be common between the primary and any of its secondary servers [RFC3974]. For example, if the primary is only IPv4 and one secondary is only IPv6, there is no common IP protocol between the two servers and the secondary will not be able to relay the hold e-mails.

23.9 Deploying IPv6 and Connecting to the IPv6 Internet

From the beginning of the IPv6 deployment to the end of the IPv4 lifetime, IPv6 will be deployed in multiple ways, far too many to describe here. However, we could group them in three different categories as described in Table 23.5.

These three categories apply to both the enterprise and the provider networks. There are other variations of these scenarios such as an IPv6-only and IPv4-only network, but only the most likely cases are discussed further.

The following sections describe detailed scenarios starting with one isolated dual-stack node, moving on to a mobile network and then to enterprise and provider networks.

23.9.1 Connecting a Single Node

A single node, where only one interface is configured, is connecting to the IPv6 Internet. Table 23.6 describes the possible scenarios:

- a node on an IPv6-enabled link (cases 1,2,3);
- a node on a network that has some far IPv6 connectivity such as inside an enterprise or in a provider network that has not fully upgraded to dual-stack (cases 4,5);
- a isolated node on the IPv4 Internet (cases 6,7,8).

Table 23.5 Deployment scenarios

Deployment scenario	Description	Implementation
Dual-stack	All devices and applications are running IPv4 and IPv6.	All applications, operating systems and network devices run IPv4 and IPv6 at the same time.
IPv4 dominant	Some devices of the network are dual-stack but an important part of the network is still IPv4-only.	Some applications, operating systems and network devices are running IPv4 and IPv6. The IPv4-only part of the network is using some IPv6 in IPv4 tuneling technique to connect the IPv6 islands in the network.
IPv6 dominant	Most devices on the network are IPv6-only while some are still dual-stack.	Most applications, operating systems and network devices are running IPv6-only. The rest of the network elements are dual-stack. An IPv4 in IPv6 tunneling technique is used to connect the IPv4 islands in the network.

Table 23.6 shows the scenarios and solutions related to the IPv6 connectivity that are available to the single node.

23.9.2 Connecting a Mobile Node

As discussed in Chapter 11, a mobile node has a permanent address used for communications. While moving, the mobile node restores the reachability of its permanent IPv6 address. Two mobile solutions are available to provide that functionality: MobileIPv6 or TSP tunnel broker. Table 23.7 lists the possible scenarios when the mobile node is visiting a network.

23.9.3 Connecting a Home Network

The home networks include the use of a gateway which is implementing IPv4 NAT and DHCP on the local network and has some WAN port connected typically to some broadband connection (see Table 23.8). If the home network is only one node, then refer to Section 23.9.1. RFC3750 and RFC3904 are also good references.

23.9.4 Connecting a Small Network

Small networks are similar to home networks, however, the stability of the IPv6 address is required to use or provide services. A typical scenario is a small remote office or a home office. If a transition mechanism is used and the mechanism embeds the IPv4 address in the IPv6 address, then the IPv4 address might change, which also makes the IPv6 address, and prefix, change. These mechanisms cannot be used in this context.

The small network (see Table 23.9) has a gateway implementing IPv4 NAT and DHCP on the local network.

Table 23.6 Scenarios for connecting a single node to IPv6

Step	IPv6 connectivity	Requirements	Example	Solution	Book reference section
1	RA on the local link.	An IPv6 router on the connected link sends router advertisements.	The node is on an IPv6-enabled link with a router sending router advertisements.	**RA**: The node receives router advertisements on its connected link and autoconfigures itself.	5.2.2.1
2	DHCPv6	The node is a DHCPv6 client and a DHCPv6 server is reachable.	The node is on an IPv6-enabled link where a DHCPv6 server is reachable.	**DHCPv6**: The node supporting DHCPv6 sends a DHCPv6 request and receives a DHCPv6 answer with information on how to configure itself.	5.5
3	PPPv6	The node supports PPPv6 and connects to a PPPv6 server.	The node is connecting to an IPv6-enabled access provider where PPPv6 is offered, either dial-up or DSL (PPPoE).	**PPPv6**: The node supporting PPPv6 connects to a PPP server and negotiates IPv6 connectivity on the PPP link and IPv6 address of its endpoint.	6.7
4	IPv6 on site but not on the connected link. No IPv4 NAT inside the site.	The node is an ISATAP client, an ISATAP router is available on the site and there is no NAT between the site's ISATAP router and the node. The node knows the address of the ISATAP router on the site.	The node is on a link with no IPv6 connectivity but the site has deployed the ISATAP service.	**ISATAP**: The node has not received any RA on its connected link nor is DHCPv6 available. The node receives a router advertisement through the ISATAP tunnel interface.	16.2.6
5	IPv6 on site but not on the connected link. IPv4 NAT or no NAT inside the site.	The node is a TSP client and a TSP tunnel broker is available on the site. The node knows the address of the TSP tunnel broker. A NAT might exist between the node and the TSP tunnel broker.	The node is on a link with no IPv6 connectivity but the site has deployed the TSP tunnel broker service.	**TSP tunnel broker**: The node has not received any RA on its connected link nor is DHCPv6 available. The node connects to the TSP tunnel broker and requests an IPv6 over IPv4 tunnel, with or without NAT traversal. The tunnel broker establishes a tunnel with the node. TSP works in both NAT and non-NAT cases.	16.2.9

(*continued overleaf*)

Table 23.6 (*continued*)

Step	IPv6 connectivity	Requirements	Example	Solution	Book reference section
6	No IPv6 connectivity on site. No NAT.	The node is a 6to4 client and a 6to4 relay is reachable. The node knows the address of the 6to4 relay. The node must have a global address and not be behind a NAT.	The node is connected directly to the IPv4 Internet and received an IPv4 global address (i.e. no NAT). A provider, either the connecting one or a far one, offers the 6to4 relay service.	**6to4**: The node has not received any RA on its connected link nor is DHCPv6 available. The node creates its 6to4 IPv6 address using its IPv4 address, configures a 6to4 interface and creates a default route to the 6to4 relay.	16.2.5
7	None. NAT in the path.	The node is a Teredo client and Teredo servers and relays are reachable. The node knows the address of the Teredo server. The node is behind a NAT.	The node is connected on the IPv4 Internet behind a NAT, such as behind corporate firewalls, home gateways, in hotel access networks, public LANs, wireless LANs, etc. A provider, either the connecting one or a far one, offers the Teredo service.	**Teredo**: The node has not received any RA on its connected link nor is DHCPv6 available. The node contacts the Teredo server to acquire its IPv6 address.	16.2.10
8	None. NAT or no NAT in the path.	The node is a TSP client and a TSP tunnel broker is reachable. The node knows the address of the TSP tunnel broker. The node may or may not be behind a NAT.	The node is connected to the IPv4 Internet behind a NAT or directly with a global address, such as behind corporate firewalls, home gateways, in hotel access networks, public LANs, wireless LANs, etc. A provider, either the connecting one or a far one, offers the TSP tunnel broker service.	**TSP tunnel broker**: The node has not received any RA on its connected link nor is DHCPv6 available. The node connects to the TSP tunnel broker and requests an IPv6 over IPv4 tunnel, with or without NAT traversal. The tunnel broker establishes a tunnel with the node. TSP works in both NAT and non-NAT cases.	16.2.9

Table 23.7 Scenarios for connecting a visiting mobile node to IPv6

Step	IPv6 connectivity	Requirements	Example	Solution	Book reference section
1	RA on the local link, DHCPv6 or PPPv6.	The mobile node supports MobileIPv6. The visiting network does not filter MobileIPv6 messaging through its border security gateways.	The mobile node is visiting an IPv6-enabled network where MobileIPv6 is permitted to be used.	**MobileIPv6**: The mobile node acquires a 'careof' address on the connected link of the visiting network and then signals its new 'careof' address to its home agent and its correspondents.	Chapter 11
2	RA on the local link, DHCPv6 or PPPv6.	The mobile node supports MobileIPv6. The visiting network filters MobileIPv6 messaging through its border security gateways. MobileIPv6 cannot be used. The mobile node supports TSP.	The mobile node is visiting an IPv6-enabled network where MobileIPv6 cannot be used. Another mobility protocol should be used.	**TSP tunnel broker**: The node connects to the TSP tunnel broker and requests an IPv6 over IPv4 tunnel, with or without NAT traversal. The TSP tunnel broker establishes a tunnel with the node. TSP works in both NAT and non-NAT cases.	11.12.4, 16.2.9
3	On site but not on connected link. No NAT inside the site.	The mobile node supports MobileIPv6. The visiting network does not filter MobileIPv6 messaging through its border security gateways. The mobile node also supports ISATAP, an ISATAP router is available on the site and there is no NAT between the site ISATAP router and the node. The node knows the address of the ISATAP router on the site.	The node is on a link with no IPv6 connectivity but the site has deployed the ISATAP service. The ISATAP router is reachable without a NAT between the node and itself.	**MobileIPv6 with ISATAP**: The mobile node has not received any RA on its connected link nor is DHCPv6 available. The mobile node acquires the ISATAP address and uses it as a 'careof' address on the visiting network. It then signals its new 'careof' address to its home agent and its correspondents.	Chapter 11, 16.2.6

(continued overleaf)

Table 23.7 *(continued)*

Step	IPv6 connectivity	Requirements	Example	Solution	Book reference section
4	On site but not on the connected link. An IPv4 NAT or no NAT is in the path.	The mobile node supports MobileIPv6. The mobile node also supports TSP and a TSP tunnel broker is available on the site. The node knows the address of the TSP tunnel broker. A NAT might be used on the site.	The node is on a link with no IPv6 connectivity but the site has deployed the TSP tunnel broker service.	**MobileIPv6 with TSP tunnel broker**: The mobile node has not received any RA on its connected link nor is DHCP available. The mobile node contacts the TSP tunnel broker, receives an IPv6 address and establishes a tunnel. The received address is used as a 'careof' address on the visiting network. The mobile node then signals its new 'careof' address to its home agent and its correspondents.	Chapter 11, 16.2.9
5	None	The mobile node supports MobileIPv6. The visiting network does not filter MobileIPv6 messaging through its border security gateways. The mobile node also supports 6to4 and a 6to4 relay is reachable. The node knows the address of the 6to4 relay. The node must have a global address and not be behind a NAT.	The node is connected directly to the IPv4 Internet and received an IPv4 global address (i.e. no NAT). A provider, either the connecting one or a far one, offers the 6to4 relay service.	**MobileIPv6 and 6to4**: The mobile node has not received any RA on its connected link nor is DHCP available. The mobile node creates its 6to4 address and uses it as a 'careof' address on the visiting network. It then signals its new 'careof' address to its home agent and its correspondents.	Chapter 11, 16.2.5

Table 23.7 (*continued*)

Step	IPv6 connectivity	Requirements	Example	Solution	Book reference section
6	None	The mobile node supports MobileIPv6. The visiting network does not filter MobileIPv6 messaging through its border security gateways. The mobile node also supports Teredo and Teredo servers and relays are reachable. The node knows the address of the Teredo server. The node is behind a NAT.	The node is connected on the IPv4 Internet behind a NAT, such as behind corporate firewalls, home gateways, in hotel access networks, public LANs, wireless LANs, etc. A provider, either the connecting one or a far one, offers the Teredo service.	**MobileIPv6 and Teredo**: The node has not received any RA on its connected link nor is DHCP available. The node contacts the Teredo server to acquire its IPv6 address and uses it as a 'careof' address on the visiting network. It then signals its new 'careof' address to its home agent and its correspondents.	Chapter 11, 16.2.10
7	None	The node supports TSP and a TSP tunnel broker is reachable. The node knows the address of the TSP tunnel broker. The node may or may not be behind a NAT. The mobile node does not support MobileIPv6 or MobileIPv6 is filtered.	The node is connected directly with a global address or to the IPv4 Internet behind a NAT such as behind corporate firewalls, home gateways, in hotel access networks, public LANs, wireless LANs, etc. A provider, either the connecting one or a far one, offers the TSP tunnel broker service.	**TSP tunnel broker**: The node has not received any RA on its connected link nor is DHCP available. The node connects to the TSP tunnel broker and requests an IPv6 over IPv4 tunnel, with or without NAT traversal. The TSP tunnel broker establishes a tunnel with the node. TSP works in both NAT and non-NAT cases.	16.2.9

Table 23.8 Home networks deployment scenarios

Step	Gateway	IPv6 connectivity	Requirements	Solution	Book reference section
1	IPv6-ready, DHCPv6-PD client.	IPv6 native is directly provided.	Gateway implements DHCPv6-PD. A DHCPv6-PD server is reachable.	**DHCPv6-PD, RA**: The gateway receives from the DHCPv6 server a prefix (PD: prefix delegation) for the home network. The gateway announces the prefix using router advertisements (RA) on the local link. All nodes on the local link are autoconfigured with the RA.	5.5.4, 5.2.2.1
2	IPv6-ready, TSP client.	The attaching network is IPv4-only. IPv6 TSP tunnel broker service is provided either by the direct provider or a far one.	Gateway is a TSP client. A TSP tunnel broker is reachable.	**TSP tunnel broker, RA**: The gateway implements TSP and receives the information from TSP for the tunnel to establish as well as the prefix for the home network. The gateway announces the prefix using router advertisements (RA) on the local link. All nodes on the local link are autoconfigured with the RA.	16.2.9, 5.2.2.1
3	IPv6-ready, 6to4 client.	No IPv6 service is provided.	Gateway is a 6to4 client. A 6to4 relay is reachable. Gateway is not behind a NAT.	**6to4, RA**: The gateway implements 6to4 and creates its 6to4 prefix based on its external IPv4 address. The gateway announces the prefix using router advertisements (RA) on the local link. All nodes on the local link are autoconfigured with the RA. A 6to4 relay address is configured on the gateway.	16.2.5, 5.2.2.1

Table 23.8 (*continued*)

Step	Gateway	IPv6 connectivity	Requirements	Solution	Book reference section
4	Not IPv6-ready.	IPv6 native is directly provided.	Node is a TSP or Teredo client.	Gateway cannot provide IPv6. Nodes inside the home network should use a technique which supports NAT traversal such as Teredo or TSP Tunnel broker.	23.9.1
5	Not IPv6-ready.	The attaching network is IPv4-only. IPv6 TSP tunnel broker service is provided either by the direct provider or a far one.	One node inside the network is acting as an IPv6 router and implements the TSP client. A TSP tunnel broker is reachable.	**TSP tunnel broker, RA**: One node inside the network is acting as an IPv6 router. It implements TSP and receives the information from TSP for the tunnel to establish as well as the prefix for the home network. The node announces the prefix using router advertisements (RA) on the local link. All nodes on the local link are autoconfigured with the RA.	16.2.9, 5.2.2.1
6	Not IPv6-ready.	No IPv6 service is provided.	Node is a TSP or Teredo client.	Nodes inside the home network should use a technique which supports NAT traversal such as Teredo or TSP Tunnel broker.	23.9.1

Table 23.9 Small networks deployment scenarios

Step	Gateway	IPv6 connectivity	Requirements	Solution	Book reference section
1	IPv6-ready, DHCPv6-PD client.	IPv6 native is directly provided.	Gateway implements DHCPv6-PD. A DHCPv6-PD server is reachable.	**DHCPv6-PD, RA**: The gateway receives from the DHCPv6 server a prefix (PD: prefix delegation) for the network. The gateway announces the prefix using router advertisements (RA) on the local link. All nodes on the local link are autoconfigured with the RA.	5.5.4, 5.2.2.1
2	IPv6-ready, TSP client.	The attaching network is IPv4-only. IPv6 TSP tunnel broker service is provided either by the direct provider or a far one.	Gateway is a TSP client. A TSP tunnel broker is reachable.	**TSP tunnel broker, RA**: The gateway implements TSP and receives the information from TSP for the tunnel to establish as well as the prefix for the network. The gateway announces the prefix using router advertisements (RA) on the local link. All nodes on the local link are autoconfigured with the RA.	16.2.9, 5.2.2.1
3	Not IPv6-ready.	IPv6 native is directly provided.	Node is a TSP client.	Gateway cannot provide IPv6. Nodes inside the network should use a technique which supports NAT traversal and a permanent address, such as the TSP Tunnel broker.	23.9.1

Table 23.9 (*continued*)

Step	Gateway	IPv6 connectivity	Requirements	Solution	Book reference section
4	Not IPv6-ready.	IPv6 TSP tunnel broker service is provided either by the direct provider or a far one.	One node inside the network is acting as an IPv6 router and implements the TSP client. A TSP tunnel broker is reachable.	**TSP tunnel broker, RA**: One node inside the network is acting as an IPv6 router. It implements TSP and receives the information from TSP for the tunnel to establish as well as the prefix for the network. The node announces the prefix using router advertisements (RA) on the local link. All nodes on the local link are autoconfigured with the RA.	16.2.9, 5.2.2.1
5	Not IPv6-ready.	No IPv6 service is provided.	Node is a TSP client.	Nodes inside the network should use a technique which supports NAT traversal and a permanent address such TSP Tunnel broker.	23.9.1

23.9.5 Enterprise and Military Networks

This section describes the deployment of IPv6 in enterprise networks, or more generally in medium to large networks within a single administration domain. Such networks include the military networks. The book Web site (http://www.ipv6book.ca) describes many scenarios. RFC 4057 also provides a good reference [RFC4057].

Most of these networks target dual-stack networks with a staged and incremental deployment. This incremental deployment requires the use of some tunneling techniques as described in Chapter 16. Some specific networks for vertical applications such as sensor networks or military ones might use IPv6-only networks at the beginning and carry IPv4 legacy traffic within IPv6 packets.

23.9.5.1 Addressing

Many enterprise network managers like to use private address space for their network. IPv6 provides a globally unique private address space, unique-local, described in Section 4.3.2.3. The enterprise should consider using this address space for its private addressing requirements while at the same time use public address space for global connectivity. In this setup, dual-stack hosts have one IPv4 private address and three IPv6 addresses: a link-local, a unique-local and a global address. When the host talks to a private server within the enterprise network, the host uses its private address to reach the server, by the source address selection mechanism as described in Section 8.6. When the host talks to an external server on the IPv6 Internet, the host uses its global address to reach the server. Private address space usage must be carefully implemented in the DNS infrastructure with private and public views, so the private address space is not leaked to the public network.

An address plan should be designed for the enterprise network. Each link is allocated a /64 prefix out of the enterprise /48 prefix. The address plan should take into consideration the optimization method for address allocation [RFC3531], as discussed in Section 23.3.1. A large enterprise has a substantial interest in optimizing the address plan to make it flexible for the future growth of its network.

The enterprise should get a /48 prefix from its IPv6 provider. The enterprise network can be first numbered using the unique local addressing. When the provider provides the /48 prefix to the enterprise, nodes receive the additional /64 prefix automatically on their link if they are using autoconfiguration.

If the enterprise has multiple providers, a multihoming technique should be used as described in Section 9.12.

23.9.5.2 Core and Distribution Routing

RIPng, OSPFv3 or IS-IS can be used for the enterprise IGP. As discussed in Chapter 9, RIPng might not scale for medium to large networks. OSPFv3 is another routing process which makes the two IP protocols independent in topology. IS-IS is IP protocol independent, making a single topology database. If the IPv4 and IPv6 topologies are different, then care must be taken with IS-IS, as discussed in Section 9.7.2.

If the core network cannot be upgraded to dual-stack but it already uses MPLS, then the IPv6 over MPLS technique can be used.

23.9.5.3 Access

If the access network cannot be upgraded to dual-stack, then the ISATAP or TSP tunnel broker can be used to tunnel IPv6 over IPv4, as discussed in Sections 16.2.6 and 16.2.9 respectively. If NATs are present within the enterprise network, then ISATAP cannot be used whereas TSP tunnel broker has NAT-traversal support.

23.9.5.4 Remote Access

For employees working at home, traveling or for remote offices, remote access to the enterprise network from the IPv4 Internet can use the TSP tunnel broker as discussed in Section 16.2.9. The TSP tunnel broker is located at the edge of the enterprise network, behaving like a VPN server. The TSP client is located in the laptop or user computer or in the home gateway, as shown in Figure 23.4.

23.9.5.5 Node, Server and Router Configuration

To fully use the IPv6 features, such as renumbering and reliability, nodes addresses should not be configured manually but by autoconfiguration or if needed by DHCPv6, as described in Chapter 5.

Given the IPv6 large address string with hexadecimal, the manual registration of node names is difficult. Instead the dynamic DNS technique should be used.

Servers, routers and security gateways are infrastructure devices that should not rely on other devices to be configured and they should not depend on the layer 2 address of their network cards. Server, router and security gateway addresses should be configured manually.

Since an interface can have multiple IPv6 addresses, the concept of secondary addresses in IPv4 no longer exists in IPv6. For network management purposes, it is now easier to make multiple addresses to make a separate network management control plane, to create virtual networks or for other purposes.

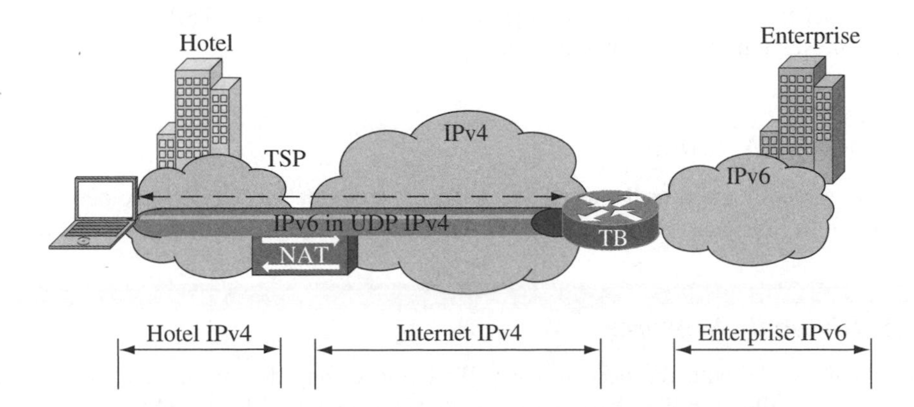

Figure 23.4 Enterprise remote access with TSP tunnel broker

23.9.6 Provider Networks

This section gives some basic hints for provider networks. The book Web site (http://www.ipv6book.ca) describes many scenarios. RFC 4029 also provides a good reference [RFC4029].

23.9.6.1 Addressing

Under the current address policy, discussed in Section 23.2, a provider initially receives a /32 from the registry. The provider assigns /48 to each of its customers, enterprise or home network. For the downstream providers, the assignment by the upstream provider could be any prefix between /32 and /48. An address plan should be designed in advance to get the maximum flexibility. The address plan should highly take into consideration the optimization method for address allocation [RFC3531], as discussed in Section 23.3.1. A provider has a substantial interest in optimizing the address plan to make it flexible for the future growth of its network.

23.9.6.2 Core, Distribution and Exchange Routing

A good architecture for the core and distribution networks is to run dual-stack networks, where all routers have all their interfaces configured for IPv4 and IPv6.

If an important part of the core and distribution networks is IPv4-only and runs MPLS, then the IPv6 in MPLS solution enables the incremental upgrade of edge routers as IPv6 demand increases, without the upgrade of the core.

Peering at exchange points can be done with dedicated IPv6 routers or dual-stack routers. As the BGP transport is independent of the address families of the routes, a single BGP peering can exchange both IPv4 and IPv6 routes. The new version of the Route Policy Specification Language (RPSL), named RPSLng [RFC4012], should be considered for managing IPv4 and IPv6 BGP peering.

23.9.6.3 Access

When the access network cannot be upgraded to dual-stack, the TSP tunnel broker, as described in Section 16.2.9, should be considered.

23.9.7 Mobile Networks

Mobile networks should consider using the enhancement of the MobileIPv6 protocol, NEMO, as discussed in Section 11.10.3, or the TSP tunnel broker, as discussed in Section 16.2.9.

23.9.8 IPv6-only Networks

IPv6-only networks with the need to carry IPv4 traffic may deploy an IPv6-only routing infrastructure while carrying IPv4 packets with IPv6 tunnels. The DSTM mechanism, as described in Section 17.1.3, with DHCPv6 or TSP should be considered.

23.10 Summary

This chapter bundles together some best practices and deployment scenarios for single nodes, home, mobile, enterprise and provider networks.

23.11 References

Blanchet, M. 'IPv6 Routing Policies Guidelines', Internet-Draft draft-blanchet-v6ops-routing-guidelines-00, July 2005.

IANA, IPv6 Global Unicast Address Assignment, http://www.iana.org/assignments/ipv6-unicast-address-assignments, August 2005.

Narten, T. et al., 'IPv6 Address Allocation to End Sites', Internet Draft draft-narten-ipv6-3177bis-48boundary-00.txt, July 2005.

[RFC1918] Rekhter, Y., Moskowitz, R., Karrenberg, D., Groot, G. and Lear, E., 'Address Allocation for Private Internets', BCP 5, RFC 1918, February 1996.

[RFC3177] IAB and IESG, 'IAB/IESG Recommendations on IPv6 Address', RFC 3177, September 2001.

[RFC3194] Durand, A. and Huitema, C., 'The H-Density Ratio for Address Assignment Efficiency: An Update on the H ratio', RFC 3194, November 2001.

[RFC3513] Hinden, R. and Deering, S., 'Internet Protocol Version 6 (IPv6) Addressing Architecture', RFC 3513, April 2003.

[RFC3531] Blanchet, M., 'A Flexible Method for Managing the Assignment of Bits of an IPv6 Address Block', RFC 3531, April 2003.

[RFC3750] Huitema, C., Austein, R., Satapati, S. and R. van der Pol, 'Unmanaged Networks IPv6 Transition Scenarios', RFC 3750, April 2004.

[RFC3904] Huitema, C., Austein, R., Satapati, S. and R. van der Pol, 'Evaluation of IPv6 Transition Mechanisms for Unmanaged Networks', RFC 3904, September 2004.

[RFC3974] Nakamura, M. and Hagino, J., 'SMTP Operational Experience in Mixed IPv4/v6 Environments', RFC 3974, January 2005.

[RFC4012] Blunk, L., Damas, J., Parent, F. and Robachevsky, A., 'Routing Policy Specification Language Next Generation (RPSLng)', RFC 4012, March 2005.

[RFC4029] Lind, M., Ksinant, V., Park, S., Baudot, A. and Savola, P., 'Scenarios and Analysis for Introducing IPv6 into ISP Networks', RFC 4029, March 2005.

[RFC4057] Bound, J., 'IPv6 Enterprise Network Scenarios', RFC 4057, June 2005.

[RIPE267] APNIC, ARIN, RIPE NCC, 'IPv6 Address Allocation and Assignment Policy', RIPE-267, ftp://ftp.ripe.net/ripe/docs/ripe-267.txt, January 22 2003.

23.12 Further Reading

American Registry for Internet Numbers(ARIN), http://www.arin.net

Asia-Pacific Network Information Center(APNIC), http://www.apnic.net

[RFC3587] Hinden, R., Deering, S. and Nordmark, E., 'IPv6 Global Unicast Address Format', RFC 3587, August 2003.

[RIPE-NCC] Réseaux IP Européens-Network Coordination Center (RIPE-NCC), http://www.ripe.net

24

Conclusion

The purpose of this book is to provide you the best knowledge of IPv6. In chapter 1, I first tried to convince you of its benefits. Chapters then describe the various IPv6 related specifications. Chapter 23 gives advice on deployment.

I spent a lot of time trying to make this the best IPv6 book, although I know that it is far from perfect. This book has a companion Web site (http://www.ipv6book.ca) to help keep the book up to date. Please refer to this Web site.

I hope you liked this book. I enjoyed writing it and learned a lot.

Marc.

Migrating to IPv6: A Practical Guide to Implementing IPv6 in Mobile and Fixed Networks Marc Blanchet

25

Quick Reference

A quick reference of key useful information for daily work is available on the book Web site (http://www.ipv6book.ca). It contains summarized information such as IPv6 headers, address assignments, neighbor cache summary, API structures and function calls, commands and configuration of various operating systems, etc.

Migrating to IPv6: A Practical Guide to Implementing IPv6 in Mobile and Fixed Networks Marc Blanchet